Composite Materials Technology

Processes and Properties

P.K. MALLICK/S. NEWMAN (Eds.)

Composite Materials Technology

Processes and Properties

With contributions from
G.B. Chapman II, B. Fisa, C.F. Johnson, M.R. Kamal,
H.S. Kliger, P.K. Mallick, S. Newman,
N. Raghupathi, M.E. Ryan, G. Slocum, C. Smith,
J. Stone, B.A. Wilson

Hanser Publishers, Munich Vienna New York

Distributed in the United States of America and in Canada by Oxford University
Press, New York

Editors:
Dr. P.K. Mallick, University of Michigan-Dearborn, Dearborn, Michigan, USA
Dr. S. Newman, Ford Motor Co. (retired), West Bloomfield, Michigan, USA

Distributed in USA and in Canada by
Oxford University Press
200 Madison Avenue, New York, N.Y. 10016

Distributed in all other countries by
Carl Hanser Verlag
Kolbergerstr. 22
D-8000 München 80

While the advice and information in this book are believed to be true and accurate at the date of going to press, neither the authors nor the editors nor the publisher can accept any legal responsibility for any errors or omissions that may be made. The publisher makes no warranty, express or implied, with respect to the material contained herein.

Library of Congress Cataloging-in-Publication Data

Composite materials technology: processes and properties/P. K. Mallick, S. Newman, eds.: with contributions from G.B. Chapman . . . [et al.].
Includes index.
ISBN 3-446-15684-4 (Hanser).
ISBN 0-19-520847-1 (Oxford University Press)
1. Reinforced plastics. 2. Polymeric composites.
I. Mallick, P. K., 1946-, II. Newman, Seymour, 1922-
TP1177,C65 1990   620.1′923--dc20   90-38921

CIP-Titelaufnahme der Deutschen Bibliothek

Composite materials technology: processes and properties/P. K. Mallick; S. Newman (Eds.).
With contributions from G. B. Chapman II . . . - Munich; Vienna; New York: Hanser; New York: Oxford Univ. Press, 1990
ISBN 3-446-15684-4 (Hanser)
ISBN 0-19-520847-1 (Oxford Univ. Press)
NE: Mallick, P. K. [Hrsg.]; Chapman, Gilbert B. [Mitverf.]

ISBN 3-446-15684-4 Carl Hanser Verlag, Munich, Vienna, New York
ISBN 0-19-520847-1 Oxford University Press
Library of Congress Catalogue Card Number 90–38921

Printed in the Federal Republic of Germany by Buch- und Offsetdruckerei Wagner GmbH, Nördlingen

Preface

The study and utilization of composite materials has been a truly interdisciplinary endeavor that has been enriched by contributions from chemistry, physics, materials science, mechanics, and process engineering. Excellent reviews of the general principles describing the mechanical, physical, and rheological properties of multicomponent systems in terms of their morphology and the properties of the constituents are available, and some of these are listed in the references (1–6) in Chapter 1. The books cited describe the pioneering work of those who have helped make composites a "leading edge" technology in the latter half of this century and have broadened its realm to include ceramic, metal, carbon, and polymer matrices.

Our particular focus in this book is a narrow but vital part of the total realm of composite materials technology, namely to provide the reader with a timely exposition of the methods and materials that have emerged as most relevant to the mass production of polymer matrix composites. A definition of a mass production technique is in order at this point. We define a mass production method as one for which production time is generally measured in the time scale of minutes rather than hours or days. Postcure heat cycles and such complex secondary operations as adhesive bonding, welding, assembly, and decorating may, in total, consume considerable periods of time. Nevertheless, the primary fabrication step to yield a shaped component should be of the order of seconds to roughly 10 minutes per part to qualify as a mass production technique.

Most commercial plastic materials contain additives and particulate fillers to modify a variety of properties (e.g., coefficient of expansion, shrinkage, electrical and thermal conductivity, flow in the liquid state, friction and wear, chemical stability, and fracture resistance). However, in this work, we are concerned primarily with materials whose mechanical and physical properties are significantly altered by fibrous materials as well as anisotropic mineral reinforcements possessing at least one long dimension; examples are mica and talc, which have platelet configurations. Consequently, some high volume production fabricating methods, such as blow molding, which qualify for inclusion on the basis of manufacturing time, are omitted because they are largely confined to the processing of resins without fibrous fillers.

The reader will note that the chapters are directed at fabricating methods and associated materials rather than at specific industries, since all composites fabricators are likely to use these methods at times. It is expected, nevertheless, that the process technologies discussed will be pertinent most often to those cost-conscious, consumer-related industries that are required to fabricate large numbers of parts and are driven by economic forces to use the most automated and least labor-intensive methods.

P.K. Mallick
Seymour Newman

Contents

I · Introduction to Composite Materials Technology: Mass Production Techniques

Seymour Newman

Contents

Seymour Newman, West Bloomfield, Michigan, USA

1.1 CONSUMPTION OF COMPOSITE MATERIALS BY INDUSTRY

An examination of the usage of composite materials by industry will suggest the industries that have the greatest long-term need for mass production techniques and are, therefore, most likely to pursue the development of methods and materials that can be adapted to their high volume product needs. The reader is referred to Table 1.1, which was released in February 1988 by the Composites Institute of The Society of The Plastics Industry. Total shipment of thermoset and thermoplastic composites for 1988 was projected to be about 1.1×10^9 kg (including the weight of the fillers and reinforcements). Roughly one-fifth of that total or 0.254×10^9 kg consists of reinforced thermoplastics.

The three industries that consume 60 % of the total are land transportation, construction, and marine; land transportation is the single largest user, taking 25 % of the total. Aircraft/aerospace/military may be one of the smallest of the composite industries as measured by quantity of materials used but is perhaps the most sophisticated in terms of its material, design, and testing needs.

TABLE 1.1   Shipment of Thermoplastic and Thermoset Composites by Market

Market	Shipments (kg $\times 10^6$)[a]		Market share 1988 (%)
	1987	1988	
Aircraft/aerospace/military	18	19	1.7
Appliances/business equipment	67	67	5.9
Construction	224	227	20.1
Consumer products	68	73	6.5
Corrosion resistant equipment	141	145	12.8
Electrical/electronic	95	98	8.7
Marine	178	180	15.9
Transportation	286	283	25.0
Other	38	39	3.5
Total	1115	1131	100

[a] Includes reinforcements and fillers.

1.1.1 Land Transportation

The main growth area for composites in land transportation is in automobiles. The potential benefits of polymer composites for automobiles include (*a*) parts consolidation, resulting in lower vehicle and manufacturing costs, (*b*) lower investment costs for plants and tooling, especially for production volumes below 200,000 vehicles, (*c*) corrosion resistance, and (*d*) the ability to modify vehicle styling at acceptable or reduced cost.

Currently, body components such as bumpers and exterior body panels represent the major growth areas. For body panels, reinforced reaction injection molding of thermoset polyurethanes and polyureas, as well as compression and injection molding of styrenated unsaturated polyester compositions (sheet and bulk molding compounds), will continue to expand in usage. Concurrently, injection molding of thermoplastics in the form of polyblends and unreinforced and mineral and/or glass fiber reinforced materials will continue to enlarge its niche. The challenge of body panels lies in conflicting requirements with respect to materials and processes. Thus, the need to generate parts free of warpage and distortion, including surface imperfections, is often at odds with other performance requirements.

The potential for structural applications is already being realized—for example, with the use of essentially unidirectional glass–epoxy composites in leaf springs for some passenger cars and light truck. Nevertheless, the long-term possibilities for a variety of chassis–frame applications and engine components remain to be fully realized. Thus, prototypes of such large chassis components as floor pans and front ends by resin transfer molding have already appeared. There are, of course, major concerns for designers in the substitution of composites for steel in structural applications; these include fracture and failure mechanisms, impact and energy absorption, fatigue, joint design and joining methods, and cost. However, it is widely held that the development of completely reliable and consistent low cost fabrication technology is the key requirement for the widespread use of composites in automobiles and trucks. In short, the benefits of high technology must be obtained at low cost, and this generalization appears to be as true for civilian and military aircraft as it is for automobiles and trucks.

1.1.2 Marine

The nation's recreational boating fleet consists of about 14 million boats and is growing, with nearly 600,000–700,000 new boats sold every year (7). Fiberglass reinforced polyester has been by far the most common construction material, based on E-glass dispensed from chopper guns with orthophthalic polyester in hand layup, open mold systems. Woven roving and chopped strand mat in roll goods form are coming into increasing use for higher performance. Specialty "knitted" reinforcements of bi- and tridirectional fiberglass rovings, S-glass, Kevlar (aramid), honeycomb cores, and vinyl ester resins are also used in high performance racing boats.

It is interesting to note, however, that to achieve higher manufacturing efficiency, some boat builders have begun to resort to faster production techniques such as resin transfer molding using vinyl ester resins and core materials in place. Extremely large molds are required to accommodate part size and injection pressures. Vacuum transfer molding has also begun to appear, using the same mold types but with airtight sealing of mold flanges.

1.1.3 Aircraft/Aerospace/Military

The aircraft/aerospace/military industry utilizes a wide variety of fabricating methods including pultrusion and filament winding, but the bulk of advanced composites processing continues to rely heavily on vacuum bag layup and autoclave cure of multilayer epoxy prepreg laminates. Automated laydown of tapes to form complex laminates has been in use for some time. Growth of composites especially in primary structures is expected to increase in military aircraft to achieve lower vehicle weight. However, since civilian aircraft such as the Boeing 757 and 767 contain only about 4% by weight of advanced composites, the major potential for growth is in commercial aircraft, provided the manufacture of such parts can be achieved at a cost consistent with the resulting increase in fuel economy. As in other industries, there are competing technologies for achieving increased performance and efficiency, and the lowest cost approach will predominate.

The use of heated molds combined with vacuum or pressure bags (eliminating autoclave processing), as well as compression molding with either "wet" resins or pre-pregs, represent important process variations. Beyond these evolutionary changes, there are serious efforts in place to develop more cost-effective fabrication techniques as well as advanced thermoplastic and thermoset materials with capabilities in the range of 100–300 °C and possessing improved damage tolerance.

In this connection, we may mention, for instance, recent efforts at NASA—Langley (8) to synthesize amorphous polyimides modified to improve flow, semicrystalline polyimides, semi-interpenetrating networks, etc. Since thermoplastic prepregs are a manufacturing need, various techniques are under investigation for coating graphite fibers. While some exchange of high volume fabrication technology between this and other industries is likely to occur, it is less likely that the specialized materials in current use or planned for future aircraft/aerospace components will find any large-scale applications in the extremely cost-conscious industries such as automotive and construction.

Some of the development activities mentioned above reflect the growing emphasis on high temperature specialized thermoplastics such as polyetheretherketone, polysulfone, and polyphenylene sulfide. These materials offer the following advantages: (a) the potential for fast cycle and low cost fabrication as by hot stamping, (b) relatively high interlaminar fracture toughness, (c) unlimited shelf life, and (d) freedom from the vagaries of the cure process. The methodology of preparing thermoplastic composites will find a broad audience even if the specialized materials themselves may not.

One high speed fabrication method mostly relevant to thermosets that has considerable potential for a variety of industries including aircraft is resin transfer molding and its variation, structural reaction injection molding, as discussed in Section

1.2.2.3 and elsewhere in this book. Thus, molding trials with this method using high performance materials such as bismaleimides and epoxy-based liquid resins have been reported (9).

1.2 PROCESSING TECHNOLOGY

1.2.1 Introduction

Rapid, economical process technology is probably the single most important factor in the growth of composites. Injection molding, a highly automated technique, is among the oldest and most firmly established composites fabrication methods. Its capability, however, is restricted to short fiber composites, which suffer further fiber degradation during the molding process. Manufacturers, recognizing this limitation, have recently attempted to pultrude "long" fiber thermoplastic molding compounds containing fibers in the range of 10–15 mm (before molding) and to establish machine and molding conditions for maximum fiber length retention. Analogously, in thermoset technology, the breakdown of fibers during the injection molding of thermoset polyesters has led to machine and material changes to minimize fiber degradation during processing as in Billion's "ZMC" process. Poor control of fiber orientation distribution and the difficulty of processing high fiber concentrations place severe limitations on this method for composites fabrication.

For more demanding structural applications, there are two essential requirements: the ability to control the orientation and placement of fibers and the ability to achieve part integration. These features allow for minimum weight designs, uniform and optimized properties, and minimum cost. Compression molding of sheet molding compounds (SMC) approaches these requirements, but several other evolving methods including stamping of fiber-reinforced thermoplastic sheet, resin transfer molding, and its variant, structural reaction injection molding (SRIM), hold considerable promise for the future. For less structural use, in situ polymerization as in reaction injection molding is now also in direct competition with thermoplastic injection molding.

1.2.2 Reactive Processing

Reactive processing, whereby the final shape and molecular structure are achieved simultaneously, is among the oldest polymer processing technologies extant; phenol-formaldehyde resins have been in use since the early part of this century. The scope of reactive processing (RP) has evolved steadily to encompass a broad range of forming methods: compression molding of prepreg composites and sheet and bulk molding compounds, injection molding of bulk molding compounds, pultrusion and filament winding of fibers wet with liquid resins, and various liquid injection molding techniques. The processing of reactive liquids to generate composites has seen a decade of feverish research and development and includes the following methods:

1. *Reinforced reaction injection molding (RRIM):* the extremely rapid impingement mixing of reactive liquid streams, containing reinforcements and fillers, which are

then injected into a closed mold. Polymerization, cross-linking and part formation occur simultaneously.

2. *Resin transfer molding (RTM)*: a method characterized by preplacement of dry reinforcement in the mold before the mold is closed and resin admitted. In most cases, a low viscosity resin must be used and either low pressure or low vacuum employed to assist resin flow and wetout of reinforcement.

3. *Structural reaction injection molding (SRIM)*: a fast resin transfer molding method that exploits the productivity advantages of standard reaction injection molding machines to achieve rapid impingement mixing and injection into the mold chamber containing the preplaced reinforcement.

1.2.2.1 Compression Molding of Sheet Molding Compounds

The compression molding of SMCs, which has seen a remarkable growth since the early 1950s, comprises a wide range of thermosetting prepreg molding materials reinforced generally with short glass fiber strands about 25–50 mm long; however, continuous glass strands may be included with conventional SMC machines or by filament winding as in "XMC" technology. Traditionally, the system has been formulated with unsaturated polyester resins; however, other compositions based on vinyl ester, epoxy, and urethane–polyester hybrids are also available. Molding and simultaneous curing of sheets of materials is generally carried out at about 150 °C with pressures of 7–10 MPa (1000–1500 psi) for periods of 1–3 minutes depending on part thickness. Normally, about 60–70 % of the mold surface is covered by the SMC charge, and filling of the mold cavity is accomplished by subsequent flow of the material. Shaping and consolidation of parts with essentially 100 % mold coverage have also been studied. For a full description of this material, its preparation and properties, see Chapter 2.

While SMC has been extremely successful in a remarkably broad range of applications, as in automobile hoods, the process does possess certain inherent limitations. Compression molding of these compounds involves relatively high molding pressures, which result in large machine costs and place a practical limit on part size. Then too, as normally practiced, SMC requires considerable flow to achieve optimum mechanical and surface properties. This flow, in turn, results in complex and difficult-to-control fiber orientations and concentrations with a considerable distribution of properties at different locations. It is not surprising that a broad range of issues have been under intense investigation in the past few years, including control of composition uniformity, control of flow and of shrinkage, the origin of surface imperfections, the influence of pressure distribution on the curing process, and the rheology of reactive systems etc. (10). With increased knowledge and faster curing chemistries, this composite manufacturing system should continue to grow through the 1990s to compete with the evolving methods discussed below and in subsequent chapters.

1.2.2.2 Reaction Injection Molding

RIM is generally considered to be an energy-efficient process, since it involves the metering of relatively low viscosity reactants and a moderate mold temperature of roughly 65 °C for most exothermic systems. Polyurethane-based polymers have been highly successful and widely used. The final compositions are block polymers of the $(A/B)_n$ type and are generally synthesized from polyether polyols ("soft" segments), glycol chain extenders, a diisocyanate such as diphenylmethane-4,4' diisocyanate (MDI), and a catalyst. The need for higher modulus and reduced thermal expansion has dictated the use of various mineral fillers (e.g., milled glass fibers, mica, flake glass, glass beads, wollastonite, clay). However, the use of fibrous, flake, and particulate reinforcements in RRIM increases the difficulty of pumping the slurries, leading as well to more rapid machine wear and to fiber orientation, with resultant part anisotropy.

The limitation on the use of mineral fillers alone to achieve the desired profile of properties and processing characteristics has led to a number of modifications and/or alternatives:

1. Modified urethane compositions containing a higher proportion of "hard" segments (reaction product of glycol chain extender and diisocyanate).
2. Use of glass flake reinforced polyureas, which generally consist of polyether polyamines, diamine chain extenders, and MDI-based isocyanates. (As a rule, catalysts are not needed, as these are faster reacting than the conventional polyurethanes.)
3. Simultaneous interpenetrating networks from a homogeneous mixture of monomers, linear polymers, and cross-linkers. This method, as discussed by Nguyen and Suh (11), appears to have the potential of providing a wide range of morphologies and synergistic behavior with a controllable degree of covalent bonding and phase separation between the networks.
4. Nonisocyanate systems such as caprolactam (nylon), acrylic, vinyl ester, polydicyclopentadiene, and various polyester–urethane hybrids.

For some typical values of the tensile properties of composites based on polyurethanes, polyureas, and other systems of interest in this chapter, see Table 1.2.

1.2.2.3 Resin Transfer Molding and Structural Reaction Injection Molding

The resin transfer method, with its ability, in principle, to achieve precise control in the placement of fibers in high concentration combined with rapid processability, should provide many of the fundamental requirements to meet the economic needs of a mass production industry. Also, the capability of producing large integrated sections and complex part geometries (including box sections using foam cores or inflatable bags) would appear to make this an ideal composites fabricating method, provided the part production time can be shortened to the time scale of minutes or less and variable costs can be kept at acceptable levels. Various techniques are

TABLE 1.2   Property Comparison of Some Representative Composites at Room Temperature[a]

Matrix	Polypropylene injection molding compound[b]	Polyurea RIM[c]	Structural RIM[d]	Polybutylene terephthalate stampable sheet[e]
Glass type	"long" glass[f]	Flake	Mat	Mat
Glass, %	40	20	54	35
Tensile strength, MPa	110	29	227	103
Flexural modulus, GPa	7.5	1.7	12.8	8.3
Notched Izod strength, KJ/m^2 [g]	20	24	74	68
Heat distortion temperature, °C at 1.8 MPa	156			218

[a] All values should be considered to be approximate; they may not correspond to current company specifications.
[b] ICI's Verton MFX 7008, "long" glass fiber injection molding compound; data from ICI's *Engineering Plastics*, issue 29, 1988.
[c] Mobay's Bayflex 150; data trade literature.
[d] Dow's polyisocyanurate; data from D. Nelson, *Plastics Engineering*, 1987, p. 41.
[e] General Electric's thermoplastic sheet based on polybutylene terephthalate and trademarked Azmet, data from D. Reinhard, SAE Congress, Detroit, February 1988.
[f] See discussion of "long" fiber in Chapter 7.
[g] $5.25 \, kJ/m^2 = 1$ ft-lb/in. of notch.

potentially available to meet this need, not the least of which is combining the technology of resin transfer molding and reaction injection molding as in structural RIM molding mentioned below.

The fiber content achievable with RTM is about 40–50 %; tooling and investment costs are relatively low because of low injection pressures (~ 6 bar): there are no storage problems of unstable B-staged prepregs, and sandwich construction can be used to optimize the efficiency of laminate parts. The further development of preform technology to optimize cycle time, fiber wetout, mechanical properties and surface appearance still represent major development needs. Currently, complex preform shapes are generated either by spraying of fibers and resins onto a perforated mold or by shaping of mats or fabrics, pretreated with thermoplastic binder or unsaturated polyester, with heat and pressure.

Since fiber orientation and concentration are particularly difficult to control at sharp edges, component design, process technology, and part performance requirements must enter into tool design decisions. A wide range of resins are available for RTM including epoxy, modified polyesters, and vinyl ester, as well as reinforcements such as continuous filament random glass mats, unidirectional cloth mats, and woven fabrics. Most fiber constructions are available in glass, aramid, or carbon fibers; but E-glass will probably be most widely used because of its price-versus-performance balance. To meet various application and design objectives, combinations of reinforcements are generally needed.

The development issues in resin transfer molding apply as well to structural reaction injection molding. Additionally, with high reactivity systems, fast fill, complete penetration of the preform, good wetout, elimination of voids, lack of movement of the reinforcement fibers, and avoidance of resin-rich areas present a formidable combination of requirements. Low initial viscosity and a precisely controlled gel time followed by a rapid cure are essential features of these systems. An interesting range of resins with these characteristics have appeared in the marketplace. Dow's polyisocyanurates and polycarbamates as well as Ashland's "acrylamates" (acrylesterol–isocyanate) are examples of such systems. For large parts weighing about 18 kg (40 lb), cycle times as short as 90 seconds have been claimed for the Dow systems (12), based on fill times of less than 20 seconds and cure times of about 45 seconds. Such production rates in a manufacturing environment appear to require the use of "shuttle presses", which accommodate one mold upper and two mold lowers, as well as internal mold release agents, both of which may be considered to be developing technologies at this time.

1.2.3 Thermoplastic Processing

As is evident from the preceding discussion, major strides have been achieved in reactive processing; indeed, the emphasis in this book is on cross-linkable polymers and advances in thermoset processing science and technology. Nevertheless, remarkable progress has also been evident in thermoplastic materials and fabrication methods. Processing based on melt flow and phase changes in chemically stable systems rather than on complex chemical reactions has many attractions. We shall report on only two methods in this chapter: novel injection molding technology and sheet stamping. Considerably more detail on thermoplastics can be found elsewhere.

1.2.3.1 Novel Injection Molding Technology

Our objective here is simply to describe several recent and promising advances in injection molding technology: hollow injection molding, coinjection molding, and injection–compression molding. By injecting a known volume of inert gas into the flow of molten plastic as it enters the mold, it is possible with suitable nozzle design to retain the gas in the middle of the thicker sections of the molding, thereby forming hollow box sections. Hence the term hollow injection molding. Gas pressure must be maintained at about 3.0–22 kPa greater than the melt pressure and may be introduced from a hydraulic cylinder; intimate mixing of gas and melt is avoided. Because suitable ribs are designed into the part, continuous channels transmit pressure without loss to the ends of the tool; sink marks are thereby minimized. Freon or nitrogen may be used; these gases vary in their "chill" effect on the molten polymer. In general, reduced clamp tonnages and lighter plasticating units may be realized because of the "lubricating" effect of the gases and the use of the gas to help complete filling the tool after near-complete packing by the injection unit.

As the name indicates, coinjection molding involves the dual injection of materials that may differ for skin and core. Two injection units are utilized, and the resultant melt streams are delivered through a single nozzle with two channels. The timing sequence of the two streams is critical. A number of advantages accrue from the potential of combining two different materials; thus, a principal opportunity lies in injecting a foamed core and a solid skin. This would appear to provide an opportunity for reduced pressure molding, as well as freedom from sink, reduced internal stresses, close tolerance dimensions, paintability, and other advantages. However, there are a number of constraints: foaming requires a minimum of 4 mm part thickness, and adhesion between skin and core is necessary to avoid delamination. Other concerns include relative shrinkage and coefficient of thermal expansion, which can cause distortion and warpage (particularly when either skin or core is reinforced and the other is not). Furthermore, skin viscosity should generally be lower than interior viscosity to maintain stable flow patterns and to avoid the tendency for core material to penetrate the skin.

Injection–compression molding refers to a two-step process that involves (a) injection molding "molten" thermoplastic plastic into a slightly open mold using specially programmed injection molding equipment, and (b) closing the tool and filling out the part by squeeze flow as in conventional compression molding. Potential benefits include reduced injection and clamp pressure and less polymer and/or filler orientation. The cost effects of more complex machine controls, possibly increased cycle time, and special tooling and tool edge requirements remain to be fully evaluated. While successful for small parts, the method remains to be exploited for large components.

1.2.3.2 Stamping

Thermoplastic sheet is becoming increasingly of interest in the fabrication of large semistructural parts, thereby providing considerable competition to SMC and structural RIM, particularly in applications that do not require metal-like appearance. Hot stamping of thermoplastic sheet is the principal fabricating method, but thermoforming and cold stamping have also been used to form parts.

An increasing range of thermoplastic matrices with a spectrum of use temperatures is available, including polypropylene, polyethylene terephthalate, polybutylene terephthalate, nylon 6,6, polyphenylene sulfide (PPS), and a variety of polymer blends. For aircraft use, in addition to PPS, a number of other specialized matrices such as polyetheretherketone, polyamide–imide (Torlon), and polyetherimide (Ultem) are under consideration. Continuous glass strand mat is widely used as the reinforcement, but sheet material with some portion of the reinforcement as continuous unidirectional fiber has become available; additionally some sheet is available from du Pont with aligned discontinuous fiber. In addition to glass fibers, cellulose, aramid, and carbon fibers have been used, the last mentioned especially in such higher temperature organic matrices as PPS.

A number of methods are available for sheet production; two of the most common include continuous lamination and various versions of paper-making technology that involve sheet formation from slurries of fiber and resin in water. Azdel, a polypropylene composite developed originally by PPG available with continuous strand mat as well as with unidirectional fiber is representative of continuous lamination technology. Endless steel belt presses are used to consolidate (13) continuously the following five-layered sandwich:

> Polypropylene film (outside skin)
> Continuous glass strand mat
> —Extruded polypropylene melt layer—
> Continuous glass strand mat
> Polypropylene film (outside skin)

This process can produce relatively thick sheets, and a variety of reinforcement types can be combined readily.

Some of the main advantages of thermoplastic sheet lie in the relatively short cycle times of about 15–60 seconds in presses (after preheating) to flow the material and cool it down; the potential for postformability; the enhanced impact properties normally encountered with longer length/diameter fiber reinforcements; and the wide range of matrices and fibers feasible.

Like all method–material combinations, sheet stamping entails disadvantages as well. Like SMC, high tonnage, costly presses are required to achieve molding pressures of 7–20 MPa, depending on resin type, reinforcement content, and flow distance. Then too, various design details as attaching devices must often be added in secondary operations; and surface finish for demanding appearance applications is generally not currently feasible, partly because of the large difference in shrinkage between reinforcement and matrix on cool-down from molding. To take full advantage of the short press time, automated methods for locating a charge in and removing parts from the press would be desirable. Overall, sheet stamping provides the designer with an increasingly versatile composites manufacturing method with its own set of unique characteristics.

1.3 MATERIAL SUBSTITUTION WITH COMPOSITES

Because of the unique characteristics of composites and the many new technical issues they pose, the full potential of composite materials to replace such traditional materials as steel, aluminum, wood and concrete remains to be fully realized. Successful applications rest on a detailed understanding of the advantages and disadvantages of composite materials as well as on a complex interplay among the design, performance, economic, and manufacturing issues that must be resolved in each instance. We next take some examples from the automotive industry, which has succeeded in making substantial use of short fiber composites but has as yet made little use of these and/or long fiber composites for heavily loaded (primary) vehicle applications.

1.3.1 Some Processing and Materials Issues

A glance at Table 1.3 quickly discloses that while composites are of lower cost than steel on a cost/volume basis, if we take modulus into account and compare materials on a modulus/cost/volume scale, steel has a considerable advantage. Consequently, in stiffness-critical applications, such as an automobile hood or other exterior body panels, a comparison of steel and various composite panels of equivalent stiffness shows that steel possesses a lower *material* cost for a given surface area. The industry has long recognized, therefore, the need to offset the material cost disadvantage by designing parts that integrate as many functions and subcomponents as possible in one forming operation. This reduces the tooling and labor costs associated with the secondary operations that otherwise would be needed. Fabricating methods that lend themselves, therefore, to this consolidation of parts and function should be favored. Then too, since tooling costs for composites are generally lower than steel stamping dies, a substantial saving in tooling costs can be realized when car volumes are less than about 200,000 vehicles per year. Regarding strength, composite materials usually have good in-plane tension and compression properties. In-plane shear and transverse shear properties, however, are relatively low, and applications that involve particularly high local shear loadings should be avoided or carefully analyzed and met by appropriate design and material measures.

TABLE 1.3   Some Material Cost Comparisons at Equivalent Stiffness[a]

Material	Glass (%)	Modulus (GPa)	Density (g/cm^3)	Approx. cost ($/kg)	Relative thickness[b]	Relative cost[c]
HSLA steel		207	7.8	0.66	1.0	1.0
SMC R50[d]	50	15.2	1.8	1.65	2.38	1.4
Polycarbonate, glass filled	40	9.7	1.5	1.65	2.78	3.3
Polypropylene, random glass mat[e]	40	5.5	1.3	2.76	3.35	2.1

[a] In bending, the stiffness is defined as EI, the product of modulus and second moment of area. For a simple rectangular beam, I is proportional to the thickness, t^3.
[b] Relative thickness for equivalent stiffness.
[c] Relative material cost per unit area at equivalent stiffness.
[d] Polyester sheet molding compound with 50% random, chopped glass.
[e] Azdel sheet.

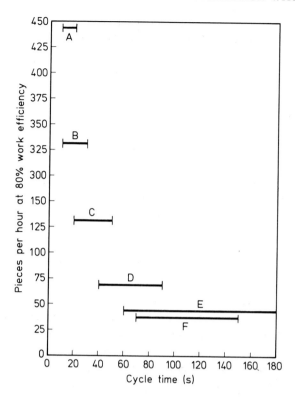

FIGURE 1.1   Productivity at 80% work efficiency versus cycle time for various fabricating processes: A, steel stamping; B, aluminum stamping; C, thermoplastic sheet compression molding; D, injection molding; E, thermoset compression molding; F, reaction injection molding of polyurea. Note that values depicted must be considered to be approximate and to depend on a variety of factors (e.g, part size, temperature, formulation).

It is of interest to note that during the 1970s a considerable effort was mounted in the automotive industry to achieve improved fuel efficiency by reducing vehicle weight through simple part-for-part substitution with composite materials. Major structural components (wheels, frame cross members, steering wheels, truck leaf springs, etc.) were identified for replacement. These programs generally were not commercialized, however, partly because of the cost penalty involved and the cost competitiveness of other fuel efficiency measures (aerodynamic design and improved driveline efficiency) at that time. Nevertheless, with improved materials, greater fabricating efficiency, and shorter cycle times, many of these and other applications will likely be realized in the next decade.

The development of rapid, completely reliable, and consistent manufacturing methods is often the major hurdle to overcome in the goal of exploiting composite materials. Figure 1.1 compares the productivity of some composites manufacturing methods with steel stamping. [This figure should be viewed with two reservations:

21

(*a*) the data depend on a number of variables and should be taken as being approximate only; and (*b*) since productivity improvements are an ongoing process, any comparison may be in error by the time it appears in print.] Productivity relates directly, of course, to the number of machines required to produce a given number of parts in a specified time span and, therefore, bears on investment costs as well as on manufacturing costs per part.

Unidirectional fiberglass composites possess a high elastic strain and about 10 times the specific strength of steel. A flexural application such as a leaf spring allows a design with the same curvature but in much thicker sections. The result is that stiff springs can be built with fewer leaves, which saves weight (and material cost) dramatically beyond simple one-for-one part replacements. Not surprisingly then, transverse leaf springs have been in use on a few sporty passenger cars for nearly 10 years. Longitudinal leaf springs for light trucks present more complex design and material issues, since they involve lateral loads and fatigue under complex stresses. These technical issues have been resolved; nevertheless, the development of foolproof, low cost fabricating methods has already spanned a decade of time or longer without yet accomplishing the objective of replacing steel springs to any substantial degree. This is illustrative of the general problem encountered—that is, the long, costly development programs that are generally required to completely validate a manufacturing process. It is not unreasonable to expect that cooperative industry programs resulting in shared technology and development costs will be required to overcome this obstacle in the future for difficult composite applications.

Other material and manufacturing issues of concern in material substitution programs include fatigue life; reliable prediction of long-term performance under adverse environmental conditions; impact damage; energy absorption; joining and adhesive bonding methodology; low cost repair procedures; rapid, nondestructive defect analysis; and education and training of design and manufacturing personnel. Each industry has its own set of development needs and priorities. Undisclosed impact damage is of particular concern to the aircraft industry, for instance. This poses a formidable material problem because of the nonlinear relation (14) of interlaminar fracture energy on matrix ductility (Fig. 1.2).

1.3.2 Role of Computers

The computer has dramatically altered the design process by providing a more rapid execution of known principles. Hardware and software are available for the following tasks: initial concepts and design illustrations; structural analysis by finite element analysis models; heat transfer and thermal analysis by similar techniques; cost, weight, and preliminary performance analysis; injection molding evaluation with mold fill and shrinkage analysis programs designed to provide information on fiber orientation, weldline location, and out-of-mold shrinkage; and tool cutting. In addition to these general programs, numerous special softwares are being developed for special tasks (e.g., fatigue and crack propagation computations).

Carried out in conjunction with prototype testing and fabrication programs, these techniques should dramatically shorten the time required to have a test-worthy,

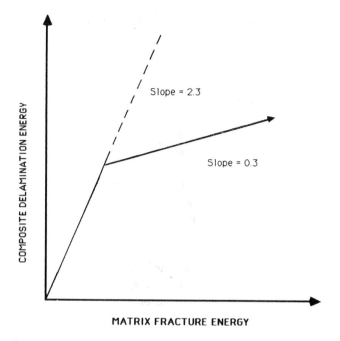

FIGURE 1.2   Schematic drawing showing the effect of matrix toughness on interlaminar fracture energy of advanced composites. (Redrawn from Fig. 7 of Ref. 14.)

manufacturable design of a particular component. In the short term, development costs and training requirements will present an obstacle to the full exploitation of these methods; over the long term, they can be expected to dramatically reduce the costly trial-and-error methods now used in the design and manufacture of composite applications.

REFERENCES

1. B. Agarwal and L. J. Broutman, *Analysis and Performance of Fiber Composites*, Wiley, New York, 1980.
2. D. Hull, *An Introduction to Composite Materials*, Cambridge University Press, London, 1981.
3. J. E. Ashton, J. C. Halpin, and P. H. Petit, *Primer on Composite Materials: Analysis*, Technomic, Westport, CT, 1969.
4. P. K. Mallick, *Fiber Reinforced Composites*, Dekker, New York, 1988.
5. L. Nielsen, *Mechanical Properties of Polymers and Composites*, Dekker, New York, 1974.
6. M. J. Folkes, *Short Fiber Reinforced Thermoplastics*, Research Studies Press, a division of John Wiley and Sons, Chichester, 1982.
7. R. Scofield, *Proceedings of the Third Annual Industry Conference of Suppliers of Advanced Composite Materials Association*, Arlington, VA, 1988.
8. T. St. Clair, N. J. Johnston, and R. M. Baucom, in "Polymer Composites for Automotive Applications," Society of Automotive Engineers, SP-748, Detroit, February 1988.

9. R. V. Wilder, *Mod Plas* April 1988; p. 54; see also L. K. English, *Mater. Eng.*, June 1987; p. 37.
10. S. Newman and D. G. Fesko, *Polym. Compos*, 5(1), 88 (1984).
11. L. T. Nguyen and N. P. Suh, *Polym. Eng Sci.*, 26(12), 781, 799, 843 (1986).
12. G. Ellerbe, In *Automotive Plastics, RETEC*, Society of Plastic Engineers, Dearborn, MI; November 1987, p. 348.
13. D. Reinhard, in "Polymer Composites for Automotive Applications," Society of Automotive Engineers, SP-748, Detroit, February 1988.
14. T. Vu-Khanh, *Poly. Comp*, 8(6), 363 (1987).

2 · Sheet Molding Compounds

P. K. Mallick

Contents

P. K. Mallick, Department of Mechanical Engineering, University of Michigan–Dearborn, Dearborn, Michigan, USA

2.1 INTRODUCTION

Sheet molding compound (SMC) is a continuous sheet of ready-to-mold composite material containing fibers and mineral fillers dispersed in a thermosetting resin. The resin is in an uncured, but highly viscous (thickened) state. It is cured (cross-linked), and transformed into a finished product by the compression molding operation.

Sheet molding compounds can be used to produce many structural composite parts of complex shapes and designs in a relatively short molding time. Following are the principal advantages of an SMC composite.

1. *Variety*: Fiber content and type in an SMC can be easily controlled to produce a wide variety of mechanical and physical properties in the finished part.

2. *Part consolidation*: SMC parts can be compression molded with ribs, bosses, curvatures, holes, and inserts. Using these design features, a number of components can be consolidated into one part and a number of secondary operations or assembly steps can be eliminated.

3. *Light weight*: SMC composites have a lower specific gravity than structural metals. As a result, SMC parts are considerably lighter than metal parts on an equal volume basis. The strength-to-weight ratio of SMC composites is comparable to that of many structural metals (Table 2.1).

4. *Dimensional stability*: Parts molded with SMC exhibit better dimensional stability than thermoplastic composites over a wide range of temperatures and other environmental conditions. The volume change on cure can also be controlled by the addition of various low shrink additives. Furthermore, the coefficient of thermal expansion of SMC composites can be controlled to match that of steel or aluminum, thus making SMC parts compatible with steel or aluminum parts.

Historically, the development of SMC started in the early 1950s after the finding that the viscosity of unsaturated polyester resins increases with the addition of Group IIA metallic oxides, hydroxides, or carbonates (1). The early applications of SMC materials were in the electrical and industrial goods (light fittings, control boxes, electrical fixtures, machine guards, tool boxes, etc.). Over the next two decades, SMC technology grew with the development of improved equipment, low profile additives, new catalyst systems, etc. In the early 1970s, the automotive industry used the SMC material for producing exterior body components, such as grille opening panels and, later, hoods. In the mid-1970s, high strength SMCs were introduced. Many structural automotive components, such as bumper beams, road wheels, crossmembers, and tailgates, were prototyped with these high strength SMC materials. Some of these components are now production parts in the automotive industry. Moreover, a wider acceptance of SMC in exterior body panels has also occurred.

The transportation industry utilizes the largest volume of SMC composites. Other applications of SMC composites are found in appliances (washing machine doors, refrigerator housings, etc.), furniture (chairs, table tops), business machines (computer housings), and construction (door panels).

TABLE 2.1   Properties of Various Structural Materials

Property	Steel SAE 1010	Aluminum alloy 2014-T6	SMC-R25[a]	SMC-R65[b]	Glass fiber reinforced nylon 6,6[c]
Specific gravity	7.87	2.7	1.83	1.82	1.40
Tensile strength, MPa	365	483	82.4	227	186
Tensile modulus, GPa	200	70	13.2	14.8	10.5
Tensile strength-to-weight ratio, 10^3 m	4.73	18.23	4.59	12.71	13.54
Tensile modulus-to-weight ratio, 10^6 m	2.59	2.64	0.73	0.83	0.76
Coefficient of thermal expansion $(10^{-6}/°C)$	11.7	23	23.2	13.7	23

[a] Contains 25 wt % E-glass fibers.
[b] Contains 65 wt % E-glass fibers.
[c] Contains 33 wt % E-glass fibers.

Manufacturing of sheet molding compound composites can be divided into three major stages, namely, compounding, maturation, and compression molding. In the compounding stage, the SMC sheet is produced by embedding fibers between two thin layers of a resin paste. The latter contains a number of ingredients, including a resin, catalyst, filler, and thickener. In the maturation stage, the resin paste viscosity is allowed to increase to high values so that the SMC sheet can be handled properly in preparation for the molding stage. Depending on the composition of the resin paste, the maturation stage can last anywhere from a few days to a few months. In the compression molding stage, SMC sheets are cut, stacked, and placed in a heated mold. With the application of heat and pressure, the sheet molding compound first flows and fills the mold cavity and then cures into a solid part. In the curing reaction that takes place in the mold., the resin molecules are cross-linked to form a three-dimensional network structure.

In the first part of this chapter, we discuss the composition of sheet molding compounds and the production of SMC sheets. The compression molding operation for manufacturing SMC parts is discussed in Chapter 3. In the second part of this chapter, we present the key mechanical properties of SMC composites which are obtained in laboratory tests and can be used at the design stages of SMC parts.

2.2 COMPOSITION

The major components in a sheet molding compound are resin, fibers, and fillers. Other components, such as catalyst, thickener, inhibitor, mold release agent, and low profile additives, are used in small quantities. However, they have important functions during the maturation and molding stages. The functions of each of these components are described in this section. A typical composition of SMC is shown in Table 2.2.

TABLE 2.2   Typical Composition of a Sheet Molding Compound

Material	Amount (wt %)
Resin paste	
Unsaturated polyester	10.50
Styrene monomer	13.40
Low profile additive	3.45
Filler	40.70
Thickener (MgO)	0.70
Catalyst	0.25
Mold release agent	1.00
Inhibitor	Trace amount
Fiber: E-glass (25 mm long)	30.00
Total	100.00

2.2.1 Resin

Principal resins used in commercial sheet molding compounds are unsaturated polyesters and vinyl esters. Epoxies are also used; however, the molding time for epoxy-based compounds is generally much longer than for compounds based on polyester or vinyl ester. Recently, phenolics have also been used in producing sheet molding compounds (2). The phenolic-based SMC has lower flammability, reduced smoke generation, and higher thermal stability than conventional polyester-based SMC.

Polyesters are obtained by the reaction of an unsaturated difunctional anhydride or acid, such as maleic anhydride (MA) or fumaric acid (FA), and an alcohol, such as propylene glycol (PG) or diethylene glycol (DG). The resulting polyester molecule contains a number of carbon–carbon double bonds (Fig. 2.1), which act as cross-link-

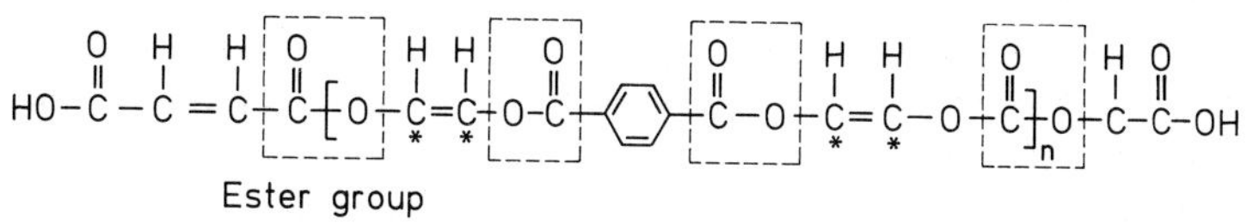

FIGURE 2.1   Schematic representation of an unsaturated polyester molecule; unsaturation sites in the molecule are indicated by asterisks.

29

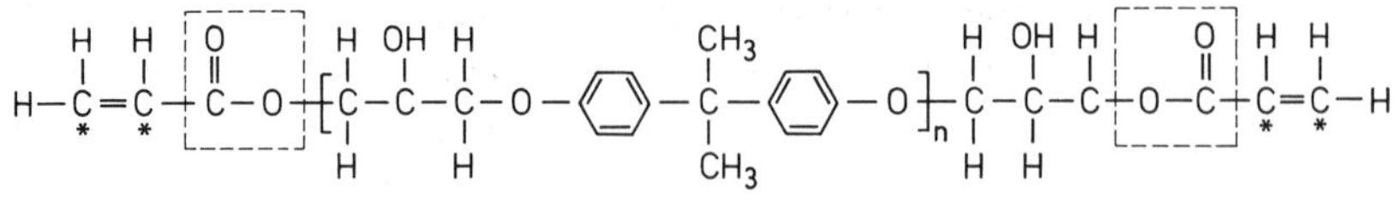

FIGURE 2.2   Schematic representation of a vinyl ester molecule; the unsaturation sites in the molecule are indicated by asterisks.

ing sites at the time of curing. Saturated acids, such as orthophthalic or isophthalic acid, are also added in various proportions to control the reactivity and mechanical properties of the resin (3).

Vinyl esters are produced by the reaction of an unsaturated carboxylic acid, such as methacrylic or acrylic acid, and an epoxy resin. Unlike polyester molecules, the carbon–carbon double bonds in vinyl ester molecules occur only at their ends (Fig. 2.2). Thus, in vinyl esters, cross-linking can take place only at the terminal double bonds. Because there are fewer cross-links, a cured vinyl ester matrix tends to be more flexible and resistant to microcrack formation than a cured polyester matrix (4).

An ingredient essential for the cross-linking of either polyester or vinyl ester resin is a reactive monomer, such as styrene or vinyl toluene, which also contains carbon–carbon double bonds in its molecules. This reactive monomer has two functions: namely, it serves as a diluent for the unreacted resin and it later coreacts with the resin, forming cross-links between the resin molecules.

Styrene is the most common monomer used in the SMC industry. Its advantages are low cost, very low viscosity, and high solvency; however, emission of styrene vapor into the shop environment is considered to be a health hazard. In the United States, government regulations currently permit only 100 ppm of styrene vapor in the shop air; in the future, this may be reduced to 50 ppm. Because of this health concern, new polyester chemistry is being developed that utilizes less styrene monomer for curing (5).

The resin is usually supplied to an SMC compounder as a mixture of styrene and polyester (or vinyl ester) with a typical styrene content of 30–50% by weight. Depending on the resin chemistry and styrene content, the room temperature viscosity of the mixture ranges from 300 to 3000 cP. High styrene content favors a better wetout of the fiber surface with the resin, since the viscosity of the mixture is lowered. However, excess styrene content tends to lower the mechanical and thermal properties of the cured resin.

One recent development in the resin technology is a hybrid of unsaturated polyester and urethane (6). In this hybrid, a low molecular weight unsaturated polyester is reacted with a diisocyanate in the presence of a styrene monomer. This reaction leads to a high molecular weight linear polymer that remains soluble in styrene. The linear polymer can be cured with a conventional catalyst. The principal advantage of the hybrid resin is its versatility, since cure time, viscosity, and mechanical properties can be tailored over a wide range by controlling both polyester and isocyanate chemistry. Another advantage of the hybrid system is its low styrene content (as low as 25 wt %).

Another recent development in the SMC resin technology (7) is an acrylesterol resin that is reacted with a polyisocyanate, such as 4,4'-diphenylmethane diisocyanate, at room temperature to form urethane linkages in its backbone. The resulting acrylamate prepolymer contains a number of unsaturation sites in its molecule. It can be transformed into a cross-linked network structure via a free radical cure mechanism.

Both acrylesterol resin and polyisocyanate have low viscosities at room temperature. When they are combined, the viscosity of the mixture builds up relatively slowly through the urethane reaction. If fibers and fillers are added during this period, good surface wetout results. However, as with conventional SMC, the viscosity levels off at 60×10^6 to 70×10^6 cP within 7 days after mixing, and a tackfree SMC sheet is obtained. Since polyisocyanate has a low viscosity (in the range of 35 cP), it serves as a diluent for the base resin and therefore, the addition of volatile monomers, such as styrene, is completely avoided.

2.2.2 Fibers

The most commonly used fiber in sheet molding compounds is E-glass. Other fibers, such as S-2 glass, carbon, and Kevlar 49, are used in limited quantities. Among these fibers, E-glass has the lowest modulus of elasticity and a relatively high specific gravity (Table 2.3). However, it costs much less than other reinforcing fibers. The wide acceptance of E-glass fibers in the SMC industry is due to this cost advantage.

Both short (discontinuous) and long (continuous) forms of fibers are used in sheet molding compounds. Short fibers (with random orientation in the plane of the sheet) are commonly used in applications requiring isotropic properties (i.e., equal properties in all directions). Long fibers, oriented unidirectionally in the sheet, provide very high strength and modulus in the fiber direction; however, the strength and modulus in the transverse direction are relatively low. To improve transverse strength and modulus, short fibers in random orientation are combined with unidirectionally oriented long fibers.

TABLE 2.3   Fibers Used in Sheet Molding Compounds

Fiber	Specific gravity	Tensile strength (GPa)	Tensile modulus (GPa)	Tensile failure strain (%)	Coefficient of thermal expansion ($10^{-6}/°C$)
E-glass	2.54	3.45	72.4	4.8	5
S2-glass	2.48	4.30	86.9	5.0	2.9
Carbon (graphite)	1.76–2.15	1.5–5.6	220–690	0.3–1.2	−0.1 to −1.2 (longitudinal) 7–12 (radial)
Kevlar 49	1.45	3.62	131	2.8	−2 (longitudinal) 59 (radial)

The conventional SMC material uses continuous fiber roving, which is a collection of 25–40 untwisted fiber strands. Each fiber strand typically contains 200–400 E-glass filaments. Short fibers are obtained by chopping (cutting) the rovings into desired lengths at the time of the SMC sheet production. The frequently used chopped fiber length is 25.4 mm (1 in.), although chopped fibers with lengths up to 50.8 mm (2 in.) are also used.

At the time of producing glass fiber strands, each filament surface is coated with a specially formulated mixture of chemicals. This chemical treatment, commonly referred to as sizing, has several important functions: it reduces the abrasion and static charge buildup between filaments, improves binding between filaments in the strand, and promotes a better coupling between the fibers and the resin matrix in SMC. The type of sizing used influences the roving characteristics that are important at the compounding stage as well as at the molding stage. For example, if a "hard" or insoluble sizing is used, the strands in a roving do not separate easily when it is chopped or when it is exposed to the styrene monomer in the resin.

Owing to this strand integrity, chopped fibers remain stiff during the molding process, giving the SMC good flow properties in the mold. However, the coupling between the fibers and the resin is generally poor, which affects the mechanical properties of the SMC composite. With a "soft" or soluble sizing, the roving tends to separate into individual strands. Filament-to-filament bonding in each strand may also degrade when the strands are exposed to styrene monomer. As a result, there is a better dispersion of filaments in the resin. However, the dispersed filaments tend to restrict the resin flow during molding, which may contribute to incomplete mold filling, air entrapment, and surface waviness.

The fiber content in SMC can be varied easily to match the application requirements. High fiber content, in the range of 50–70 % by weight, is used for structural applications requiring high strength and modulus. Fillers are not added in sheet molding compounds containing more than 60 wt % fibers. In general, high viscosity of these sheet molding compounds results in reduced flow in the mold and, therefore, poor moldability.

Sheet molding compounds are designated according to the form of fibers used in them (Fig. 2.3). These designations are:

1. "SMC-R", in which R represents randomly oriented short fibers in the sheet.
2. "SMC-CR", in which C represents continuous parallel fibers on one side of the sheet and R represents randomly oriented short fibers on the other side of the same sheet.
3. "XMC", in which X represents criss-crossed continuous fibers in the sheet.

In both SMC-R and SMC-CR, the fiber content is shown by the weight percent at the end of each letter designation. For example, SMC-R40 contains 40 wt % of randomly oriented short fibers; SMC-C30R10 contains 30 wt % of continuous parallel fibers and 10 wt % of randomly oriented short fibers. The fiber content in XMC is in the range of 70 wt %, with a majority of the fibers in the continuous form. It may also contain small amounts of randomly oriented short fibers for improving

 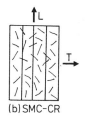 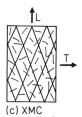

(a) SMC-R   (b) SMC-CR   (c) XMC

FIGURE 2.3   Various forms of sheet molding compounds: (a) SMC-R, (b) SMC-CR, and (c) XMC; L and T refer to longitudinal and transverse directions, respectively.

the transverse properties. The angle between the criss-crossed fibers in XMC is usually in the range of 5–7°.

2.2.3 Fillers

Fillers have several functions in a sheet molding compound. They reduce the volumetric shrinkage of the resin, improve the moldability by promoting better flow, and enhance the surface quality of the molded part. They also reduce the material cost, since they are generally much less expensive than the resin they replace in the resin paste.

Calcium carbonate ($CaCO_3$) is the most commonly used filler in sheet molding compounds. Kaolin clay, talc, glass spheres (both solid and hollow), and alumina trihydrate ($Al_2O_3 \cdot 3H_2O$) are also used as fillers. Among these fillers, hollow glass spheres have the lowest specific gravity (Table 2.4). They are useful in reducing the weight of the sheet molding compound; however, they tend to crush at the high pressures used during molding operations. Alumina trihydrate acts as a filler as well as a fire retardant, since the water of hydration in its molecule is released at elevated temperatures (ca. 220 °C). It is used in many electrical, appliance, and construction applications, where flammability and smoke generation are critical material selection criteria.

The filler-to-resin ratio in a sheet molding compound influences the resin paste viscosity, which controls the fiber surface wetout and resin flow in the mold. In

TABLE 2.4   Mineral Fillers Used in Sheet Molding Compounds

Filler	Specific gravity	Average particle size (μm)
$CaCO_3$	2.71–2.93	3
Kaolin clay	2.60	4
$CaSO_4$	2.96	4
Alumina trihydrate	2.42	9
Solid glass sphere	2.50	10
Hollow glass sphere	0.40	20
Talc	2.70	2–20

33

general, the higher the fiber content, the lower the filler-to-resin ratio. For example, typical filler-to-resin ratios for SMC-R30, SMC-R50, and SMC-R65 are 1.5, 0.5, and 0, respectively.

Electrically conductive fillers are sometimes used in glass fiber reinforced sheet molding compounds to reduce the accumulation of static charge, which may otherwise interfere with electromagnetic or radio frequencies. Examples of these fillers are carbon particles (carbon black), carbon fibers, short aluminum or nickel fibers, powders or flakes, and aluminized chopped glass fibers. These fillers are added in small quantities in conjuntion with the primary filler, such as $CaCO_3$ to enhance both electrical and thermal conductivities of SMC composites that are based on glass fiber materials.

2.2.4 Catalyst

The function of a catalyst is to initiate the curing reaction at elevated temperatures. Common catalysts are organic peroxides, such as t-butyl perbenzoate (TBPB), which decomposes readily at the mold temperature to form free radicals. These free radicals react with styrene and polyester molecules to break up the carbon–carbon double bonds, thus initiating the curing reaction.

The rate at which a catalyst decomposes into free radicals increases with increasing ambient temperature. The rate of catalyst decomposition is measured by its half-life $t_{1/2}$, which is the time required for the catalyst concentration to reduce to half its initial value. For convenience, the catalyst reactivity is listed according to the temperature at which the half-life is either 1 hour or 10 hours. The 1-hour half-life temperature is considered to be the minimum practical operating temperature for the catalyst. The 10-hour half-life temperature is frequently used in comparing the reactivities of various catalysts.

Table 2.5 lists the 10-hour half-life data for various commercial catalysts used in sheet molding compounds. Among these, t-butyl perbenzoate and peroxyketals, such as 1,1-di(t-butylperoxy)-3,3,5-trimethyl cyclohexane and 1,1-di(t-butylperoxy)cyclohexane, are the most commonly used catalysts in sheet molding compounds. Both are in the category of organic peroxides ($ROOR''$) and are used at mold temperatures of 138–160 °C. Peroxyketals are slightly more reactive than TBPB, but in an SMC formulation, both provide room temperature shelf life of up to 6 months. More reactive peroxides, such as t-butyl peroctoate (TBPO), are used if shorter molding cycles or lower molding temperatures are desired. However, with such catalysts, the shelf life of the SMC sheet is greatly reduced.

Another type of catalysts used for SMC is in the category of azo compounds ($R—M=N—R'$). They are generally less hazardous than peroxides of comparable reactivity. Since their decomposition rates are much less sensitive to contaminants, such as acids, bases, and transition metals, they tend to have a higher pot life than peroxide initiators.

The cure characteristics of a resin–catalyst combination are frequently determined by the SPI gel time test. In this test, 10 g of a resin–catalyst mixture is poured into a standard sized test tube. The test tube is suspended in a 82 °C water bath, and the

TABLE 2.5 Half-Life Data for Various Catalysts Used in Sheet Molding Compounds (8)

Catalyst	Half-life at 32 °C (days)	Temperature (°C) at 10-hour half-life	Mold cure time at 149 °C (min)[a]	Pot life (days)[b]	Maximum storage temperature (°C)
t-Butyl perbenzoate	24,068	105	0.80	8	38
t-Butyl peroctoate	159	72		2–3	18
1,1-di-(t-Butylperoxy) cyclohexane[c]		93	0.62		32
2-t-Butyl azo-2-cyano-4-methyl pentane		70			
1-Cyano-(1-t-butylazo) cyclohexane	10,954	96	0.73	71	29
2-t-Butylazo-2-cyanopropane	600	79	0.49	43	27

[a] Based on 1 wt % commercial resin formulation.
[b] Resin formulation at molar equivalent of 1 wt % t-butyl perbenzoate.
[c] 80 wt % solution in butylbenzyl phthalate.

temperature rise in the mixture is monitored by means of a thermocouple inserted into the test tube. A typical temperature–time curve (referred to as an exotherm) obtained in a gel time test (Fig. 2.4) indicates the time at which the curing reaction initiates and the mixture begins to transform into a gel-like mass. The heat generated by the curing reaction increases the temperature of the mixture which, in turn, accelerates the catalyst decomposition. As a result, curing proceeds at a progressively increasing rate and a rapid temperature rise follows, since the heat generated by the curing reaction does not dissipate effectively into the surrounding water bath. The SPI gel time curve shows an exothermic peak temperature, which indicates that the curing reaction is completed. The slope of the exotherm is a measure of the cure rate, which depends primarily on the catalyst reactivity. The time at which the temperature rise is 5.5 °C above the bath temperature is considered to be the gel time.

One way of reducing the cure time of a resin is to mix a small percentage (up to 20 mol %) of low temperature catalyst, such as TBPO, with a high temperature catalyst, such as TBPB. A dual-catalyst system can cause a significant reduction in both gel time and cure time, often with a slight decrease in pot life (Table 2.6).

2.2.5 Inhibitor

Inhibitors are added in trace amounts to an SMC resin formulation to prevent or reduce any curing reaction that may occur during mixing, maturation, and storage. This, in turn, improves the shelf life of SMC rolls prior to molding. When the SMC sheet is placed in a heated mold, the curing reaction is first slowed by the inhibitor; but its effect is quickly overcome as a result of rapid generation of free radicals from

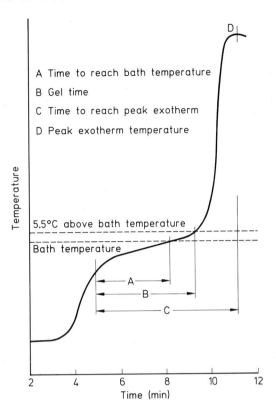

A Time to reach bath temperature
B Gel time
C Time to reach peak exotherm
D Peak exotherm temperature

FIGURE 2.4   Typical temperature–time curve obtained in an SPI gel time test.

TABLE 2.6   SPI Exotherm Data for Curing Polyesters at 121 °C (9)

Catalyst		Amount of catalyst 2 (wt %)	SPI exotherm data			Pot life at 32 °C (days)
1[a]	2[b]		Gel time (min)	Cure time (min)	Peak temperature (°C)	
TBPB		0	4.2	5.60	217	40
TBPB	TBPO	17	2.3	3.13	209	12
Peroxyketal		0	2.71	3.71	209	81
Peroxyketal	TBPO	10	2.20	3.07	204	40
Peroxyketal	Azo compound	10	1.85	2.79	199	56

[a] Peroxyketal = 1,1-di(t-butylperoxy)cyclohexane.
[b] Azo compound = 2-t-butyl azo-2-cyano-4-methylpentane.

the catalyst. As the inhibitor concentration decreases, the reaction rate accelerates in the mold.

Hydroquinone and parabenzoquinone (PBQ) are two common inhibitors used in SMC formulations. Since inhibitors reduce the cure rate and the corresponding increase in viscosity at the early stages of molding, the flow of SMC in the mold can be controlled by selecting a proper combination of catalyst and inhibitors. Minor flow problems, such as short shots or voids, can often be corrected by adjusting the inhibitor concentration in the resin paste.

2.2.6 Internal Mold Release Agent

The function of a mold release agent is to prevent adhesion between the resin and the mold surface, thus facilitating part ejection from the mold. Internal mold release agents, blended in the resin paste, are generally effective. However, smoother part ejection from the mold may require periodic spraying of the mold surface with a mold release agent.

Mold release agents are selected on the basis of their melting points, which must be lower than the mold temperature. Commonly used mold release agents are zinc stearate (mp 133 °C) and calcium stearate (mp 150 °C). They can be used at mold temperatures up to 155 and 165 °C, respectively. Both are in powder form at room temperature but dissolve easily in the resin formulation. They are added in small concentrations, usually less than 2 % by weight of the total compound. Excessive amounts of internal mold release agents can reduce the tensile strength of the molded composite.

2.2.7 Thickener

The role of a thickener is to increase the viscosity of the resin so that the SMC sheet can be handled, cut, stacked, and draped on the mold surface. In the thickened state, the resin is uncured, but its viscosity is in the range of 30×10^6 to 100×10^6 cP. In this condition, the SMC sheet is dry and feels nontacky, but it is still pliable.

Common thickeners used for polyester and vinyl ester SMCs are alkaline Group IIA metal oxides and hydroxides, such as MgO, CaO, and $Mg(OH)_2$. Addition of these oxides or hydroxides sets off a two-step reaction with the resin (10). An example is shown below.

Step 1:

$$HOOC — COOH \; + \; \underset{\text{Magnesium oxide}}{MgO} \; \rightarrow \; \underset{\text{Basic salt}}{HOMgOOC — COOH}$$
$$\underset{\text{Resin}}{}$$

Step 2:

$$\underset{\text{Resin}}{HOOC — COOH} \; + \; \underset{\text{Basic salt}}{HOMgOOC — COOH}$$

$$\rightarrow \; \underset{\text{Neutral salt}}{HOOC — COOMgOOC — COOH} \; + \; H_2O$$

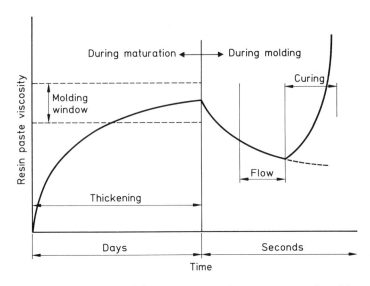

FIGURE 2.5   Viscosity variation of the resin paste at the maturation and molding stages.

The neutral salt formed in this reaction has a higher molecular weight than the resin. It can, in turn, react with other basic salts, creating large increases in the molecular weight, hence the viscosity of the resin paste.

The metal oxide/hydroxide thickeners react with the resin immediately upon mixing with the resin formulation. The viscosity of the mixture increases rapidly in the beginning and tends to level off after a period of time (Fig. 2.5) called the maturation time. The amount of thickener determines the ultimate viscosity of the resin paste as well as the maturation time. The maturation rate (i.e., how quickly the viscosity reaches the plateau) can be increased by adding various amounts of maturation control agents. It also increases with increasing moisture content in the resin. However, if the viscosity increases very rapidly after mixing, the fibers may not be properly wetted by the resin, which may cause air entrapment at the fiber surface and poor mechanical properties for the molded part.

When an SMC sheet is placed in a heated mold, the resin viscosity is reduced (Fig. 2.5), since the high molecular weight neutral salt molecules in the thickened SMC dissociate into resin molecules at the molding temperature. In the initial stages of molding, particularly before the curing reaction begins, the resin must attain low viscosities for good mold flow. Otherwise, the mold may not be completely filled (resulting in a "short shot"), or there may be unusually high air entrapment (resulting in voids) in the molded part. It may also require high molding pressures to spread the SMC in the mold.

The molding pressure and the extent of viscosity reduction during molding depend on the final viscosity of the thickened SMC just before molding. This final viscosity depends on the thickener type and concentration. In the example shown in Table 2.7, MgO and $Mg(OH)_2$ are used for thickening a low profile SMC resin system. Magnesium oxide, even at a lower concentration, increases the initial resin viscosity at a higher rate than magnesium hydroxide. However, the latter tends to increase the viscosity much more slowly during maturation, thus keeping the thickened SMC in the moldable range over a longer period of time.

It is important to note that the molding of good quality parts requires the final viscosity of the SMC sheet to be within a molding window span between 50×10^6 and 130×10^6 cP. Final viscosities below 50×10^6 cP may produce resin separation from fibers during molding, and viscosities higher than 130×10^6 cP are likely to produce unfilled parts (short shots).

A second approach (12) to thickening of an SMC resin paste is known as interpenetrating thickening process (ITP). It involves the reaction of a polyisocyanate with a polyol, which is incorporated into the styrenated polyester or vinyl ester resin. This results in an in situ formation of a polyurethane network that mechanically thickens the system by resin chain entanglement. In contrast to the conventionally thickened system using MgO, ITP-thickened systems absorb less moisture, which significantly reduces the variability of mechanical and electrical properties of SMC composites. The impact energy of ITP-thickened SMC composites is also higher than the MgO-thickened SMC composites (13).

2.2.8 Low Profile Additives

Low profile additives are thermoplastic powders that are mixed with polyester and vinyl ester resins to control the shrinkage of SMC composites. Both resins exhibit 5–9 % reduction in linear dimensions because of a combination of polymerization (curing) shrinkage in the mold and thermal shrinkage while cooling outside the mold. Besides causing dimensional changes of the molded part, shrinkage may create a number of surface defects, such as long- or short-term waviness, sink marks, and voids.

Shrinkage in SMC composites is reduced significantly by the addition of 10–20 wt % of a thermoplastic material, such as polyvinyl acetate, polycaprolactone,

TABLE 2.7   Thickening Effects of MgO and Mg(OH)$_2$ on SMC at 24–27 °C (11)

Thickener type	Amount of thickener based on resin (wt %)	SMC viscosity ($\times 10^6$ cP) at:				
		4 hours	1 day	2 days	7 days	14 days
MgO	0.5	0.85	6	11	19	20
	0.55	1	10	17	32	33
	0.60	3	12	25	46	48
	0.65	3	19	30	52	60
	0.75	4	28	47	73	80$^+$
MgO/CM-201[a]	0.65/0.30	6	30	47	56	54
	0.65/0.85	14	39	48	46	37
Mg(OH)$_2$	2.5		8.5	25	42	49

[a] CM-201 is a proprietary maturation control agent.

39

polyacrylate copolymers, cellulose acetate butyrate, polystyrene, and polyethylene. Polyvinyl acetate is the most effective shrinkage control additive among these thermoplastics (Table 2.8); however, it does not accept color very well. Polyethylene and polystyrene are less effective in shrinkage control, but they accept pigmentation very well. Some of these thermoplastic powders are dissolved in the styrene monomer before being mixed with the resin. The others are mixed directly with the resin. The resins containing low profile additives are sometimes called low profile resins.

According to Atkins et al. (14), thermoplastic additives separate and form a dispersed second phase in the resin at the onset of curing. This thermoplastic phase absorbs small quantities of unpolymerized resin and styrene. The higher thermal expansion of the thermoplastic phase and the vapor pressure of the absorbed styrene resists the polymerization (curing) shrinkage of the resin. Microvoids are formed in the thermoplastic phase as the residual resin and styrene polymerize. When the molded part begins to cool outside the mold, both matrix and dispersed thermoplastic phase shrink. However, since the glass transition temperature of the cured polyester resin is higher than that of the thermoplastic additive, the thermal contraction rate of the cured matrix is much less than that of the thermoplastic phase. The difference between the two shrinkage rates causes void and microcrack formation at the interface of the two phases, which, in turn, compensates for the thermal shrinkage of the cured matrix.

Ross et al. (15) observed that low profile additives are more effective with large amounts of styrene and a highly reactive resin. The styrene content is important because its expansion within the thermoplastic phase is responsible for partial compensation of the polymerization shrinkage. High resin reactivity assures the gelation of the continuous phase at low degrees of cure. A rapid temperature rise occurs at the onset of gelation, which accelerates the thermal expansion and microvoid formation in the thermoplastic phase.

TABLE 2.8   Linear Shrinkage of SMC[a] with Low Profile Additives (14)

Low profile additive	Amount (parts by weight)	Linear shrinkage (%)
None	0	0.7
Polyvinyl acetate	10	0.07
Polymethyl methacrylate	10	0.14
Polystyrene	10	0.29

[a] Containing 10 wt % of 6.35 mm long E-glass fibers in a polyester resin based on 0.25 mol of isophthalic acid, 0.75 mol of maleic anhydride, and 1.1 mol of propylene glycol.

2.2.9 Pigments

Organic or inorganic pigments, either in powder form or in a carrier resin, are added to a sheet molding compound to produce color in the molded part. Inorganic pigments, such at titanium dioxide or iron oxide, have higher specific gravities, lower

TABLE 2.9 Pigments Used in Sheet Molding Compounds (1)

Pigment	Color	Form	Secondary effect
Cadmium salts	Greenish yellow to orange	Concentrate	Accelerator
Carbon black	Black	Powder	Inhibitor
Iron oxide	Red	Powder	Accelerator
Titanium dioxide	White	Powder	Slight accelerator
Phthalocyanine green	Green	Concentrate	Slight inhibitor
Kaolin clays		Concentrate	Slight accelerator
Organic dyes and pigments		Concentrate	Inhibitors

oil absorption, and better heat stability than organic pigments. Some pigments may have inhibiting or accelerating effects on the cure characteristics of the resin (Table 2.9).

2.2.10 Other Additives

Other additives frequently included in the SMC resin paste formulation are flame retardants, ultraviolet (UV) absorbers, and impact modifiers. Most notable among these additives are the elastomeric impact modifiers, such as acrylonitrile–butadiene copolymer and styrene–butadiene copolymer, which improve the impact energy absorption, fracture toughness, and damage resistance of the SMC composite. These modifiers usually are added to the resin paste in liquid form. However, in the cured composite, they remain as discrete rubber particles ranging in size from 5 to 25 μm and toughen the matrix resin in a manner similar to that found in high impact polystyrene (16).

2.3 PRODUCTION OF SMC SHEET

The principal steps in the production of SMC sheet are mixing, compounding, compaction, and maturation. The resin paste is prepared in the mixing step. Fibers are added to the resin paste layers in the compounding step, which is followed by resin impregnation (wetting) of fibers in the compaction step. Both compounding and compaction for SMC-R and SMC-CR compounds are carried out on an SMC machine, shown schematically in Figure 2.6. For XMC, these two steps are carried out on a filament winding machine. The XMC sheet is removed by slitting the filament-wound roll along its length. Finally, the maturation step is used to increase the resin paste viscosity in the SMC sheet to a uniform and reproducible value suitable for the molding operation.

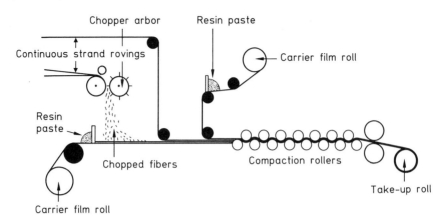

FIGURE 2.6   Schematic of an SMC machine.

2.3.1 Mixing

Various ingredients in an SMC resin formulation are divided into two parts:

1. The A side or the resin side, which contains the resin, low profile additive, catalyst, inhibitor, mold release agent, and a majority of the filler content.
2. The B side or the thickener side, which contains the thickener in an unthickenable carrier resin, pigment, and the balance of the filler.

Ingredients in each part are thoroughly blended in separate mix tanks by high intensity mixers. On the A side, thermoplastic powders are added to the resin first and the mix is agitated for an amount of time sufficient to ensure their uniform dispersal in the resin. The B side mix is blended with the A side mix immediately before the resin paste is transferred to the SMC machine. Application of vacuum during the mixing step reduces air entrapment in the resin paste and, consequently, the porosity in the molded part.

Typical resin paste formulations for three different sheet molding compounds are listed in Table 2.10. Among these, SMC-R50 and SMC-R65 are considered to be structural-grade sheet molding compounds. Fiber content in such compounds is equal to or higher than 50 wt %, while thermoplastic additives and fillers either are eliminated or are added only in small quantities.

2.3.2 Compounding

The resin paste is either poured or pumped onto two 0.05 mm thick polyethylene carrier films at two locations just behind the metering (doctor) blades (Fig. 2.6). The resin paste thickness drawn by each carrier film is controlled by the vertical adjustment of the metering blade. Continuous fiber rovings are pulled from various roving packages and fed into the chopper assembly, which is located 400–800 mm above the horizontal machine bed.

The function of the chopper assembly is to cut the continuous rovings into the desired length and sprinkle the chopped fibers uniformly on the resin paste just

TABLE 2.10   Typical SMC Formulations

Ingredient	SMC-R30		SMC-R50		SMC-R65	
	Wt by parts	Wt %	Wt by parts	Wt %	Wt by parts	Wt %
Resin	100	30.96	100	34.48	100	31.85
Thermoplastic additive	15	4.64	15	5.17	0	0
Catalyst	1	0.31	1	0.345	1	0.32
Mold release agent	5	1.55	3	1.035	3	0.95
Filler	100	30.96	20	6.90	0	0
Thickener	5	1.55	6	2.07	6	1.91
Fiber	97	30.03	145	50.00	204	64.97
Total	323	100	290	100	314	100

underneath the assembly. It includes a cutter roll, an elastomeric arbor, an oscillating comb, and static charge eliminator bars. The number of blades in the cutter roll can be varied to adjust the chopped fiber length. The oscillating comb moves the row of continuous rovings back and forth across the chopper assembly. This allows all areas of the cutting roll to be used evenly. The static charge eliminator bars, located just below the cutter roll, serve to ionize the surrounding air, which reduces the static charge buildup and the clumping of glass fibers.

Chopped fibers fall onto the bottom resin paste layer in a random fashion and are pulled forward by the lower polyethylene film. For producing SMC-CR type material, collimated continuous fiber rovings are placed on top of the random fiber layer. The top polyethylene film coated with a second thin layer of resin paste is combined with the bottom layer of resin paste and fibers. This sandwich construction is pulled into the compaction section, where fibers are impregnated with the resin paste.

2.3.3 Compaction

The compaction is usually performed with a roll-type impregnator that contains a series of hollow rolls with decreasing gap between each pair toward the take-up section of the machine. The rolls are made from Teflon-coated steel or aluminum alloys and may be externally serrated or spiral fluted to improve the impregnation process. The pressure from the compaction rolls forces the resin paste layers into the fibers to achieve a uniform wetout. Some of these rolls, especially toward the machine end, are internally heated to control their surface temperatures. This increases the sheet temperature, which lowers the resin paste viscosity and reduces the maturation time. Lower resin paste viscosity improves the fiber wetout. One or more of the rolls in the compaction section can also be used to prick holes in the upper polyethylene film and allow entrapped air to escape. Pulling the sheet assembly through large calender rolls at the end of the compaction section can further reduce the air entrapment by squeezing the air out through the sides of the sheet. This extra step produces a less porous surface in molded parts and a better fiber wetout, which increases the tensile strength of the molded parts.

The SMC sheet is wound on a take-up roll located at the end of the compaction line. The SMC roll is then wrapped with a barrier film, such as an aluminized Kraft paper, cellophane, or Mylar film, which prevents styrene evaporation from the sheet and moisture absorption into the sheet while it is stored for maturation.

2.3.4 Maturation

The SMC roll is stored after production in a temperature-controlled environment, referred to as the maturation room. The maturation temperature is maintained at 29–32 °C. Storage time may vary from 1 to 7 days, depending on the resin–thickener combination. The maturation of most SMC sheets requires approximately 3 days. After the proper viscosity level in the sheet has been achieved, the SMC roll can be removed from the maturation room for immediate use or stored in the maturation room for several days or months. The shelf life in storage depends on the resin–catalyst–inhibitor combination and the storage temperature.

2.3.5 Quality Control Methods

Principal batch-to-batch quality control measures performed during the production of SMC sheet are listed below.

1. *Resin paste viscosity*: The resin paste viscosity is measured in a Brookfield viscometer using a small sample of the resin mix. After its initial viscosity has been measured, the sample is stored at the same temperature as the SMC roll and its viscosity is monitored periodically throughout the maturation step.

2. *Weight per unit area*: The weight per unit area is measured on a 300 mm × 300 mm sample cut from the SMC sheet. Any variation in the weight per unit area of the SMC sheet will cause a corresponding change in the number and size of SMC plies in the compression molding charge (stack). It may also be necessary to add make-up strips to the charge to compensate for the reduced weight per unit area of the SMC sheet. In either case, a change in the charge assembly can cause flow variations in the mold that may affect the part performance or the surface quality.

3. *Fiber content*: Fiber content is measured after stripping the polyethylene films and removing the resin paste from the SMC sheet in an ultrasonic bath containing methylene chloride. Changes in fiber content can affect the mechanical performance of the molded part.

4. *Cure characteristic*: In addition to the tests above, the SPI gel time test is frequently performed to obtain information on the cure characteristics of the resin-catalyst combination.

5. *Spiral flow test*: The spiral flow test is performed to measure the flow characteristics of the sheet molding compound. In this test, a measured quantity of SMC is transfer

molded into a standard spiral cavity. The spiral length filled by the SMC is considered to be a measure of its moldability.

2.4 MECHANICAL PROPERTIES

The mechanical properties of sheet molding compound composites are determined with specimens machined from compression-molded flat plaques. The molding of these plaques is carefully controlled to exclude the gross molding defects, such as knit lines and undercure. Thus, the properties measured with flat plaque specimens may not always reflect the variability observed in a compression-molded production part. However, they provide information and guidelines for preliminary part design and material selection. A study of these properties may also help one to understand the effects of various material variables in an SMC formulation.

2.4.1 Static Properties

Static properties measured in tensile, flexural, compressive, and shear tests are discussed in this section. Tensile and flexural properties are routinely measured and reported by the material suppliers. Compressive and shear property data are not usually available from the material suppliers.

2.4.1.1 Tensile Properties

Tensile stress–strain diagrams of SMC-R, SMC-CR, and XMC composites are shown in Figures 2.7–2.9. Unlike many ductile metals and plastics, SMC composites do not exhibit any yielding and their strain-to-failure is low (< 2 %). However, at low fiber contents, the tensile stress–strain diagrams are nonlinear, which indicates the appear-

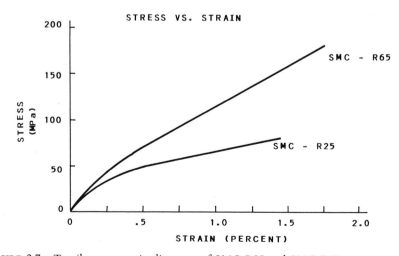

FIGURE 2.7   Tensile stress–strain diagrams of SMC-R25 and SMC-R65.

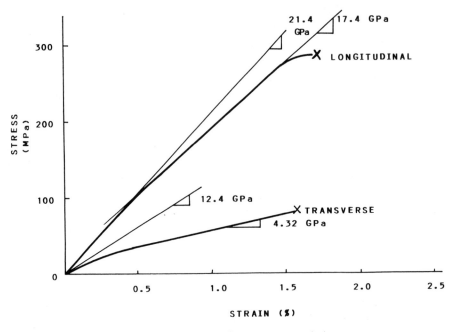

FIGURE 2.8   Tensile stress–strain diagram of SMC-C20R30 (17).

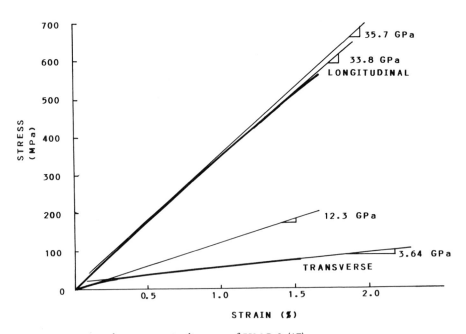

FIGURE 2.9   Tensile stress–strain diagram of XMC-3 (17).

ance of microdamage (resin cracking or debonding at the fiber–resin interfaces) in the material. At high fiber contents, the nonlinearity decreases, but the stress–strain diagram exhibits a bilinearity with a break point (knee) close to the matrix failure strain. At stress levels higher than the break point, the slope of the stress–strain diagram is lower than the initial slope.

In most flat plaque data concerning SMC-R composites, the tensile properties are observed to be independent of the specimen orientation. Thus, SMC-R composites are ideally planar isotropic materials. However, with long or obstructed flow in the mold, there may be preferred orientation of fibers in the molded part. This, in turn, will cause differences in tensile or other properties in various directions of the molded part.

For continuous fiber SMC composites, such as SMC-CR or XMC, the tensile properties are much higher in the longitudinal (L) direction than in the transverse (T) direction. The difference between the longitudinal and transverse properties is reduced with increasing proportion of random chopped fibers in the composite (Table 2.11).

Other observations on the tensile properties of SMC composites are summarized below.

1. Tensile strength of SMC-R composites increases significantly with increasing fiber content (Table 2.12); however, the tensile modulus is affected only moderately. Watanabe and Yasuda (19) developed the following empirical relationships for the

TABLE 2.11 Variation of Tensile and Flexural Properties with Fiber Type and Random Fiber Content in SMC-CR Composites[a] (18)

Property	SMC-R65		SMC-C60R5		SMC-C45R20	
	E-glass	S-glass	E-glass	S-glass	E-glass	S-glass
Tensile strength, MPa	225	271	768 (L) 22.1 (T)	1059 (L) 30 (T)	675 (L) 44 (T)	909 (L) 59 (T)
Tensile modulus, GPa	14.5	15	35.7 (L) 8.9 (T)	41 (L) 9.9 (T)	24.7 (L) 12 (T)	40.1 (L) 12.9 (T)
Poisson's ratio	0.26		0.31 (LT) 0.16 (TL)	0.29 (LT) 0.04 (TL)	0.30 (LT)	0.28 (LT)
Flexural strength, MPa	381	433	825 (L)	886 (L)	872 (L)	862 (L)
Flexural modulus, GPa	14.9	17.4	35.8 (L)	38.4 (L)	35.6 (L)	41.0 (L)

[a] Composition (wt %): polyester resin (31.8%), catalyst (0.32%), MgO (1.6%), zinc stearate (1.28%); total glass fiber content = 65%.

TABLE 2.12 Mechanical Properties of Various SMC Composites (17)

Property	SMC-R25	SMC-R50	SMC-R65	SMC-C20R30	XMC-3
Specific gravity	1.83	1.87	1.82	1.81	1.97
Tensile strength, MPa	82.4	164	227	289 (L) 84 (T)	561 (L) 69.9 (T)
Tensile modulus, GPa	13.2	15.8	14.8	21.4 (L) 12.4 (T)	35.7 (L) 12.3 (T)
Poisson's ratio	0.25	0.31	0.26	0.30 (LT) 0.18 (TL)	0.31 (LT) 0.12 (TL)
Strain-to-failure, %	1.34	1.73	1.63	1.73 (L) 1.58 (T)	1.66 (L) 1.54 (T)
Flexural strength, MPa	220	314	403	645 (L) 165 (T)	973 (L) 139 (T)

effect of fiber volume fraction on the tensile strength and tensile modulus of SMC-R composites:

$$S_u = 0.33\, S_f v_f + 0.31\, S_m (1 - v_f) \qquad (2.1)$$

$$E = 0.59\, E_f v_f + 0.71\, E_m (1 - v_f) \qquad (2.2)$$

where S_u = ultimate tensile strength of the SMC-R composite
S_f = ultimate fiber strength
S_m = ultimate matrix strength
E = initial tensile modulus of the SMC-R composite
E_f = fiber modulus
E_m = matrix modulus
v_f = fiber volume fraction

Note that the matrix in Eqs. (2.1) and (2.2) refers to the resin–filler combination, and not just the resin.

2. Tensile strength of SMC-R composites increases as the chopped fiber length is increased (Table 2.13); however, fiber length has virtually no effect on the tensile modulus.

3. The fiber type (e.g., E-glass vs. S-glass) has a significant effect on the tensile strength and modulus of the composite. The effect is more pronounced for the fiber-dominated properties of the composite, such as the longitudinal tensile strength and modulus of SMC-CR composites, as shown in Table 2.11.

TABLE 2.13   Effect of Chopped Fiber Length on Tensile and Flexural Properties of SMC-R25

Chopped fiber length (mm)	Room temperature properties		
	Tensile strength (MPa)	Flexural strength (MPa)	Flexural modulus (GPa)
12.5	63.4	191.7	14.1
25	89.7	213.1	14.3
37.5	109.0	245.5	14.3
50	109.0	276.6	17.0

4. The resin type may also influence the tensile properties of SMC composites, particularly at low fiber contents.

5. Even with carefully molded flat plaques, a wide variation in tensile strength is observed in SMC-R composites (20). Such variation is attributed to the presence of dry fiber, nonuniform filler dispersion, and fiber bundle separation near the surface (21).

6. Stress concentration effects on the tensile strength of SMC composites have been studied by a number of investigators (22–24). All these studies used circular holes as the source for stress concentration. In general, the tensile strength of SMC-R composites is relatively insensitive to the presence of holes, since microcracks formed at the naturally occurring flaws in these composites (fiber ends, voids, etc.) tend to reduce the stress-concentration at the hole boundary.

Mallick (23) as well as Shirrell and Onachuk (24) have shown that the Whitney–Nuismer point stress failure criterion (25) can be used to predict the net tensile strength of SMC-R composites. For isotropic composites, this failure criterion can be written in the equation form as follows:

$$\sigma_n = \left[\frac{2}{2 + \lambda^2 + 3\lambda^4} \right] \sigma_0 \qquad (2.3)$$

where σ_n = net tensile strength of a specimen containing a central hole
  σ_0 = unnotched tensile strength
  λ = $r/(r + r_0)$
  r = hole radius
  r_0 = a characteristic distance from the hole edge at which the tensile stress due to the applied load is equal to the unnotched tensile strength.

The characteristic distance r_0 in Eq. (2.3) depends on the fiber content in SMC-R composites.

2.4.1.2 Flexural Properties

Flexural properties of SMC composites are determined by three-point flexural tests of rectangular beam specimens in accordance with ASTM D 690. These tests are easy to perform and require simple fixtures. However, if relatively high span-to-depth ratios are not used, the measured flexural modulus value may be low because of the influence of shear deflection in the beam (26).

In general, flexural strengths are higher than the tensile strengths (Table 2.12), but the flexural modulus values are similar to the tensile modulus values. Flexural load–deflection diagrams are nonlinear, which indicates the occurrence of microcracking even at the early stages of loading.

In general, flexural properties follow the same trend as the tensile properties. They are affected by the fiber content, length, type, and orientation in the same way as the tensile properties are affected by these parameters.

2.4.1.3 Compressive and Shear Properties

As with tensile properties, both compressive strength and modulus depend on the fiber content and fiber orientation (Table 2.14). For SMC-CR and XMC composites, both values are higher in the longitudinal direction than in the transverse direction.

In-plane shear strength and shear modulus are measured by rail shear tests of flat plaque specimens. A description of the rail shear test method can be found in Reference 26. In-plane shear properties of SMC-R, SMC-CR, and XMC composites are presented in Table 2.14. For SMC-R composites, the following equation relating

TABLE 2.14   Compressive and Shear Properties of Various SMC Composites (17)

Property	SMC-R25	SMC-R50	SMC-R65	SMC-C20R30	XMC-3
Compressive strength, MPa	183	225	241	306 (L) 166 (T)	480 (L) 160 (T)
Compressive modulus, GPa	11.7	15.9	17.9	20.4 (L) 12.2 (T)	36.8 (L) 14.5 (T)
In-plane shear strength, MPa	79	62	128	85.4	91.2
In-plane shear modulus, GPa	4.48	5.94	5.38	4.09	4.47
Interlaminar shear strength, MPa	30	25	45	41	55

the shear modulus G_{xy}, tensile modulus E, and Poisson's ratio v can be used to predict the shear modulus within 10 % of the experimental values:

$$G_{xy} = \frac{E}{2(1 + v)} \qquad (2.4)$$

It should be noted that Eq. (2.4) is valid for isotropic materials. For design purposes, SMC-R composites are considered to be planar isotropic.

Interlaminar shear strength (ILSS) is a measure of the shear strength in the thickness direction. It is determined by three-point flexural testing of beams with short span-to-depth ratios. It is considered to be a quality control test for molded composites. In general, vinyl ester SMC composites exhibit a higher interlaminar shear strength than polyester SMC composites, particularly at elevated temperatures. Typical ILSS values of various SMC composites are given in Table 2.14.

2.4.2 Fatigue Properties

Most of the data available on the fatigue properties of SMC composites are based on tension–tension cyclic loading of unnotched specimens. The S-N diagrams of SMC-R25, SMC-R50, SMC-R65, SMC-C20R30, and XMC-3 are shown in Figures 2.10–2.14.

Unlike low carbon steels, no fatigue limit is observed in any of these S-N diagrams, and the number of cycles to failure increases with decreasing fatigue stress level. Another point to observe from these S-N diagrams is that the lower the fiber content in SMC, the greater the scatter in fatigue life at any given stress level. Fatigue strengths at 10^6 cycles obtained from the best-fit lines of the S-N data are listed in Table 2.15.

In general, the fatigue strength of SMC-R composites increases with increasing fiber content. For SMC-CR and XMC composites, a large difference in fatigue strength is observed between the longitudinal and transverse directions. The lon-

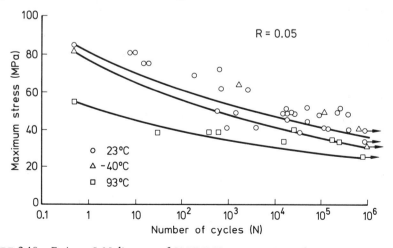

FIGURE 2.10   Fatigue S-N diagram of SMC-R25 at − 40, 23, and 93 °C.

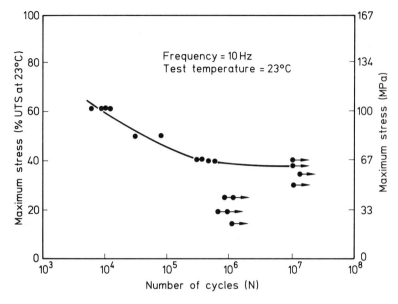

FIGURE 2.11   Fatigue S-N diagram of SMC–R50 at 23 °C (28).

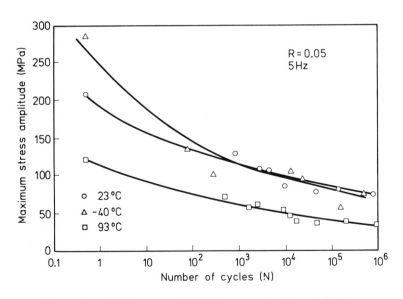

FIGURE 2.12   Fatigue S-N diagram of SMC-R65 at − 40, 23, and 93 °C.

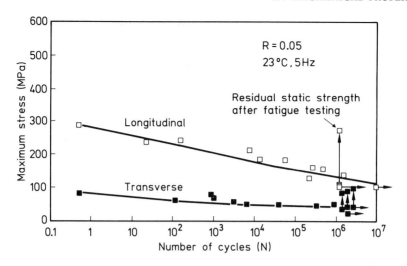

FIGURE 2.13   Fatigue *S-N* diagram of SMC–C20R30 (17).

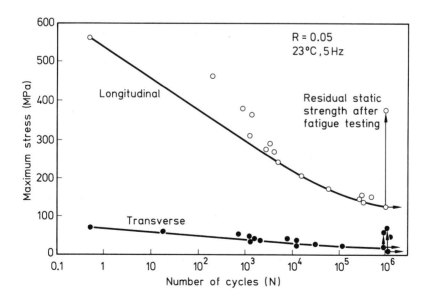

FIGURE 2.14   Fatigue *S-N* diagram of XMC-3 (17).

TABLE 2.15 Fatigue Strengths at 10^6 Cycles for Various SMC Composites (17, 27)

Material	R	Fatigue strength (MPa)		Ratio of fatigue strength to ultimate static strength	
		At 23 °C	At 90 °C	At 23 °C	At 90 °C
SMC-R25	0.05	40	25	0.49	0.55
SMC-R50	0.05	63	53	0.38	0.41
SMC-R65	0.05	70	35	0.31	0.28
SMC-C20R30 (L)	0.05	130	110	0.45	0.44
SMC-C20R30 (T)	0.05	44	35	0.52	0.43
XMC-3 (L)	0.05	130	85	0.23	0.18
XMC-3 (T)	0.05	19	27	0.15	0.26
SMC-C60R5 (L)	0.05	338		0.43	
SMC-C45R20 (L)	0.05	262		0.47	

gitudinal fatigue strength of SMC-CR composites decreases significantly as the continuous fiber content in the material is reduced.

Wang and his coworkers (29–31) have studied the fatigue damage development in SMC-R50 composites. The microscopic fatigue damage appears in the forms of matrix cracking and fiber–matrix interfacial debonding. The matrix cracking is influenced by the stress concentrations at poorly dispersed clusters of filler as well as at the chopped fiber ends. Matrix cracks are generally normal to the loading direction. However, in fiber-rich areas, with fibers oriented parallel to the loading direction, the matrix crack lengths are small and limited by the interfiber spacing. If the fibers are

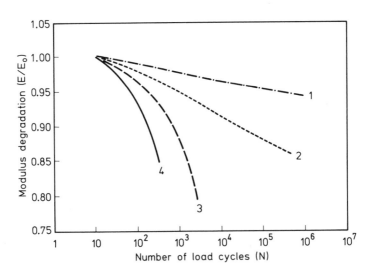

FIGURE 2.15 Modulus degradation of SMC-R50 during a tension fatigue test. The maximum stress levels for curves 1–4 are 20, 50, 70 and 85%, respectively, of the static ultimate tensile strength (29).

oriented at an angle with the loading direction, microcracks generally grow along the fiber–matrix interface. With increasing fatigue stress level, the microcrack density tends to increase, but the crack length tends to decrease.

Mallick (32) has shown that the microcrack density also depends on the type of the matrix material. For example, at any given stress level, the microcrack density is higher in a polyester matrix SMC-R65 than in a vinyl ester matrix SMC-R65, possibly because of higher chain flexibility of vinyl ester molecules and better adhesion to glass fibers of vinyl ester resin than polyester resins.

One consequence of the microdamage development and growth is that the dynamic modulus of the material is decreased (Fig. 2.15) with increasing number of cycles. Increasing the fatigue stress level results in significant increase in the rate of modulus reduction. Similar behavior is also observed for SMC-CR composites; however, the modulus reduction is less pronounced in the longitudinal direction than in the transverse direction.

As a result of the accumulated fatigue damage in SMC composites, the residual tensile strength of fatigued specimens is reduced (27, 32). Other evidence of fatigue damage accumulation consists of an increase in damping factor in fatigued specimens (27). Both phenomena are observed to depend on the fatigue stress level as well as the number of cycles endured. Additionally, in SMC-R composites, the reduction in residual static strength depends on the matrix type, as demonstrated in Table 2.16.

TABLE 2.16   Residual Flexural Strength of SMC-R65 at Room Temperature After Completely Reversed Flexural Fatigue Test at 54 °C (32)[1]

Matrix	Alternating stress, σ_a (MPa)	σ_a/σ_f (%)	Cycles tested	Residual flexural strength, σ_{res} (MPa)	σ_{res}/σ_f
Polyester	62	20	1.72×10^6	269	0.70
	68.9	22.5	1.63×10^6	397.4	1.03
	86.2	28	1.30×10^6	280.7	0.73
	103.4	34	2,000	367.4	0.96
	103.4	34	18,000	309.22	0.80
	103.4	34	48,000	223.3	0.58
	103.4	34	367,000	254.4	0.66
	137.9	45	40,000	248.1	0.645
	172.4	56	4,110	251.4	0.654
Vinyl ester	103.4	38	895,000	232	0.89
	137.9	51	15,000	296.2	0.99
	137.9	51	55,020	304	1.02
	137.9	51	376,000	232.1	0.78
	172.4	64	40,000	270.8	0.91
	206.8	76	10,000	276.9	0.93

[1] σ_f = flexural strength

Wang et al. (30) used a damage parameter D to represent the fatigue damage accumulation in SMC-R composites. The damage parameter in terms of the tensile modulus is defined as follows:

$$D = 1 - \left(\frac{E}{E_0}\right) \tag{2.5}$$

where E = tensile modulus after N cycles
E_0 = initial tensile modulus

It is found that the fatigue damage parameter follows a power law equation given by:

$$\frac{dD}{dN} = AN^{-B} \tag{2.6}$$

where B is a constant and A is a function of the damage state, which depends on the loading history, stress amplitude, mode of loading, temperature, and other variables.

2.4.3 Impact Properties

A number of different test methods have been used to determine the impact properties of SMC composites. The following observations can be made from these test results.

1. Conventional Izod or Charpy impact tests (ASTM D 256-72) are not useful in assessing the impact toughness of SMC composites. For example, Owen (33) performed notched Izod impact tests on various SMC composites containing 27–34 wt % E-glass fibers in low profile polyester resins and found that the Izod impact energy was within a narrow range of 1.11–1.19 kJ/m, even though these SMC composites exhibited considerable difference in toughness in other impact tests. Denton (28) reported a notched Izod impact energy of 1 kJ/m for a vinyl ester SMC-R50.

2. The drop dart impact test (ASTM D 3029-72) is capable of detecting changes in resin flexibility as well as fiber content (33). Typical drop dart impact energy values of various SMC-R composites are given in Table 2.17.

TABLE 2.17   Various Impact Properties of SMC-R Composites (33)

Impact test	Glass fiber content (wt %)		
	27	30	34
Dart drop impact energy, J/m	370	420	540
Rheometrics impact test			
Ultimate force, kN	4.1	5.35	6.15
Yield energy, J	0.82	1.04	0.91
Ultimate energy, J	10	16	17
Total energy, J	30	46	44

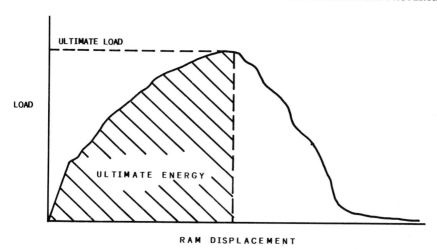

FIGURE 2.16   Typical load–displacement diagram obtained in a Rheometrics constant speed impact test of SMC specimens (17).

3. Several investigators have used a Rheometrics constant speed impact test to determine the impact properties of SMC composites. A typical load–displacement diagram obtained in such tests is shown in Figure 2.16. Test results reported by Owen (33) indicate that the ultimate impact energy of SMC-R composites depends on the glass fiber content (Table 2.17). Riegner and Sanders (17) used a similar test to determine the impact properties of SMC-R25, SMC-C20R30, and XMC-3 composites at −35 and 23 °C. Different impact velocities ranging from 0.5 to 10 m/s were used. In general, temperature and impact velocity had no significant effect on the impact load and energy values for these composites. The ultimate impact energy of XMC-3 composites was slightly higher than that of SMC-C20R30 or SMC-R25 composites.

4. The instrumented drop-weight impact tests conducted by Golovoy et al. (34) indicate that the fracture initiation energy of XMC composites is independent of the impact velocity; however, the fracture propagation energy increases with increasing impact velocity. In Golovoy's tests, failure in impacted beam specimens took place through shear delamination, since the beam span-to-thickness ratio was relatively small. Similar tests by Myers (35) show that the drop-weight impact energy of SMC-R composites, tested with plate specimens, increases with increasing glass fiber length. The drop weight impact energy corresponding to the maximum load in the load-deflection diagram of SMC-R composites containing 25 mm long chopped fibers was 9.8 J at 2.1 m/s impact velocity.

   Damage caused by low energy impact and postimpact residual strength of SMC composites has been studied by several investigators (36–38). The impact damage is induced by firing a steel ball or a striker against a properly supported SMC plate. The incident energy levels are low so that the striker does not penetrate the impacted plate. The damaged plate is then examined nondestructively for surface or internal damage

TABLE 2.18   Tensile Properties of Impact-Damaged SMC-R Composites (36)

Average impact velocity (m/s)	SMC-R50		SMC-R65	
	Tensile strength (MPa)	Tensile modulus (GPa)	Tensile strength (MPa)	Tensile modulus (GPa)
0 (undamaged)	157.3	14	228	17.8
31	137.6	12.9	209	16.5
46	119	10.9	167	17.3
54	120	11.1	151	14.0
68	94	9.7	124	12.4

and tested for residual tensile strength. The macroscopic damages observed in SMC-R composites consist of a small local indentation (dent), a ring of cracks on the front surface surrounding the dented area, and a cluster of cracks on the back surface. The surface damage area as well as the damage intensity increase with increasing impact energy. Tensile specimens cut from the damaged area exhibit nearly linear strength and modulus reduction with increasing impact energy (Table 2.18).

2.4.4 Damping Characteristics

Vibration damping characteristics of various SMC materials have been reported by Riegner and Sanders (17), who performed a forced vibration test on three-point flexural specimens and measured both complex modulus and loss coefficient over the 0.1–40 Hz frequency range. The complex modulus is found to remain constant in this frequency range. The loss coefficient does not also vary significantly with test frequency.

Table 2.19 lists the loss coefficient values of SMC-R25, SMC-R65, SMC-C20R30, and XMC-3. The loss coefficients of all these materials are much higher than those of steel and aluminum alloys. This is an indication of better noise and vibration damping in SMC structures than in steel or aluminum alloy structures.

TABLE 2.19   Vibration Damping Loss Coefficients of Various SMC Composites (17)

Material	Loss Coefficient	
	Frequency = 0.1 Hz	Frequency = 10 Hz
SMC-R25	0.037	0.035
SMC-R65	0.039	0.034
SMC-C20R30 (L)	0.034	0.029
SMC-C20R30 (T)	0.051	0.049
XMC-3 (L)	0.028	0.025
XMC-3 (T)	0.063	0.053

2.4.5 Effects of Environmental Conditions

The test environment may have a significant effect on the properties of SMC composites. For example, both tensile strength and tensile modulus decrease at elevated temperatures (Fig. 2.17). Similar effects are also observed for flexural properties (Fig. 2.18). These effects are related to the softening of the matrix at increasing temperatures. The fatigue properties of SMC composites are also affected

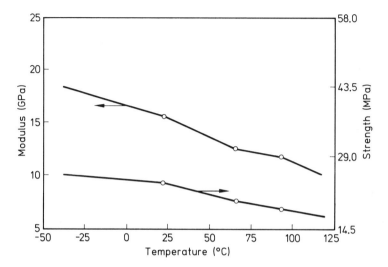

FIGURE 2.17   Tensile strength and modulus variation of SMC-R50 with increasing ambient temperature (28).

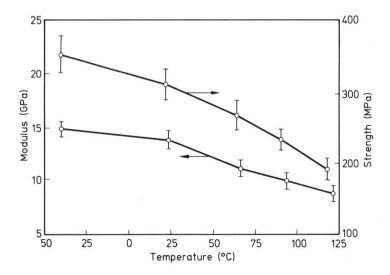

FIGURE 2.18   Flexural strength and modulus variation of SMC-R50 with increasing ambient temperature (28).

TABLE 2.20   Environmental Effects on the Tensile Strength of SMC-R Composites After Continuous Exposure for 6 Months (39)

Humid air	Antifreeze
Virtually no effect at 23 °C, 50% RH 20–25% reduction at 23 °C 100% RH 35–45% reduction at 93 °C, 100% RH	Slight reduction at 23 °C 55–75% reduction at 93 °C
Salt water	Gasoline
20–35% reduction at 23 °C 47–55% reduction at 93 °C	Slight reduction at 23 °C 15–30% reduction at 93 °C
Motor oil 0–20% reduction at 23 and 93 °C	

by the test temperature (Figs. 2.10 and 2.12). As indicated in Table 2.15, there is a significant reduction in fatigue strength at 10^6 cycles as the test temperature is increased from 23 to 90 °C.

Properties of SMC composites may also be affected by moisture and liquid environments. Molecules of these liquids diffuse into the resin matrix and may act as plasticizers. Furthermore, liquid molecules that diffuse into the fiber–matrix interface may cause debonding of fibers and deteriorate their mechanical strength. Springer and his coworkers (39) soaked samples of various SMC composites in humid air, salt water, motor oil, and several other liquids and tested them for static properties after 3 and 6 months of soaking. Their test results are summarized in Table 2.20. In most cases, a vinyl ester matrix exhibited a lower reduction in properties than a polyester matrix.

Moisture absorption into SMC composites under ambient conditions depends on the relative humidity of the surrounding air. Experiments by Springer et al. show that the tensile properties of SMC composites may deteriorate with moisture absorption, particularly at 100 % relative humidity and elevated temperatures. However, the interlaminar shear strength was not significantly affected by moisture absorption. Hosangadi and Hahn (40) observed that a combination of high humidity (e.g., 98 % RH) and elevated temperature (e.g., 75 °C) can cause discoloration, surface blistering, and chemical degradation of the matrix resin in SMC-R65. In their study, interlaminar shear strength was reduced by as much as 40 % from the unexposed value.

Ngo et al. (41) studied the fatigue behavior of SMC-R65 composites after immersing the test specimens in water and isooctane for 40 days at room temperature. Their data indicate that water immersion may be more detrimental than isooctane immersion in terms of fatigue at any given stress level. Isooctane immersion does not cause any significant reduction in fatigue behavior from that observed in air.

Devine (42) reported the effects of boiling water, antifreeze, gasoline, and dry heat on the flexural strengths of five SMC composites. These materials were identical in all aspects except for the base resin. The resins investigated were orthophthalic polyester, isophthalic polyester, terephthalic polyester, vinyl ester, and bisphenol A epoxy. In 3 week exposures to boiling water and antifreeze, epoxy exhibited the highest strength retention (> 50 %), followed by vinyl ester and terephthalic polyester. Orthophthalic and isophthalic polyesters retain only 30 % of their initial strengths under the same conditions. In refluxing gasoline, the situation is effectively reversed, with ortho-, iso-, and terephthalic polyesters showing nearly 90 % strength retention after 3 weeks, and vinyl ester and epoxy showing less than 30 % strength retention over the same time period. All SMC composites were subjected to dry heat aging for 12 months at two temperatures, 130 and 180 °C. At 130 °C, all the composites retained more than 80 % of their initial strengths. However, at 180 °C, the standard orthophthalic SMC lost its strength completely. The remaining SMC composites were much superior to orthophthalic SMC, ranging from 52 % strength retention for epoxy to 70 % retention for vinyl ester.

Gruenwald and Walter (43) reported the effect of long-term outdoor weathering on the physical and mechanical properties of a polyester-based SMC-R. The average E-glass fiber content in the test panels was 25–30 wt %. The time of exposure varied from 8 to 15 years. The most notable physical effect was color fading, which depends strongly on the type of pigment used. In general, there is less color fading with inorganic pigments, while organic pigments tend to be chemically unstable to weathering.

Another physical effect was the reduction of surface gloss, which was followed by the erosion of surface fibers. The gloss is influenced by the resin type as well as the filler content. High filler content accelerates both gloss reduction and color fading. The flexural strength is reduced in the first 2–3 years of exposure and levels off to a constant value. Depending on resin type, filler content, and fiber content, the final flexural strength may be between 60 and 100 % of the initial value.

2.4.6 Creep Properties

Creep is a viscoelastic phenomenon manifested by an increase in strain with increasing time even though the stress level is constant. Figure 2.19, which shows the creep response of SMC-R50 at four different test temperatures, demonstrates that significant creep strains may occur at elevated temperatures even when the stress level is relatively low. At a given temperature, the creep strain increases as the applied stress level is increased. Among the material variables, the fiber content has the greatest influence on the creep strain. At a given temperature and stress level, creep strain is higher if the fiber content is reduced.

Yen, Hiel, and Morris (44) used the Findley equation to calculate the creep strain of SMC-R50:

$$\varepsilon = \varepsilon_0 + mt^n \tag{2.7}$$

where ε = time dependent creep strain
ε_0 = instantaneous strain
t = time
m, n = experimentally determined constants

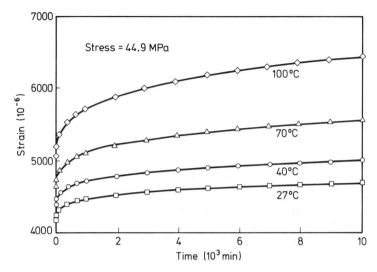

FIGURE 2.19   Creep response of SMC-R50 at different test temperatures (44).

Their experimental data in both single-step and multistep loading tests fit well with the Findley equation.

2.5 OTHER MOLDING COMPOUNDS

This section briefly describes bulk molding compound (BMC) and thick molding compound (TMC). Both are ready-to-mold composite materials containing short random fibers in a thermosetting resin. Parts produced from BMC are either compression molded, injection molded, or transfer molded, while TMC parts are compression molded.

2.5.1 Bulk Molding Compound

Bulk molding compound, also known as dough molding compound (DMC) in Europe, is produced in the form of a thick rope or log. It generally contains 15–20 wt % of 6–12 mm long E-glass fibers in a polyester resin. The resin formulation is similar to that used for sheet molding compounds. Chopped glass strands are compounded with the resin formulation in an intensive mixer and then extruded into the form of a continuous log. The extruded log is cut into desired lengths by means of a pneumatic cutter located outside the extruder. In general, thickeners are not used in the resin formulation if the BMC log is molded into the finished product just after compounding it in the extruder. Otherwise, thickeners, such as MgO powder, are added to the resin formulation and proper maturation time is allowed for the BMC to increase in viscosity.

Mechanical properties, such as tensile strength, tensile modulus, and impact energy, are lower for BMC composites than those for SMC composites. The

TABLE 2.21   Typical Properties of Bulk Molding Compound (BMC) Composites

Property	Value
E-glass fiber content, wt %	10–25
Specific gravity	1.8–2.0
Tensile strength, MPa	27–54
Tensile modulus, GPa	3.4–10.3
Flexural strength, MPa	69–172
Flexural modulus, GPa	5.5–8.3
Coefficient of thermal expansion, $\times 10^6/°C$	14.4–27

difference in mechanical properties is attributed to shorter fiber length and lower fiber content in BMC. Some typical mechanical properties of BMC composites are presented in Table 2.21.

2.5.2 Thick Molding Compound

Thick molding compound is a form of sheet molding compound in which the sheet thickness may range up to 50 mm, versus 6 mm or less for conventional sheet molding compounds. It is produced by metering the premixed resin paste and chopped fibers directly onto two large impregnating rolls (Fig. 2.20). The compound is deposited between two moving polyethylene carrier films.

The chopped fibers in a TMC sheet are dispersed randomly in a three-dimensional form instead of a two-dimensional form in a conventional SMC sheet. Also, since

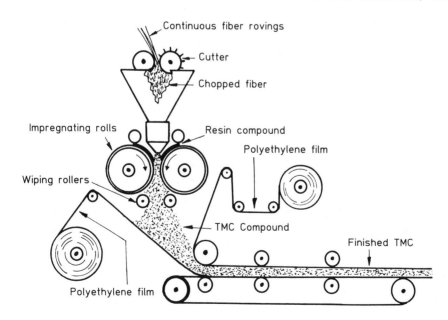

FIGURE 2.20   Schematic of a TMC machine.

fewer layers of TMC are required to compression mold a given part thickness, the possibility of interlaminar cracks in the molded part is reduced. Another advantage of the TMC process is its superior impregnation efficiency and control, which provide good fiber wetout even at high filler contents.

REFERENCES

1. R. W. Meyer, *Handbook of Polyester Molding Compounds and Molding Technology*, Chapman & Hall, New York, 1987.
2. M. K. Gupta, D. W Hoch, and J. F. Keegan, *Mod. Plast*, 64, 70 (1987).
3. "How Ingredients Influence Unsaturated Polyester Properties", Amoco Chemical Corporation, Chicago, Bulletin IP-70.
4. "DERAKANE Vinyl Ester Resins for Corrosion Resistance", Dow Chemical Company, Midland, MI, 1975.
5. L. Walewski and S. Stockton, *Mod. Plast*, 62, 78 (1985).
6. H. R. Edwards, *Mod. Plast*, 64, 66 (1987).
7. F. J. Butwin and W. C. Howes, *Proc., 41st Annual SPI Technical Conference*, 1986.
8. G. A. Harpell, J. P. Cassoni, and O. L. Mageli, *Proc., 30th Annual SPI Technical Conference*, 1975.
9. E. R. Kamens, R. B. Gallagher, and V. R. Kamath, *Proc., 34th Annual SPI Technical Conference*, 1979.
10. K. S. Gandhi and R. Burns, *J. Polym. Sci., Polym. Chem. Div.*, 14, 793 (1986).
11. G. E. Forsyth and F. J. Ampthor, *Plast. Eng.*, (1974).
12. J. Ferrarini, D. M. Longnecker, N. N. Shah, J. Fletzin, and G. G. Greth, *Proc., 33rd Annual SPI Technical Conference*, 1978.
13. A. T. Hurst, S. A. Smith, and J. A. Reitz, *Proc., 36th Annual SPI Technical Conference*, 1981.
14. K. E. Atkins, J. V. Koleske, P. L. Smith, E. R. Walter, and V. E. Matthews, *Proc., 31st Annual SPI Technical Conference*, 1976.
15. L. R. Ross, S. P. Hardebeck, and M. A. Bachman, *Proc., 43rd Annual SPI Technical Conference*, 1988.
16. F. J. McGarry, E. H. Rowe, and C. K. Riew, *Polym. Eng. Sci.*, 18, 78 (1978).
17. D. A. Riegner and B. A. Sanders, *Proc., SPE National Technical Conference*, 1979.
18. J. F. Kay, *Compos. Tech. Rev*, 4, 110 (1982).
19. T. Watanabe and M. Yasuda, *Composites*, 13, 54 (1982).
20. C. D. Shirrell, *Polym. Compos.*, 4, 172 (1983).
21. C. D. Shirrell, in ASTM STP 873, 1985, p. 3.
22. S. V. Hoa, *Proc., 39th Annual SPE Technical Conference*, 43 (1981).
23. P. K. Mallick, *Composites*, 19, 283 (1988).
24. C. D. Shirrell and M. G. Onachuk, in ASTM STP 907, 1986.
25. J. M. Whitney and R. J. Nuismer, *J. Compos. Mater.*, 8, 253 (1974).
26. P. K. Mallick, *Fiber-Reinforced Composites: Materials, Manufacturing and Design*, Dekker, New York, 1988.
27. R. Kundrat, S. Joneja, and L. J. Broutman, *Proc., 37th Annual SPI Technical Conference*, 1982.
28. D. L. Denton, SAE Paper 790671, 1979.
29. S. S. Wang and E. S. M. Chim, *J. Compos. Mater.*, 17, 114 (1983).
30. S. S. Wang, E. S. M. Chim, and H. Suemasu, *J. Appl. Mech.*, 53, 347 (1986).
31. S. S. Wang, E. S. M. Chim, and N. M. Zahlan, *J. Compos. Mater.*, 17, 250 (1983).

32. P. K. Mallick, *Polym. Compos.*, 2, 18 (1981).
33. G. E. Owen, *Polym. Eng. Sci.*, 21, 467 (1981).
34. A. Golovoy, M. F. Cheung, and H. Van Oene, *Proc., 39th Annual SPI Technical Conference,* 1984.
35. F. A. Myers, *Proc., 37th Annual SPI Technical Conference,* 1982.
36. S. K. Chaturvedi and R. L. Sierakowski, *Compos. Struct.*, 1, 137 (1983).
37. R. P. Khetan and D. C. Chang, *J. Compos. Mater.*, 17, 182 (1983).
38. S. K. Chaturvedi and R. L. Sierakowski, *J. Composite Mater.*, 19, 100 (1985).
39. G. S. Springer, B. A. Sanders, and R. W. Tung, *J. Compos. Mater.*, 14, 213 (1980).
40. A. B. Hosangadi and H. T. Hahn, in ASTM STP 873, 1985, p. 103.
41. A. Ngo, S. V. Hoa, and T. S. Sankar, in ASTM STP 873, 1985, p. 65.
42. F. E. Devine, *Composites*, 14, 353 (1983).
43. R. Gruenewald and O. Walter, *Proc., 30th Annual SPI Technical Conference,* 1975.
44. S. C. Yen, C. Hiel, and D. H. Morris, in ASTM STP 873, 1985, p. 131.

3 · Compression Molding

P. K. Mallick

Contents

P. K. Mallick, Department of Mechanical Engineering, University of Michigan—Dearborn, Dearborn, Michigan, USA

3.1 INTRODUCTION

Compression molding is one of the oldest manufacturing techniques in the plastics industry. Traditionally, it has been used for molding thermosetting (such as phenolic and alkyd) powders and rubber compounds. However, compared with the injection molding process, it has found only limited use for either thermoplastic or thermosetting resins. With the recent development of high strength sheet molding compounds and a greater emphasis on the mass production of composite materials, the compression molding process is receiving a great deal of attention from both industry and research community. A recent industry-wide projection shows that the volume of compression molded SMC used in the United States will grow nearly threefold in the next 5 years. Much of this potential growth is expected to take place in the automotive and appliance markets.

Compression molding has a number of advantages over the injection molding process. It is performed in relatively simple tools with no sprues, runners, or gates. Consequently, very little material is wasted in a compression molding operation. High fiber volume fractions or long fiber lengths can be easily accommodated in compression molding, while injection molding is limited to low fiber volume fractions and fiber lengths of 3 mm and less. Thus, in general, better physical and mechanical properties can be achieved in compression-molded parts. The molding pressure in the compression molding process is lower than that in the injection molding process. This means that for comparable part surface areas, a lower capacity press is required for compression molding than for injection molding. For this reason, compression molding is more suitable for producing parts with large surface areas than injection molding.

In this chapter, we discuss the compression molding process as it relates to the manufacturing of sheet molding compound composite parts. Compression molding is also used for bulk molding compounds. However, from the standpoint of current usage, SMC composites have a larger market share than BMC composites. Recently, a large number of research works have been published on the cure mechanism, flow characteristics, and other fundamental aspects of SMC compression molding. Some of these research works are reviewed here. Other topics in this chapter include defects in compression-molded parts, molding parameters, and new process developments in the compression molding technology. A few application examples are given, as well.

3.2 COMPRESSION MOLDING PROCESS

The process of compression molding can be divided into three basic steps (Fig. 3.1).

1. *Charge preparation and placement*: The stack of SMC plies placed in the preheated mold is referred to as the charge. The plies are die cut in the desired shape and size from a properly matured SMC roll, polyethylene carrier films are peeled off from each ply, and the plies are stacked into a charge outside the mold.

Rectangular ply patterns are commonly used in the charge; however, circular, elliptical, or any other ply pattern can also be used. The ply dimensions are selected

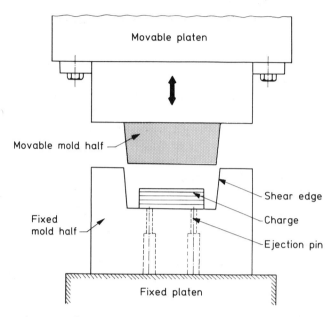

FIGURE 3.1   Schematic of a compression molding press.

to cover 60–70 % of the mold surface area with the charge; the charge weight is measured just before placing it in the mold. Since the charge weight is left constant for each part, any variation in SMC sheet weight per unit area is made up by adding small strips cut from the SMC roll. The location of the charge placement in the mold is one of the key factors in determining the quality of the molded part, since it influences the fiber orientation, void content, knitline formation, etc. The importance of proper charge placement is discussed later.

2. *Mold closing*: After the charge has been placed in the bottom mold half, the top mold is quickly moved down to contact the top surface of the charge. Then the top mold is closed at a slower rate, usually at 5–10 mm/s. With rising temperature in the charge, the viscosity of SMC is reduced. As the molding pressure increases with continued mold closure, the SMC flows toward the cavity extremities, forcing the air in the cavity to escape through the shear edges or other vents. If the mold closing is too slow, the charge surface layers may gel before mold filling is completed. On the other hand, if the mold is closed too fast, air may be entrapped by the advancing flow front. Thus, mold closing speed is another key molding parameter that influences the quality of a compression-molded part.

The molding pressure based on the projected part area ranges from 1 to 40 MPa depending on the part complexity, flow length, and SMC viscosity at the mold temperature used. The common mold surface temperature is approximately 149 °C. Both top and bottom molds are externally heated to maintain the mold surface temperatures within ±5 °C of the desired mold temperature. The amount of flow in a compression molding process is relatively small compared with an injection molding process. However, it is still a critical parameter, which determines the quality of the part. It controls the void content due to air entrapment and fiber orientation in the part and thus affects the mechanical properties of the part.

3. *Curing*: After the cavity has been filled, the mold remains closed for a predetermined period of time to assure a reasonable level of curing and ply consolidation throughout the part. The curing time, which may vary from one to several minutes, depends on several factors, including the resin–catalyst–inhibitor reactivity, part thickness, and mold temperature. At the end of the curing time, the top mold is opened and the part is removed from the bottom mold with the aid of ejector pins. The part is then allowed to cool outside the mold, while the mold surfaces are cleaned of any remaining debris and sprayed with an external mold release agent in preparation for the molding of the next part.

As the part cools outside the mold, it continues to cure and shrink. Since the pressure restraint is not present at this postmolding stage, the part may exhibit distortions or generate residual stresses as a result of differential cooling at various sections in the part. The temperature distribution during cooling is important in determining the residual stresses in the final part; however, it is very much influenced by the part design.

In recent years, process developments have occurred that improve either the productivity or the quality of the compression-molded part. Three of these are described briefly.

1. *Charge preheating*: Preheating the charge (just before placing it in the mold) to a temperature below the gel point reduces the molding time and thus improves the productivity. The preheating can be done in dielectric heaters outside the mold so that the entire charge volume is heated up quickly to the same temperature.

2. *Vacuum-assisted molding*: The application of vacuum during the flow of SMC in the closed mold reduces air entrapment in the part, decreases the surface porosity, and improves the strength of the molded part.

3. *In-mold coating*: In-mold coating is used to mask the surface defects in a molded part, such as surface waviness, porosity, and sink marks. The most common method of in-mold coating involves opening the mold by a small amount (0.2–0.5 mm) part way through the molding cycle, injecting a flexible coating of a polyester or a polyester–urethane hybrid to cover the entire surface, reclosing the mold, and restoring the curing operation to the normal molding pressure. Curing of the coating may require additional time in the mold, thus increasing the overall mold cycle time. The second method of in-mold coating uses a high pressure injection of the coating and does not require opening and closing of the mold. Instead, the coating is injected at the precise moment when SMC exhibits the maximum curing shrinkage in the mold.

3.3 CURE CYCLE

The curing reaction that takes place in the mold transforms the uncured SMC into a solid part through the establishment of cross-links between the resin molecules. The degree of cure determines the number of cross-links, which, in turn, influences the physical and mechanical properties of the cured SMC part. In general, tensile strength, tensile modulus, heat resistance, and chemical resistance of SMC composites increase with increasing degree of cure, while the strain-to-failure and impact resistance show a decreasing trend. A high degree of cure is desirable for good quality parts.

The length of time required to produce a predetermined degree of cure is called the cure cycle. In a compression molding operation, the cure cycle comprises a major part of the molding cycle. Thus, any reduction in cure cycle would mean an increase in productivity. Such a reduction is achieved by changing either the material formulation, such as fast curing resins and highly reactive catalyst–inhibitor system, or the molding parameters, such as the mold temperature.

Figure 3.2 shows how the degree of cure of a vinyl ester resin varies with time at different cure temperatures. The rate of cure is very rapid at the beginning of the reaction and slows down considerably as the degree of cure asymptotically reaches a maximum value. Both cure rate and cure time can be controlled by adjusting the cure temperature.

Several investigators (2, 3) have shown that the cure rate of a thermosetting resin fits the following empirical relationship:

$$\frac{d\alpha_c}{dt} = (k_1 + k_2 \alpha_c^{\,m})(1 - \alpha_c)^n \tag{3.1}$$

where $\dfrac{d\alpha_c}{dt}$ = cure rate

α_c = degree of cure at time t

k_1, k_2 = rate constants

m, n = constants describing the order of reaction

A second-order reaction (i.e., $m + n = 2$) is commonly assumed for unsaturated polyester and vinyl ester resins. Lem and Han (3) have shown that the rate constants k_1 and k_2 increase with increasing temperature, while m and n are not much affected by the temperature (Table 3.1). Han and his coworkers have also shown that the low profile thermoplastic additives, which are used for controlling shrinkage, decrease the rate of cure as well as the final degree of cure.

As the cure progresses, the resin viscosity increases so rapidly that the flow of SMC in the mold can be severely affected. Poor flow may result in unfilled mold (short shot) and voids in the molded part due to air entrapment and incomplete wetout. As shown in Figure 3.3, the viscosity depends on the degree of cure as well as the curing temperature.

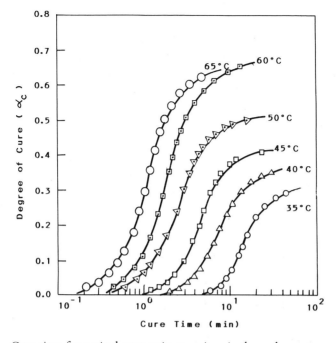

FIGURE 3.2   Cure time for a vinyl ester resin at various isothermal cure temperatures (1).

TABLE 3.1   Kinetic Parameters for Unsaturated Polyester and Vinyl Ester Resins (3)

Resin	Temperature (°C)	Constants			
		k_1 (min^{-1})	k_2 (min^{-1})	m	n
Unsaturated	50	0.0278	0.793	0.35	1.65
polyester	60	0.0924	1.570	0.40	1.60
Unsaturated polyester	50	0.0064	0.458	0.29	1.71
with 10 wt % polyvinyl acetate	60	0.0500	0.780	0.35	1.65
Vinyl ester	50	0.0151	0.998	0.35	1.65
	60	0.0624	1.590	0.49	1.51

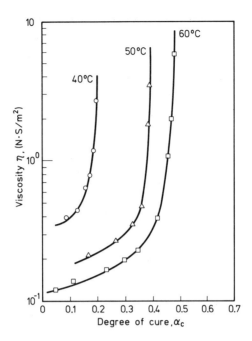

FIGURE 3.3   Viscosity versus degree of cure for a vinyl ester resin at various isothermal cure temperatures (1).

3.4 THERMAL EFFECTS

3.4.1 Temperature Distribution and Cure Time

When a charge is placed in the mold, the temperature in its surface layers increases rapidly to the mold surface temperature, whereas the temperature in the interior layers increases relatively slowly. Thus, in general, curing begins at the outer layers and progresses inward. The heat generated due to the curing reaction in each layer accelerates the initiation of curing in the next interior layer and increases the enthalpy content of the charge. However, as the curing reaction continues, the heat of reaction from the interior is not efficiently removed to the mold surface because of the poor thermal conductivity of the SMC material. The imbalance between the heat generation and the heat removal causes a temperature rise in the interior layers that may exceed the mold temperature. When the curing reaction nears completion, the heat generation is reduced and the temperature in the interior layers slowly approaches the mold temperature.

Mallick and Raghupathi (4) measured the temperatures at the outer surface, subsurface, and centerline of a 12.5 mm thick charge. The time–temperature curves (Fig. 3.4) for this charge thickness exhibit exothermic temperature peaks in the interior layers with the maximum peak observed at the centerline. The temperature in the surface layers remains close to the mold temperature. For a thinner charge, the temperature profiles may be more uniform than those shown in Figure 3.4.

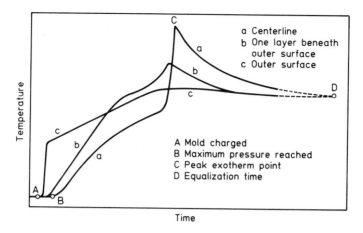

FIGURE 3.4 Schematic representation of temperature–time curves during a compression molding process (4).

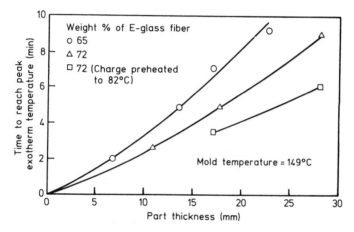

FIGURE 3.5 Time to reach peak exotherm temperature as a function of part thickness (4).

The time to reach the peak exotherm at the centerline is commonly considered to be the minimum cure time in the mold. It has been shown that nearly 90 % of the curing reaction is completed at this time. Removal of a part from the mold earlier than this time leaves the center of the charge uncured and, since the pressure is relieved on mold opening, interlayer cracks may generate near the center of the part. Mallick and Raghupathi (4) have shown that the mechanical properties of the molded part may be severely affected if the time in the mold is less than the minimum cure time.

The cure time, which controls the production rate, depends on the part thickness, the mold temperature, the resin–catalyst–inhibitor reactivity, and the thermal characteristics of the SMC material. It increases almost linearly with part thickness and may be as high as 10–15 minutes for part thicknesses greater than 12 mm (Fig. 3.5). One

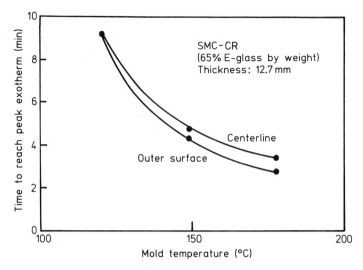

FIGURE 3.6 Time to reach peak exotherm temperature as a function of mold temperature (4).

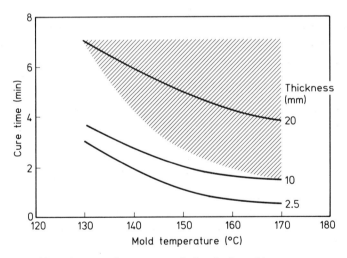

FIGURE 3.7 Molding diagram depicting a safe (hatched) mold temperature zone (5).

efficient method of reducing the cure time is to preheat the charge dielectrically outside the mold and quickly transfer it to the mold, where the curing reaction is completed. If the charge is preheated, the temperature distribution in the charge is more uniform in the thickness direction, which reduces the possibility of generating high residual stresses in the molded part.

TABLE 3.2   Effects of Initiator and Inhibitor on the Time to Reach Peak Exotherm (6)

Initiator		Inhibitor[a] concentration (parts by wt)	Time to reach peak exotherm temperature (seconds)	
Type[b,c]	Concentration (parts by wt)		Mold temperature = 125 °C	Mold temperature = 145 °C
TBP	0.49	0	268	90
TBP	0.98	0	178	81
PDO	5.5	0.55	113	65
PDO + TBP	2.75 + 0.25	0.28	120	60
PDO	5.5	0.28	80	54
PDO	5.5	0	47	43

[a] Inhibitor = benzoquinone.
[b] TBP = t-butyl perbenzoate; PDO = t-butyl peroxy-2-ethylhexanoate.
[c] The resin formulation contains 57 parts of an unsaturated polyester resin in styrene, 43 parts of polymethyl methacrylate in styrene, 138 parts of $CaCO_3$, and 9.6 parts of $Mg(OH)_2$.

The cure time in compression molding can be reduced by increasing the mold temperature (Fig. 3.6); however, the peak exotherm temperature is increased at high mold temperatures. If the maximum peak exotherm temperature exceeds 200 °C, the resin near the centerline may char and decompose, creating a weak interlaminar zone in the interior of the part. Based on this observation, Panter (5) has generated molding diagrams (Fig. 3.7) that can be used to determine the safe mold temperature zones for various part thicknesses.

The effect of the resin–catalyst–inhibitor reactivity on the cure time was studied by Fan, Marinelli, and Lee (6). Their data are presented in Table 3.2, which shows that the catalyst concentration, dual-catalyst combination, and inhibitor concentration in the SMC resin paste can be controlled to reduce the time to reach peak exotherm.

The thermal characteristics of the SMC material that influence the cure time in compression molding are its heat capacity and thermal conductivity. In general, these properties do not depend strongly on the type of the resin; however, they can be controlled by varying either the filler content or the fiber content. Increasing the filler content in SMC reduces the maximum peak exotherm temperature and allows higher mold temperatures to be used, which in turn, reduces the cure time. Increasing the fiber content has a similar effect on the cure time of SMC.

3.4.2 Cure Kinetics

The curing reaction of sheet molding compounds is basically a free radical chain growth copolymerization between the styrene monomer and the unsaturated polyester or vinyl ester molecules. In recent years, Lee (7) has developed the following kinetic equations for the curing reactions of SMC:

77

1. Initiation

$$\frac{dC}{dt} = 2k_d I \tag{3.2}$$

2. Inhibition

$$qZ_0 = 2f(I_0 - \bar{I}_0) \tag{3.3}$$

3. Propagation

$$\frac{d\alpha}{dt} = 2f\bar{I}_0 k_p \left\{ (1 - \alpha)\left[1 - \exp\left(\int_{t_z}^{t} k_d\, dt\right)\right]\right\} \tag{3.4}$$

where C = free radical concentration
     I = catalyst concentration
     I_0 = initial catalyst concentration
     Z_0 = initial inhibitor concentration
     $\bar{I}_0$ = catalyst concentration after all inhibitors are consumed
     f = catalyst efficiency
     q = inhibitor efficiency
     t = time
     t_z = induction time before propagation
     k_d = rate constant for catalyst decomposition
     k_p = rate constant for monomer propagation
     α = fractional conversion

The rate constants k_d and k_p depend on the surrounding temperature:

$$k_d = A_d \exp\left(\frac{-E_d}{RT}\right) \qquad k_p = A_p \exp\left(\frac{-E_p}{RT}\right)$$

where A_d, A_p = constants
     E_d, E_p = activation energies
     R = universal gas constant
     T = temperature

Unlike the empirical nature of Eq. (3.1), Eqs. (3.2)–(3.4) are based on the cure mechanism envisioned in a compression molding process. Three major simplifying assumptions are involved in developing the kinetic equations (3.2)–(3.4). They are:

1. Homopolymerization of the polyester molecules is negligible.
2. The monomer reaction begins only after the catalyst radicals have consumed all inhibitor molecules initially present in the system.
3. The propagation rate is not affected by the monomer diffusion at the early stages of conversion.

In spite of these assumptions, Lee and his coworkers (6, 7) have found a reasonable agreement between the reaction time predicted by Eq. (3.2)–(3.4) and the data obtained in isothermal differential scanning calorimetry (DSC) experiments. These equations are useful in predicting the resin–catalyst–inhibitor reactivity and optimizing resin formulations for maximum productivity.

3.4.3 Heat Transfer in the Mold

Heat transfer accompanying the curing process is modeled by the following one-dimensional heat conduction equation:

$$\rho c_p \frac{\delta T}{\delta t} = k \frac{\delta^2 T}{\delta z^2} + H_R M_0 \frac{\delta \alpha}{\delta t} \tag{3.5}$$

where T = temperature at time t and location z

ρ, c_p, k = density, specific heat, and thermal conductivity of SMC, respectively

H_R = heat of reaction

M_0 = initial monomer concentration

$\dfrac{\delta \alpha}{\delta t}$ = conversion rate given by Eq. (3.4).

Equation (3.5) has been solved using a finite difference method for a part of uniform thickness, h. The mold temperature is assumed to be constant. The solution of this heat transfer equation generates considerable amount of pertinent molding information, such as the time to reach peak exotherm (Fig. 3.8) or the temperature profiles during molding (Fig. 3.9).

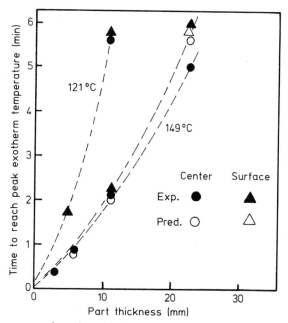

FIGURE 3.8   Time to reach peak exotherm temperature versus part thickness at various mold temperatures (7).

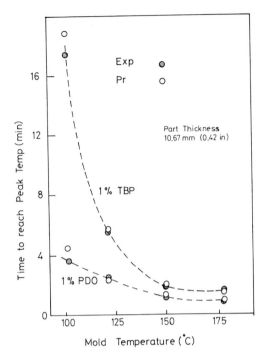

FIGURE 3.9   Time to reach peak exotherm temperature versus mold temperature for two different catalysts (7).

3.5 FLOW BEHAVIOR

The flow of SMC in the mold is a complex phenomenon; however, it is the most important factor in determining the quality of a molded part. It not only determines the extent of cavity filling, but also influences the fiber orientation, fiber distribution, porosity, and surface defects in a molded part. This section discusses various experimental and theoretical studies undertaken to provide insight into the flow behavior of SMC.

3.5.1 Experimental Observations

The first experimental study on the flow behavior of SMC was conducted by Marker and Ford (8). They used an instrumented mold with a disc-shaped cavity to examine the pressure distribution, temperature profile, and flow during the compression molding process. A round charge pattern with layers of white, black, and gray SMC was employed. The mold closing speed was approximately 1 mm/s. From the cross-sectional views of the molded disc, the investigators concluded that the charge surface layers flow out to the edges of the mold before the interior layers begin to deform. This was attributed to the rapid reduction in the viscosity of the surface SMC layers after they contact the mold surface. Since the hotter surface layers have moved out to the edges, the curing reaction begins at the edges and propagates toward the center of the charge. Thus, the curing reaction does not occur simultaneously over the entire part.

Several years later, Barone and Caulk (9) used an experimental approach similar to that of Marker and Ford, but used a larger cavity and two different mold closing speeds, namely 1.75 and 10 mm/s. Circular charge patterns with alternate layers of black and white SMC were used. The amount of mold closing was controlled to produce parts with various degrees of mold flow. From examinations of cross sections of these parts (Fig. 3.10), Barone and Caulk drew the following conclusions.

1. At the faster mold closing speed, all layers in a charge extend uniformly, with slip occurring only at the mold surface. The outer layers do not flow any further than the interior layers. The charge thickness does not affect the uniform flow pattern at this mold closing speed.

2. At the slower mold closing speed, the flow pattern depends on the charge thickness. For a charge 5 or 6 layers thick, the surface layers adjacent to the hot mold surfaces undergo a greater extensional deformation than the interior layers. This is similar to the flow pattern observed by Marker and Ford for a 10-layer charge at a

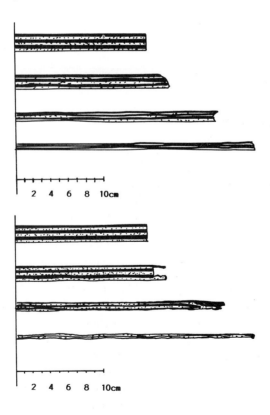

FIGURE 3.10   Stages of deformation for a six-layered charge at fast (top) and slow (bottom) mold closing speeds (9).

mold closing speed of 1 mm/s. However, as the charge thickness is reduced to only 3 layers, the flow pattern becomes uniform, with no relative motion between the layers. For both thick and thin charges, slip occurs at the mold surface; for thick charges only, however, slip also occurs between the surface and interior layers at slow mold closing speeds.

3.5.2 Theoretical Models

In recent years, attempts have been made to develop theoretical models for the flow pattern of sheet molding compounds in a compression molding process. The ultimate goal of these models is twofold: to predict the occurrence of molding defects, such as short shots (partial filling of the cavity) and knit lines, and to provide guidelines for selecting the optimum charge shape as well as its location in the mold. Two theoretical approaches for the mold flow study are briefly described in this section. In both cases, the SMC charge is assumed to be thin, so that an isothermal condition exists through its thickness. The flow takes place in the xy plane and the mold closing is in the negative z direction (Fig. 3.11).

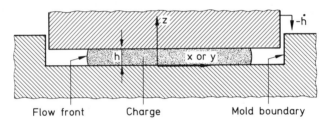

Flow front   Charge   Mold boundary

FIGURE 3.11   Nomenclature for compression mold filling analysis (11).

The first approach is due to Tucker and his coworkers (10–12), who have used the following assumptions in their models:

1. The sheet molding compound is an incompressible, isotropic, Newtonian fluid at the molding conditions used.
2. No-slip conditions exist between the charge surface layers and the mold surfaces.
3. The dominant stresses are the transverse shear stresses across the narrow gap between the mold surfaces; and the in-plane stresses are negligible.

Based on these assumptions, they developed the following equation, which is solved for the pressure distribution across the charge:

$$\frac{\delta^2 p}{\delta x^2} + \frac{\delta^2 p}{\delta y^2} = -\frac{12\,\mu\dot{h}}{h^2} \tag{3.6}$$

where p = pressure (which is a function of x and y)
μ = viscosity (assumed to be a constant)
h = instantaneous charge height
$\dot{h}$ = mold closing speed (assumed to be a constant)

Equation (3.6) is known as the generalized Hele–Shaw model. It can be solved for the pressure distribution using a finite difference (13), finite element (10, 11), or boundary element method (12). If the pressure distribution is known, the velocity field can be determined across the entire flow area. Tucker and his coworkers calculate the average velocities u and v in the x and y directions, respectively, using the following two equations:

$$\bar{u} = - \frac{h^2}{12\,\mu} \left(\frac{\delta p}{\delta x} \right) \tag{3.7}$$

$$\bar{v} = - \frac{h^2}{12\,\mu} \left(\frac{\delta p}{\delta y} \right) \tag{3.8}$$

After their experimental observations (see Section 3.5.1), Barone and Caulk (14, 15) proposed that a thin resin layer between the mold surface and the charge surface layer acts as a lubricant and provides a condition for slippage (relative motion) at the mold surface. Furthermore, uniform extension of each layer suggests negligible shear stresses in the thickness direction.

Barone and Caulk (14) found that the hydrodynamic friction model is appropriate for describing the slippage at the mold surface. Assuming that the hydrodynamic friction forces f_x and f_y are proportional to the relative velocity between the SMC and the mold surface, that is,

$$f_x = - k_H\, u$$

$$f_y = - k_H\, v$$

where k_H is the coefficient of hydrodynamic friction, they developed the following equation for the flow of SMC:

$$\frac{\delta^2 p}{\delta x^2} + \frac{\delta^2 p}{\delta y^2} = \mu h \left(\frac{\delta^3 u}{\delta x^3} + \frac{\delta^3 u}{\delta x\, \delta y^2} + \frac{\delta^3 v}{\delta x\, \delta y^2} + \frac{\delta^3 v}{\delta y^3} \right)$$
$$- 2k_H \left(\frac{\dot{h}}{h} \right) \tag{3.9}$$

The first term on the right-hand side of Eq. (3.9) represents the resistance of the material to extensional deformation, and the second term represents the frictional resistance generated by the slip of SMC surface layer at the mold surface. Equation (3.9) has been solved by Barone and Osswald (16) using the boundary element method.

For a thin charge, material resistance is neglected, which reduces Eq. (3.9) to a simpler form:

$$\frac{\delta^2 p}{\delta x^2} + \frac{\delta^2 p}{\delta y^2} = - 2k_H \left(\frac{\dot{h}}{h} \right) \tag{3.10}$$

Although Eqs. (3.6) and (3.10) are apparently similar, the difference in their right-hand sides should not be overlooked. In Eq. (3.6), the right-hand side represents

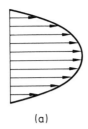 

(a)   (b)

FIGURE 3.12   Velocity profile through the thickness of a charge with (a) no slip at the mold wall and (b) slip allowed at the mold wall.

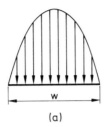 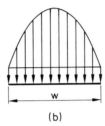

(a)   (b)

FIGURE 3.13   Pressure distribution according to (a) model proposed by Tucker and coworkers and (b) model proposed by Barone and Caulk.

the resistance to transverse shear deformation, and in Eq. (3.10) it represents the hydrodynamic frictional resistance at the mold surfaces.

The difference in the two approaches becomes more clear as one compares the velocity profiles and pressure distributions predicted by Eqs. (3.6) and (3.10). With the no-slip condition assumed in Eq. (3.6), the velocity profile is parabolic through the charge thickness (Fig. 3.12a). Slip is allowed in Eq. (3.10) and, therefore, the velocity profile becomes more uniform (Fig. 3.12b). Figure 3.13 compares the pressure distributions across the surface of a circular charge pattern predicted by the two approaches. According to Tucker and his coworkers, the normal pressure variation is parabolic with zero values at the free boundary. In the approach used by Barone and Caulk, the normal pressure distribution has two components—a uniform pressure due to viscous resistance and a parabolic variation due to frictional resistance. As the flow progresses, both h and μ decrease. This increases the contribution of the friction and decreases the contribution of the viscous resistance. Eventually, the viscous resistance becomes insignificant compared to friction and the pressure distribution becomes parabolic.

Tucker and his coworkers as well as Barone and Caulk have applied their respective models to predict the flow front progression using various charge patterns. Figures 3.14 and 3.15 show some of these results and compare them with the experimental observations. Although such comparisons are limited, it appears that the Barone–Caulk model works better for a thick SMC charge. For a relatively thin SMC charge, the difference between the two models becomes small; however, the Tucker model is easier to implement, since it is difficult to determine the hydrodynamic friction coefficient k_H in the model of Barone and Caulk.

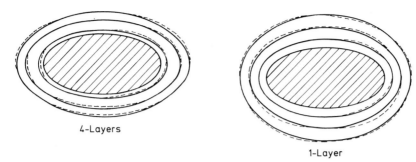

4-Layers

1-Layer

FIGURE 3.14   Experimental flow front progression (dashes) in four- and one- layered elliptical charges at three different stages of deformation compared with the thin cavity solution (solid lines) proposed by Barone and Caulk (14).

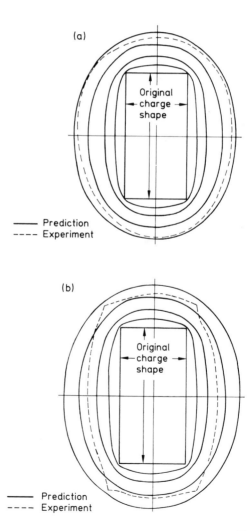

FIGURE 3.15   Flow front progression for a rectangular charge (75 mm × 150 mm) predicted by the model of Tucker and coworkers for charges 3 mm thick (a) and 9 mm thick (b) (11).

85

In the theoretical models described above, an isothermal condition is assumed to exist in the thickness direction. This assumption may be considered to be valid for a relatively thin charge. However, for a thick charge, a temperature gradient is observed across the thickness during the flow and early stages of curing. This temperature gradient may also cause a large viscosity gradient in the thickness direction with the viscosity increasing from a low value near the cavity surface to the maximum value at center of the charge. Metzner and his coworkers (17) modeled this condition through their analysis of squeeze molding of a disc of two Newtonian fluids having different viscosities. They have shown analytically that the existence of viscosity gradient across the thickness can result in a preferential flow of the lower viscosity fluid, and, therefore, a failure of the parallel squeezing assumption. They defined a nondimensional term S to divide the flow regime into two distinct flow patterns:

$$S = \frac{\mu_{max} h^2}{4 \mu_{min} r^2} \tag{3.11}$$

where μ_{max} = viscosity of the high viscosity fluid
μ_{min} = viscosity of the low viscosity fluid
h = gap between the parallel cavity surfaces
r = disc radius

When S is small, parallel squeezing occurs and the maximum velocity is in the high viscosity fluid at the center. For large values of S, parallel squeezing does not occur and the maximum velocity is in the low viscosity fluid, which is squirted out ahead of the high viscosity fluid. It is interesting to note that these analytical conclusions match the experimental observations cited earlier.

3.6 FIBER ORIENTATION

The ideal fiber orientation in an SMC-R composite is planar random, since it produces an isotropic behavior with equal physical and mechanical properties in all directions in the plane of the composite part. However, the complex flow pattern in a compression molding process may induce localized fiber orientation in a preferential direction, which makes the part stronger in that direction, but weaker in other directions. Such preferential orientations are found if the flow path is excessively long or obstructed by inserts or mold edges, or if the flow takes place across convergent or divergent channels.

3.6.1 Measurement Techniques

The average fiber orientation in the plane of a thin composite part (18) is described by orientation parameters f and g, which are defined as follows:

$$f = \tfrac{1}{2} [3 \langle \cos^2 \theta \rangle - 1] \tag{3.12}$$

$$g = \tfrac{1}{4} [5 \langle \cos^4 \theta \rangle - 1] \tag{3.13}$$

where $\langle \cos^m \theta \rangle = \int_0^{\pi/2} n(\theta) \cos^m \theta \sin\theta \, d\theta$

$\quad n(\theta) = $ distribution of the fiber orientation angle θ

The fiber orientation angle θ is usually measured with respect to the principal loading direction. For completely aligned fibers, $f = g = 1$, and, for a completely random orientation of fibers, $f = g = 0$. For compression-molded SMC composites, both f and g are between 0 and 1. Experimental studies by several investigators show that these orientation parameters may vary significantly from part to part even if the same charge pattern is used for molding all the parts. Furthermore, orientation parameters may differ significantly across part thickness or from one region of a part to another.

Fiber orientation in a compression-molded part can be examined using conventional X-ray radiography. However, the X-ray absorption of E-glass fibers is not sufficiently greater than the surrounding resin and fillers to provide high contrast radiographs. The contrast can be improved by adding to the SMC sheet small quantities of tracer fibers, such as glass fibers containing high concentrations of lead or metal-coated glass fibers. The X-ray images can be analyzed by using computer-assisted digitized techniques. From the digitized recording of the end-point coordinates of each fiber, orientation parameters f and g can be easily determined. Kau (19) has described another technique in which the top and bottom surfaces of a compression-molded plaque specimen are photographed by a video camera. These photographs are used for determining the orientation parameters with the help of a computer based data acquisition system.

Denton and Munson-McGee (20) used the Fraunhofer light diffraction technique for determining the fiber orientation pattern in short fiber composites. The diffraction pattern was obtained by passing light rays from a laser source through a high contrast photographic slide of an X-ray image. A round diffraction pattern is obtained if the fibers are randomly oriented. A two-lobed diffraction pattern appears if a preferential fiber orientation exists in the molded part. With increased degree of fiber orientation, these lobes are elongated in the direction transverse to the preferential orientation. Through comparison of these diffraction patterns with those associated with known fiber orientations, the orientation parameter of a molded part can be estimated.

3.6.2 Model

Jackson, Advani, and Tucker (21) have developed a model for the flow-induced fiber orientation in a thin, compression-molded flat part assuming no velocity variation across the thickness. The generalized Hele–Shaw equations were used for predicting the flow and deformations during mold filling. The velocity fields obtained from the solutions of the Hele–Shaw equations were input into a differential equation describing the fiber orientation distribution in a concentrated suspension of fibers. Figure 3.16 shows the theoretical fiber orientation distributions for 33 and 67 % initial mold surface coverages. These theoretical predictions compare reasonably well with the experimental measurements.

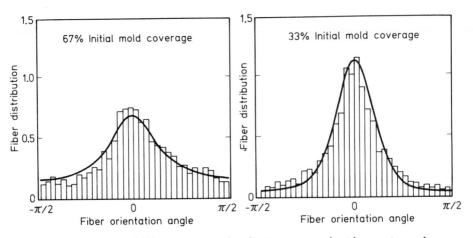

FIGURE 3.16   Theoretical fiber orientation distribution compared with experimental histograms for 67 and 33% initial mold surface coverage (21).

3.6.3 Effects on Mechanical Properties

Flow-induced fiber orientation causes direction-dependent (anisotropic) mechanical properties in a compression-molded SMC-R part. This results in higher strength and modulus values in the direction of fiber orientation than in the transverse direction. To determine the effects of fiber orientation on the mechanical properties of SMC-R composites, Berthelot (22) used charge surface areas ranging from 22 to 76 % of the mold surface area in a flat plaque mold cavity. Only the charge width was varied (Fig. 3.17); thus the flow was one-dimensional. The mechanical properties were measured on specimens cut from various locations in the flow direction of the molded plaques. Figure 3.17 shows the variation of flexural strength in specimens molded with 31 % charge surface area. Although the flexural strength is nearly independent of the specimen location, there is a significant difference between the longitudinal and transverse directions. Similar observations were made by Denton (23), who used a 38 % mold surface coverage and measured the tensile strength variation across the molded plaque. His data are given in Table 3.3.

TABLE 3.3   Tensile Properties of SMC-R50 Parallel and Normal to Flow (23)

Direction or location	Tensile strength (MPa)	Tensile modulus (GPa)
Parallel to flow	215	17.4
Normal to flow	75.2	10.9
Inside the charge area	155	14.1
Outside the charge area	135	14.3

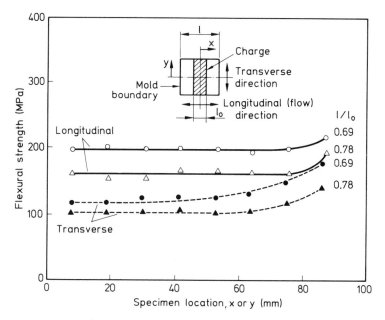

FIGURE 3.17 Variation of flexural strength in the longitudinal (flow) and transverse directions as a function of position (*x* for the longitudinal specimens; *y* for the transverse specimens) (22).

3.7 DEFECTS

Compression molding may produce a variety of surface and internal defects in SMC parts (Table 3.4). The surface defects, such as waviness, pinholes and sinkmarks, create unacceptable surface appearance for some applications, particularly for highly visible automotive body panels. The surface appearance can be improved by buffing and painting outside the mold or by in-mold coating the surface during part molding. The internal defects, such as voids, knit lines, and fiber orientation, may affect the mechanical performance of the molded part. They can be detected by nondestructive

TABLE 3.4 Defects in Compression Molded SMC Parts

Surface Defects	Internal Defects
Pinhole	Voids
Waviness	Blisters
Craters	Delamination
Sink marks	Knit lines
Surface roughness	Preferential fiber orientation
Dark areas	Fiber buckling (in SMC-CR or XMC)
Pop-up blisters in painted parts	Undercure
	Resin-rich areas
	Warpage
	Residual stress

testing techniques, such as radiography or ultrasonic testing. Many of these molding defects can be eliminated or reduced through proper material selection, part design, and molding techniques. The source and remedies for some of the major molding defects are discussed next.

3.7.1 Porosity

Porosity results from the air entrapment in a molded part. Air bubbles generated at the mixing and compounding stages of SMC sheet production remain in the resin paste, at the fiber–resin interfaces, and between the fiber bundles. Air is also trapped between the stacked layers in the SMC charge as well as in the closed mold. A substantial amount of trapped air is carried away by the flow of SMC toward the vents and shear edges in the mold. The use of vacuum in the mold also helps remove air. However, if the viscosity of SMC at the molding temperature is too high or the proper mold closing speed is not used, air entrapment in the part is increased. Blocked or inadequate venting in the mold also increases the air entrapment in the part.

The amount of entrapped air in the SMC sheet can be reduced by several means, such as mixing the SMC resin formulation in vacuum, controlling the viscosity increase of the resin paste before compounding, and improving the resin–fiber impregnation at the compaction stage. Himebaugh and Newman (24) have shown that a larger proportion of air is entrapped at the compounding and impregnating stages than at the mixing stage. Since much of this air is at the fiber–resin interfaces, it is important to improve the fiber surface wetout, which increases the displacement of air from the interfacial zones. This can be achieved by adding active additives to the resin paste (25), which tends to lower the interfacial surface energy; it also improves impregnation and increases the air release from the fiber surfaces. As shown in Figure 3.18, the interlaminar tensile strength of an SMC part is improved significantly with the inclusion of these additives.

3.7.2 Blisters

Blisters are interlayer cracks formed close to the surface of a molded part as a result of the internal pressure of volatiles (such as styrene monomer) or entrapped air. They are often manifested as visible dome-shaped bulges on the part. They usually form at the time of mold opening when the external pressure is released. If the molded part is exposed to elevated temperatures—for example, in paint bake ovens—the expansion of the entrapped gas may also cause blisters.

Blister formation can be reduced in two ways (26):

1. By minimizing the entrapment of volatiles and air by vacuum-assisted molding as well as good flow of SMC in the mold.
2. By increasing the interlaminar shear strength by using coupling agents, improving resin properties, reducing contaminants, and ensuring proper cure before the mold is opened.

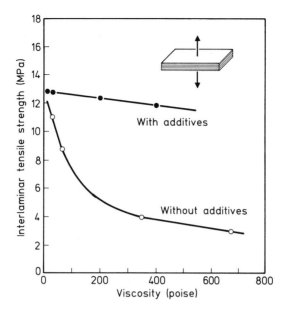

FIGURE 3.18 Interlaminar tensile strength of compression-molded plates as a function of viscosity of the SMC compound (25).

3.7.3 Delamination

Delamination or separation of layers is usually observed near the centerline of thick parts. Possible causes of delamination are excessive thermal stresses, internal stresses due to surface pregelling, incomplete curing, and poor fiber wetting by the resin. Lynskey and Robertson (27) have shown that a proper combination of catalyst, inhibitor, and retarder is beneficial in reducing the delamination. In their study, massive delamination was observed in 12 mm thick SMC-R50 plaques when molded with a standard unsaturated polyester resin and *t*-butyl perbenzoate. Additions of stilbene (a secondary monomer), parabenzoquinone (inhibitor), and orthodinitrobenzene (retarder) reduced the delamination problem up to 15 mm thickness. However, changing to a slower catalyst, such as *t*-butyl hydroperoxide, had the most dramatic effect, since it enabled plaques to be molded up to 33 mm thickness without delamination. The principal effects of these modifications are an increase in time to reach peak exotherm and a reduction in the peak exotherm temperature, both resulting from a slower or delayed curing reaction. Since the curing reaction is more uniform through the thickness, internal stresses are reduced.

3.7.4 Sink marks

Sink marks are small dimples or depressions that appear on the surface of a molded part directly opposite to the location of ribs and bosses (Fig. 3.19). They do not affect the performance of the part; however, they are visible to the eye and may not be acceptable from an appearance standpoint. The appearance problem becomes more

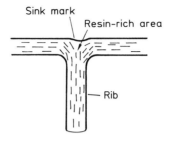

FIGURE 3.19 Schematic representation of sink marks in the surface of a compression-molded part.

unacceptable when a relatively flat surface such as an automobile hood is painted to a high gloss.

Jutte (28) as well as Marker and Ford (8) have shown that the SMC flows into ribs or bosses only after the adjacent flat areas in the mold have been filled. The material entering the ribs and bosses is elongated and folded, which tends to orient the fibers parallel to the ribs and bosses (Fig. 3.19). The process of filling the ribs and bosses creates a resin-rich zone near the opposite flat surface. Since this resin-rich zone has a higher shrinkage than the surrounding material, slight depressions are created on the flat surface as the part cools from the molding temperature.

A number of steps can be adopted to reduce the sink depth. They are: (a) use a low profile additive to reduce the resin shrinkage, (b) use SMC layers of varying fiber lengths with longer fiber (50–75 mm) layers placed near the flat surface and shorter fiber (12–25 mm) layers placed near the ribbed surface, (c) avoid thick sections directly above the ribs and use a small corner radius (0.12–0.25 mm) at the rib juncture, and (d) if possible, use a protruded corner (29) instead of a square or rounded corner at the rib juncture (Fig. 3.20).

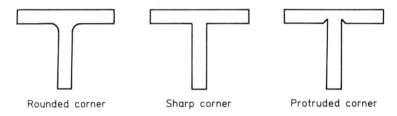

Rounded corner   Sharp corner   Protruded corner

FIGURE 3.20 Various rib corner designs (29).

3.7.5 Knit Lines

Knit lines are formed when two or more separate flow fronts join to form a single flow front. The fibers in the knit line tend to be preferentially oriented parallel to the knit line. Thus, the strength of the part in a direction normal to the knit line is reduced, creating the possibility of early cracking because of molded-in stresses or applied stresses.

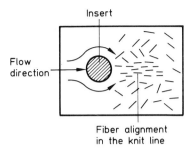

FIGURE 3.21   Knit line behind a core pin.

A common location of knit lines is behind the molded-in holes, which are produced using core pins in the mold (Fig. 3.21). If these holes are located close to the edge of a part or if two or more holes are too close together, cracks generated in the knit lines behind them may easily extend to the edge or join the adjacent holes, making the part potentially weaker than its design level. In many applications, it may be desirable to drill the holes after the molding operation. Whenever possible, molded-in holes should be located in the low stress areas of the part.

3.7.6 Warpage and Residual Stresses

Warpage is caused due to nonuniform cooling at various sections in the part. It frequently occurs in thin sections, which are located adjacent to thicker sections. For this reason, parts should be designed with uniform thickness to the extent possible. Another source of warpage is the flow-induced fiber orientation, which creates anisotropic coefficients of thermal expansion. This can cause differential shrinkage in various directions of the part and ultimately may result into warpage or residual stresses.

One source of residual stresses in a thicker part is the nonuniform curing either through its thickness or over its surface area. The areas that cure first have compressive residual stresses, while the areas that cure later are left with undesirable tensile residual stresses.

3.8 MOLDING PARAMETERS

Principal molding, material, and tooling parameters that affect the quality of a compression molded part are listed in Table 3.5. Several investigators have studied the effects of the molding parameters on the mechanical properties of compression-molded flat plaques. The results of these studies are summarized below.

1. *Cure time*: Mallick and Raghupathi (4) have shown that inadequate cure time in the mold leaves uncured and partially consolidated layers in the interior of thick parts. These, in turn, cause a strength variation across the part thickness, as demonstrated in Figure 3.22, which shows that the flexural strength at the center of a part that is 12.5 mm thick is considerably lower than that at its outer surfaces for short mold cure

93

TABLE 3.5   Principal Parameters in a Compression Molding Process

Molding Parameters	Tool Parameters
Mold temperature	Mold design (radius, ribs, etc.)
Molding pressure	
Mold closing speed	Shear edges
Charge specification	Vents
(geometry, placement, and size)	Ejection system
	Parting line
Material Parameters	Draft
Resin paste formulation	Tool material
Resin-catalyst-inhibitor reactivity	Surface finish
Maturation time	
Sheet thickness	
Sheet temperature	

times. The difference in flexural strengths will be reduced as the mold cure time is increased.

2. *Mold closing speed*: Burns et al. (30) have shown that the mold closing speed has a profound effect on the tensile strength of a compression molded part. Fast closing speeds lead to low strengths (Fig. 3.23) because inefficient removal of air from the SMC charge results in porosity.

3. *Molding pressure*: The effect of molding pressure was also studied by Burns and his coworkers (30). In their experiments, the tensile strength increased by as much as 20 % as the molding pressure was increased from 3.5 to 15 MPa.

4. *Charge size*: Charge size determines the initial mold surface coverage and the length of flow in the mold. As discussed earlier, excessively long flow lengths may produce flow-induced fiber orientation which, in turn, causes direction-dependent properties

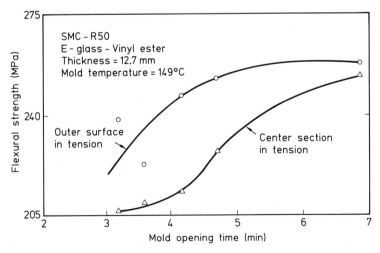

FIGURE 3.22   Flexural strength variation as a function of mold cycle time (4).

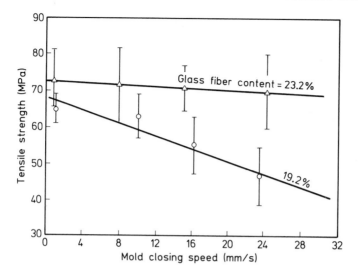

FIGURE 3.23   Effect of mold closing speed on the tensile strength of a compression-molded panel (30).

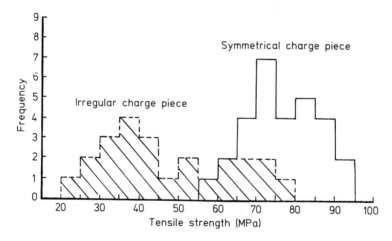

FIGURE 3.24   Tensile strength distribution as affected by the charge shape (30).

in the part. On the other hand, very limited flow may not carry the air out of the charge, thus producing voids and blisters.

To maintain a constant charge weight, small strips of SMC are often added, which compensates for the SMC thickness variation. Burns et al. (30) have shown that the addition of such small strips can produce a lower average strength as well as a wider variation of strength in the part (Fig. 3.24).

For SMC-CR and XMC containing continuous fibers, the charge should cover nearly 90 % of the mold surface area. Otherwise, resin flow in the transverse direction of continuous fibers can cause severe fiber buckling and a consequent reduction in mechanical properties in the longitudinal direction (31).

5. *Charge placement*: Proper charge placement in the mold is important for a number of reasons. If the charge is placed unevenly, it can produce eccentric loading on the press platens and tilt them, which, in turn, causes unwanted thickness variation in the part. Uneven charge placement makes the resin flow first on one side of the cavity; then, with continued pressure, it fills the entire cavity. This backflow may result in nonuniform fiber distribution, resin–rich areas, uneven curing, surface waviness, etc. If the charge is divided into two pieces and placed separately in the mold, a knit line may result at their joining (Fig. 3.25).

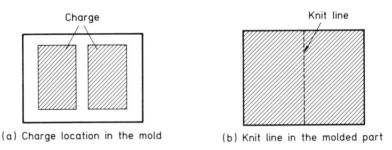

Charge

Knit line

(a) Charge location in the mold   (b) Knit line in the molded part

FIGURE 3.25   Knit line formed at the joining of two separated charges.

3.9 PROCESS CONTROL

Hydraulic presses are commonly used for compression molding operations. However, conventional hydraulic presses lack the controls and sophistication required for molding good quality parts. There have been a number of recent developments in the area of press design and process controls that are responsible for improving quality, reducing rejection rates, and increasing the productivity. These developments are now discussed.

3.9.1 Platen Parallelism

When the mold is closed, the upper platen may not remain parallel to the lower platen because of eccentric loading due to uneven charge placement or pressure distribution in the mold. The clearance in the guidance of the platens and the nonuniform elastic deformation of the press frame tend to increase the problem. Nonparallel platens cause poor thickness tolerance in the molded part. Furthermore, they result in nonuniform cure rate and flow instability, both of which affect the quality of the part. To correct this problem, the press is fitted with four independent hydraulic cylinders, one at each corner of the moving platen. Position sensors located at these corners monitor the platen movement and transmit the parallelism error signal to the hydraulic cylinders, which activate differentially to maintain platen parallelism throughout the molding operation.

3.9.2 Mold Heat Management

The molds are heated by pumping oil or superheated steam through channels drilled in the platens. The usual practice is to space these channels evenly across each platen and locate them at a uniform distance from the cavity surface. However, this is not the optimum design, since heat is removed nonuniformly from the mold at the mold closing stage. The surface area covered by the original charge becomes a heat-deficient zone, while the areas in which heated SMC flows become a heat-rich zone. As a result, a nonuniform temperature distribution exists in which the cavity surface temperature near the centerline of the mold is lower than the preset mold temperature and that near the outer perimeter exceeds the preset mold temperature. Since this condition creates a nonuniform cure rate and affects the part quality, considerable attention should be given to the optimum thermal design of the mold. Barone and Caulk (32) have developed analytical procedures that can be used to predict the heat removed or added per cycle as a function of the position on the mold surface. These calculations can be used to locate the heating channels appropriately so that a uniform cavity surface temperature exists throughout the molding cycle. An example is shown in Figure 3.26.

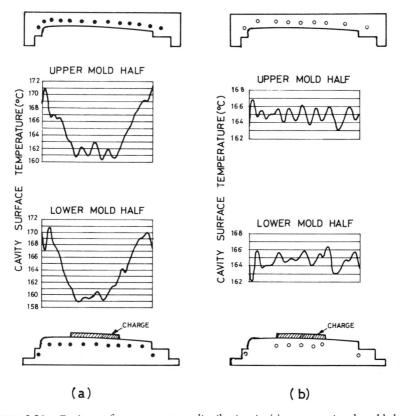

FIGURE 3.26   Cavity surface temperature distribution in (a) a conventional mold design and (b) an optimized mold design (33).

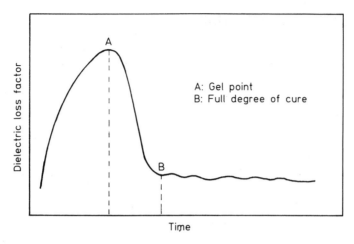

FIGURE 3.27   Dielectric loss factor as a function of cure time.

3.9.3 Cure Monitoring

In-process cure monitoring is important to ensure that each part has properly cured before the mold is opened. One method being developed for this purpose measures the dielectric loss factor (34), which increases rapidly in the beginning of the cure cycle, attains a peak, and then reduces to a constant value as the curing reaction is completed (Fig. 3.27).

The technique for monitoring dielectric loss factor is relatively simple. It uses two metal electrodes placed opposite each other at critical locations on two sides of the mold. When the SMC charge is placed between these electrodes, a capacitor is formed. Through proper instrumentation, the dielectric loss factor is continually monitored throughout the molding cycle. Although this technique has not found much commercial use, it has the potential of automatically controlling the press operation.

3.9.4 Process Automation

Considerable progress has been made in recent years in the area of compression molding process automation (35, 36). For example, auxiliary machines are now available for automatic cutting, weighing, and stacking of SMC sheets. The stacked SMC charge is then compressed outside the press to squeeze out the air between the layers and is fed into the press by means of a computer-operated robot manipulator. The use of the feeding manipulator assures exact positioning of the charge in the mold in a relatively short feeding cycle. Computer-controlled manipulators have also been developed to unload the molded part from the press automatically, deflash it around the desired contour, and perform postmolding secondary operations, such as drilling and painting. All these developments have helped improve the quality as well as the productivity of the compression molding operation.

3.10 APPLICATION EXAMPLES

Compression-molded SMC composites are used in many automotive, business machine, appliance, construction, and industrial applications. Four such applications are briefly described.

3.10.1 Computer Enclosures

Computer enclosures are commonly fabricated from sheet steel panels, which are expensive to finish and paint to an exact color, gloss, and texture. They are now being compression molded using integrally pigmented sheet molding compounds. The surface is textured in the mold to produce the desired finish. The material meets the Underwriters Laboratory (UL) requirements for flame spread and smoke generation. It also has a continuous use temperature of 130 °C.

The enclosures are molded as thin as 1.5 mm. The flatness of the molded panels is assured by designing proper ribs on the nonappearance side of the panels. For example, it is observed that X-patterned ribs are better in maintaining flatness than circular patterned ribs (Fig. 3.28). Molded-in inserts at the bases are used to secure the panels to the computer frame.

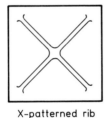 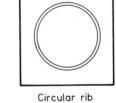

X-patterned rib   Circular rib

FIGURE 3.28   Rib patterns used in computer enclosures.

3.10.2 Dishwasher Inner Doors

Compression molded SMC composites are being used in many appliances, such as dishwashers, refrigerators, laundry equipment and air conditioners, to replace porcelainized and painted steel. An example is the inner door of a dishwashing machine (37), which is compression molded using a thick molding compound (TMC). The material is pigmented to produce a white color, which eliminates the need for painting. The TMC door is molded in one piece with mounting bosses and reinforcing ribs and has fewer seams and fasteners than a steel door. The molding time is reported as 47 seconds.

The principal advantage of compression molding the inner door is the parts consolidation, which reduces the production cost significantly over the steel door. The coefficient of thermal expansion of TMC is close to that of steel. As a result, the assembly of the inner TMC door panel and the outer steel door panel keeps a tight seal at varying temperatures. The TMC door has been tested to withstand hostile environments of highly alkaline, grease-laden water at temperatures as high as 82 °C for two or three cycles daily for more than twenty years.

3.10.3 Light Truck Tailgate

The SMC tailgate used in Ford light trucks is a two-panel design replacing a steel production tailgate that required seven spot welds. The materials used are a vinyl ester SMC-R50 for the inner panel and a polyester SMC-R50 for the outer panel. The inner panel is ribbed to provide stiffness for the tailgate.

The two panels are compression molded separately and bonded together with a urethane adhesive. Two steel hinge cups and two latch tapping plates are also included in the assembly. Excluding the cups and plates, the SMC tailgate is nearly 27 % lighter than a similar steel tailgate (38).

3.10.4 Automotive Road Wheels

The standard automotive road wheels are made of low carbon steel, such as SAE 1010 or 1020 alloy. The rim and disc sections of these wheels are manufactured separately in a series of stamping operations and then spot welded to form the wheels. Recently, high strength sheet molding compounds have been used to prototype automotive road wheels. The SMC wheels are compression molded in one step in a four-piece mold consisting of the top mold, the bottom mold, and two slides (Fig. 3.29).

The material used in SMC wheels is a combination of SMC-R50 and XMC (39). Both compounds are based on a vinyl ester resin that provides a better fatigue performance at elevated temperatures than polyester resins. The SMC-R50 charge is placed in the bottom mold and forms the disc section of the wheel. The XMC charge is placed around the bottom mold and forms the rim section of the wheel. After the two charges in the mold have been placed, the slides are drawn inward and the mold is closed. The molding pressure is exerted primarily on the SMC-R50 charge, which flows into the rim section and joins with the XMC charge.

The SMC wheels provide 40–50 % weight savings over the standard steel wheels. Other advantages of SMC wheels are their uniform weight distribution (therefore,

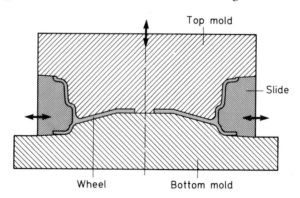

FIGURE 3.29   Molding arrangement for compression-molded SMC road wheels.

less difficulty in balancing) and their corrosion resistance. It has been reported in the literature that the SMC wheels perform well in the accelerated fatigue tests as well as the simulated road tests that are used for standard steel wheels. The major concern with the SMC wheels is the bolt torque loss at the wheel studs, which is due to the stress relaxation in SMC at elevated temperatures. One solution to this problem is the use of molded-in steel inserts in the bolt hole areas.

REFERENCES

1. C. D. Han and K. W. Lem, *J. Appl. Polym. Sci.*, 29, 1879 (1984).
2. M. R. Kamal and S. Sourour, *Polym. Eng. Sci.*, 13, 59 (1973).
3. K. W. Lem and C. D. Han, *Polym. Eng. Sci.*, 24, 175 (1984).
4. P. K. Mallick and N. Raghupathi, *Polym. Eng. Sci.*, 19, 774 (1979).
5. M. R. Panter, *Proc., 36th Annual SPI Technical Conference,* 1981.
6. J. D. Fan, J. M. Marinelli, and L. J. Lee, *Polym. Compos.*, 7, 239 (1986).
7. L. J. Lee, *Polym. Eng. Sci.*, 21, 483 (1981).
8. L. Marker and B. Ford, *Proc., 32nd Annual SPI Technical Conference,* 16-E, 1977.
9. M. R. Barone and D. A. Caulk, *Polym. Compos.*, 6, 105 (1985).
10. C. L. Tucker and F. Folgar, *Polym. Eng. Sci.*, 23, 69 (1983).
11. C. L. Tucker, "Compression Molding of Polymers and Composites," in A. I. Isayev, Ed., *Injection and Compression Molding Fundamentals*, Dekker, New York 1987.
12. T. A. Osswald and C. L. Tucker, *Polym. Eng. Sci.*, 28, 413 (1988).
13. R. J. Silva-Nieto, B. C. Fisher, and A. W. Birley, *Polym. Compos.* 1, 14 (1980).
14. M. R. Barone and D. A. Caulk, *J. Appl. Mech.*, 53, 361 (1986).
15. M. R. Barone and D. A. Caulk, "Mechanics of Compression Molding", in T. G. Gutowski, Ed., *The Manufacturing Science of Composites*, American Society of Mechanical Engineers, New York, 1988, p. 63.
16. M. R. Barone and T. A. Osswald, *Polym. Compos.*, 9, 158 (1988).
17. S. J. Lee, M. M. Denn, M. J. Crochet, and A. B. Metzner, *J. Non-Newtonian Fluid Mech.*, 10, 3 (1982).
18. R. L. McCullough, R. B. Pipes, D. Taggart, and J. Mosko, in *Composite Materials in the Automobile Industry*, American Society of Mechanical Engineers, New York, 1978, p. 141.
19. H. T. Kau, *Polym. Compos.*, 8, 82 (1987).
20. D. L. Denton and S. H. Munson-McGee, in ASTM STP 873, 1985, p. 23.
21. W. C. Jackson, S. G. Advani, and C. L. Tucker, *J. Compos. Mater.*, 20, 539 (1986).
22. J. M. Berthelot, *Fibre Sci. Technol.*, 18, 1 (1983).
23. D. L. Denton, *Proc., 36th Annual SPI Technical Conference,* 1981, 16-A.
24. D. C. Himebaugh and S. Newman, *Proc., 38th Annual SPI Technical Conference,* 1983.
25. D. Slotfeldt-Euingsen, E. Magnus, E. Ekern, E. Holtmon, and L. Corneliussen, *Polym. Compos.*, 7, 431 (1986).
26. R. M. Griffith and H.J. Shanoski, *Plast. Design and Process.*, 17, 10 (1977).
27. B. M. Lynskey and F. C. Robertson, *Proc., 40th Annual SPI Technical Conference,* 1985, 9-D.
28. R. B. Jutte, SAE Paper 730171, Society of Automotive Engineers, 1973.
29. K. L. Smith and N. P. Suh, *Polym. Eng. Sci.*, 19, 829 (1979).
30. R. Burns, A. G. Hankin, and D. Pennington, *Plast. Polym.*, 235 (1975).
31. P. K. Mallick, *Polym. Compos.*, 7, 14 (1986).
32. M. R. Barone and D. A. Caulk, *Polym. Eng. Sci.*, 21, 1139 (1981).

33. M. R. Barone, D. A. Caulk, and M. R. Panter, *Polym. Compos.*, 7, 141 (1986).
34. M. L. Bromberg, *Proc. SPE ANTEC*, 403 (1986).
35. J. Stückrad and D. Therolf, at Third Annual ASM/ESD Advanced Composites Conference, Detroit, 1987.
36. D. H. Anderson and A. F. Lawrence, SAE Technical paper 830487, Society of Automotive Engineers, February 1983.
37. A. Stuart Wood, *Modern Plast.*, 63, 44 (1986).
38. E. G. Trudeau and M. W. Lindsay, *Proc., 36th Annual SPI Technical Conference,* 1981, 21-B.
39. M. Dewan and T. N. Coppock, *Proc., 36th Annual SPI Technical Conference,* 1981, 16-B.

4 · Reaction Injection Molding

Greg Slocum

Contents

Greg Slocum, Mobay Corporation, Pittsburgh, Pennsylvania, USA

4.1 INTRODUCTION

In a book on reaction injection molding (RIM) that he edited a decade ago (1), Walter Becker cited the difficulty of analyzing or even describing such a rapidly developing field at any "point in time." Review of that book, with a mind to the state of the technology and the industry today, makes that difficulty clear. The chemistry that was standard 10 years ago is almost completely replaced. The list of applications of the late 1970s has expanded severalfold with the advances in both chemical and processing technology. The chemistry that is standard today promises to be replaced in the coming years, perhaps more than once. Today's list of applications of the RIM process also promises to continue to grow rapidly with future developments. Even so, another "stop action photograph" of this increasingly important technology, particularly as it applies to the production of composite parts, is appropriate at this time.

Becker listed in his introduction a number of reasons for the great interest in RIM, including several cost factors and the capability for molding large, complex parts. All those factors still exist today, and with the technological developments that have been made in the past decade, they are even more important. When molding large, complex parts, there are significant advantages to injecting essentially monomeric materials (i.e., those that are relatively low viscosity liquids) into the mold where they polymerize, rather than injecting already polymerized materials. When making a large number of parts, the development of the RIM process for high volume production has further advantages over competitive processes.

Discussion of these advantages and comparison of reaction injection molding with other processes requires some definition of the basics of the various molding techniques. The details of RIM processing are described later in this chapter, but the essential character of the process involves high speed mixing of two or more reactive chemicals as an integral part of the injection of those chemicals into the mold. RIM equipment allows the injection of essentially all the mixed chemicals into the mold, obviating the need for flush waste, either material or solvents. The injection mixture flows into the mold under relatively little pressure and at low temperatures, resulting in savings in energy and equipment. Because of the chemistries involved in RIM, the cure time usually is short and is not significantly dependent on part cross section. Trimming demolded parts is typically an easy operation, as is painting, at least compared with most other polymeric-based parts.

Injection molding is discussed in detail elsewhere in this book. The main difference, though, between injection molding and reaction injection molding is in the mixing and reaction of the chemicals that eventually make the polymeric material in the part. In injection molding all this takes place before the injection, instead of during the injection and molding as with RIM. This means that to be injected into the mold, the already formed polymer must be heated to the point where it flows and then cooled back to the point where it has sufficient integrity for demolding. Thus, unlike RIM, the injection molding process involves temperature cycling rather than actual chemical polymer formation. Since the viscosities of the melted thermoplastics are much higher than the essentially monomeric materials injected in the RIM process, the pressures necessary for injection through a comparable cross section and distance are much higher in injection molding. These factors lead to increased energy and equipment costs for injection molding.

Thermoset compression molding, discussed in Chapter 3, lies somewhat between RIM and injection molding conceptually. In this process, the polymer is partially formed and mixed with the reinforcement (as the reinforcement is premixed in conventionally reinforced RIM) before the molding operation. Rather than being injected into a closed mold, the material is placed in an open mold and squeezed throughout the mold cavity as the mold is closed. Final polymer formation (curing) occurs in the mold when the heat of the mold kicks off the cross-linking reactions in the material mixture. While the pressures required are somewhat less than with injection of completely formed polymer into a mold, they are still considerably higher than those involved with RIM. Mold temperatures for compression molding are necessarily high enough to drive the final cross-linking of the polymer. Energy and equipment costs are, again, higher for compression molding that for RIM.

A variation on compression molding is hot stamping of long fiber reinforced thermoplastic sheet. Here the polymer is completely formed before the molding, so that, again both temperature cycling and high molding pressures are required. Both these factors lead to higher equipment and energy costs.

Resin transfer molding (RTM; see Chapter 5) is conceptually similar to RIM in that an essentially monomeric material is injected into the mold, where it reacts to form the polymer of the finished part. Until recently, though, RIM did not use the preplaced reinforcement typical in RTM. Additionally, the chemistries usually used in the two processes have been based on different types of polymerization. With the slow injection rates typically used with RTM, molding pressures can be kept extremely low, allowing low cost tooling for low volume applications. Such low injection rates, though, require slow reaction and cure rates, making RTM less suitable for high volume production.

The foregoing considerations, which are summarized in Table 4.1, might make one wonder why anything other than reaction injection molding is used. During the development of all the molding processes for polymerics, there have been times when competitive advantages existed for a particular process in a particular application. In the continuing development of all the processes, this will undoubtedly continue to be true. For relatively flat, simple parts, the material cost advantages will continue to favor compression molding for appearance parts. For non appearance parts of similar shape, the cycle time advantages may favor stampable, long fiber reinforced thermoplastics, depending on the material cost differential. For parts of relatively complex shape, though, many factors tend to favor reaction injection molding.

In the RIM industry there are a number of terms used to describe different "types" of RIM, which are really differences in reinforcement. Despite some variation in usage, some relatively standard definitions are given and used throughout this chapter. RIM, itself, is typically used to refer to the process or the material when no reinforcement is used. Reinforced RIM (RRIM) conventionally means the material reinforced with relatively short fibers or flakes (or the process of producing this type of material) by injecting the chemicals and the reinforcement through the mixhead

TABLE 4.1 Temperature and Pressure Requirements for Various Molding Methods

Molding technique	Processing temperatures (°C)	Molding pressures (bar)
Rection injection molding	60 – 120	5 – 15
Injection molding	120 – 350	70 – 2000
Compression molding	130 – 175	20 – 140
Long fiber reinforced stampable thermoplastic	200	100
Resin transfer molding	20 – 120	4 – 10

(necessitating the relatively short fibers). Structural RIM (SRIM) or mat molding RIM (MMRIM) generally refers to the process of preplacing a long fiber reinforcing mat in the mold and then injecting a mixture of reacting chemicals through the mat. Since this book primarily concerns composites, unreinforced RIM is discussed only for historical perspective and for comparison in applications in which both RIM and RRIM are widely used. Because of the significant differences between RRIM and SRIM, the materials, the properties, the processing, and the flow of each are described separately and then contrasted. Finally some applications of each are discussed.

4.2 MATERIALS USED IN RIM

4.2.1 Historical Development of Commerical Resins in RIM and RRIM

The reaction injection molding process was initially developed in the mid-1960s as a means to decrease cycle times of cast urethanes to allow them to compete in industries geared to high speed production. While some other materials have been used in the process (e.g., caprolactam and dicyclopentadiene), polyurethane-based systems received far and away the most intensive development and have been used in nearly all commercial applications. The basic components of early RIM systems were the same as the components of cast urethane systems: a polyol, a diisocyanate, and a glycol chain extender, as in Eq. (4.1a):

$$R(\!-\!OH)_n \ + \ R'(\!-\!OH)_m \ + \ OCN\!-\!R''\!-\!NCO \tag{4.1a}$$

$$\downarrow$$

$$\begin{matrix} O & H & & H & O & & & O & H & & H & O \\ \| & | & & | & \| & & & \| & | & & | & \| \\ -\!C\!-\!N\!-\!R''\!-\!N\!-\!C\!-\!O\!-\!R'\!-\!O\!-\!C\!-\!N\!-\!R''\!-\!N\!-\!C\!-\!O\!-\!R\!-\!O\!- \end{matrix} \tag{4.1b}$$

where R was typically a repeating propylene oxide–ethylene oxide (PO/EO) unit from around 50 to 100 units long, R' was a repeating methylene unit from 2 to 6 units long, and R'' was a methylene diphenyl group or some variation thereof (MDI).

The components had some basic requirements themselves, such as relatively low viscosity ($< 1500\,cP$) and compatibility sufficient to ensure that mixing was not extremely difficult and that reaction would be essentially complete before separation into two phases could occur. The exact nature of the components depended on the application, but the function of each in the final polymer can be generalized. The PO/EO repeating units provide a long flexible chain in the polymer, giving flexibility and impact resistance. the reaction product of the glycol and the diisocyanate forms a hard, more or less crystalline material, which is incompatible with the long flexible chain. During and after reaction, this incompatibility causes separation into very small regions of two distinct phases. This "microphase separation" is what gives the RIM materials their balance of properties (2,3).

A number of factors affect specific properties of RIM materials, but the separation into small domains or segments (some composed of flexible polyether chains; others of hard, relatively polar segments) most strongly influences their overall properties. The polyether chains typically have a glass transition temperature of around $-50\,°C$, providing flexibility and allowing for absorption of the energy of an impact down to nearly that temperature. The portions of the polymer that make up the hard segment are attracted to each other, forming hard glassy or crystalline regions that maintain their integrity up to near their glass transition temperature. The attraction between these hard domains dispersed in the soft polyether phase is what gives the polymer its integrity as a solid material in a particular modulus range. This temperature can be as high as $100\,°C$, depending on the nature and quality of the hard segment.

Although there was tremendous development in RIM materials and processing based on the chemistry described above, it was the nature and quality of the hard segment, along with processing problems related to the chemicals used in their formation, that prevented wider application of the process until new chemistries were developed. The modulus of all materials varies with temperature, with most polymeric materials showing significant change in modulus within the range of -50 to $+150\,°C$. Between -30 and $+65\,°C$ the hard segment in RIM materials based on polyether polyols, glycols, and modified MDI softens to the point that the modulus of the bulk material normally drops by a factor of 6–10. In addition to this extreme variation of physical properties with temperature, the glycols that were normally used in RIM are incompatible with the long chain polyether polyols. This factor led to mix quality problems, as well as other problems associated with separation of the glycol and polyol components before mixing with the diisocyanate. Reactivity as well was a significant problem with these early RIM systems. In spite of significant levels of both amine and organometallic urethane catalysts, the development of properties (i.e., green strength) was so slow that typical demold times were around 1 minute, and longer for parts with thick cross sections. These problems, as well as a number of others, called for a solution, which came in the form of the development of "amine-extended" RIM systems.

4.2.2 RIM Systems in Current Commercial Use

In the mid-1970s Mobay Chemical Corporation and its West German parent, Bayer AG, introduced RIM systems (4) based on a substituted aromatic diamine developed at Bayer's Leverkusen laboratories. The new systems still used polyether polyols and a modified MDI, but the glycol chain extender was replaced by the aromatic diamine. The polymer that was formed is still called a polyurethane in commercial literature, but actually it is a hybrid polyurethane–urea as shown in Eq. (4.2a):

$$R—(OH)_n \ + \ H_2N—Ar—NH_2 \ + \ OCN—R''—NCO \qquad (4.2a)$$

$$\downarrow$$

$$\overset{O \ \ H}{\underset{\|\ \ \ |}{}} \ \ \overset{H \ \ O \ \ H}{\underset{|\ \ \|\ \ |}{}} \ \ \overset{H \ \ O \ \ H}{\underset{|\ \ \|\ \ |}{}} \ \ \overset{H \ \ O}{\underset{|\ \ \|}{}}$$
$$—C—N—R''—N—C—N—Ar—N—C—N—R''—N—C—O—R—O— \qquad (4.2b)$$

where R and R″ have the same meaning as Eq. (4.1) and Ar is a substituted phenyl ring. These amine-extended systems have numerous advantages over glycol-extended systems and are still used extensively today. Foremost among the advantages in terms of the material properties is the relative insensitivity of the physical properties to temperature in the normal use range of RIM materials. The ratio of the modulus at -30 and at $+65\,°C$ is normally about half that of a glycol-extended system. This can be rationalized in terms of the quality of the hard segment. While the hard segment in amine-extended systems is essentially glassy (i.e., noncrystalline) because of the asymmetry of the amine extender, the replacement of urethane with urea groups doubles the number of hydrogen bonds that can be formed. These bonds are the factor most responsible for maintaining the integrity of the hard segment, so it should be no surprise that the urethane–urea hybrid is more thermally stable.

Probably more important, though, to the widespread replacement of glycol-extended systems by amine-extended systems, as well as the continued growth in their use, are the processing advantages associated with the amine-based chemistry. Because these systems are so widely used today, including nearly all RRIM systems, the typical components are listed in some detail in Table 4.2. The ratio of the polyol to the sum of the chain extender and cross-linker, which itself affects the ratio of isocyanate to polyol blend, determines the stiffness of the polymer. The higher the chain extender content, the higher the stiffness of the resulting polymer. The components listed above the double line in Table 4.2 make up what is normally called the B side (or polyol blend), which is mixed in the mixhead with the diisocyanate, or A side.

One of the first processing advantages of amine-based systems is that the B side is completely compatible; that is, it is a one-phase mixture. This avoids any problems caused by separation into two phases before injection. Additionally, since the more compatible blend is also more compatible with the diisocyanate, better mix quality can be achieved. The amine-extended systems are more reactive than analogous glycol systems; thus green strength is better and cycle times are faster. The whole reaction profile of the amine-extended systems is different from that of glycol-extended systems (see Section 4.4.1), which helps to explain some further processing advantages. While the viscosity of an amine-based B side is approximately equivalent to that of glycol systems, the initial viscosity build out of the head is much more rapid. This leads to a much less turbulent flow of the amine-based system through the mold and fewer defects due to entrapped air. These and associated advantages led to the almost

TABLE 4.2   Typical Components in Amine-Extended Polyurethan–Urea RIM Systems

Component	Approximate amount (parts by weight)	Chemical description
Polyol	70 – 90	Polyether polyol, typically 3000 – 6000 molecular weight and EO tipped
Chain extender	10 – 30	Substituted aromatic diamine, typically DETDA
Additional cross-linker	0 – 10	Relatively small (MW < 400) polyol with higher functionality, or glycols, sometimes used to promote green strength
Amine catalyst	0.01 – 0.2	Typically triethylene diamine
Metal catalyst	0.01 – 0.2	Typically an organotin salt
Diisocyanate	Slight stochiometric excess	Typically a glycol-modified MDI

complete replacement of glycol systems by amine-extended systems over a 4 year period after the introduction of the amine-based chemistry in 1979.

Another advantage of amine-extended RIM systems was not readily apparent when they were introduced: namely, the capability for internal mold release (IMR: see Section 4.2.7). Once developed, though, the same sort of rapid switch to IMR systems was observed, with an almost complete changeover in the 5 years between the first extended production demonstration of the technology by Mobay Corporation and Dow Chemical Company in 1984 and the present. All through the 1970s, part of the normal cycle of making each RIM part was the spraying of the mold with an external release agent. Numerous attempts were made to incorporate into RIM formulations, without adversely affecting other characteristics (such as paint adhesion), materials that functioned as internal release agents in other plastic molding techniques. Amine-based systems allowed the successful incorporation of just such an agent. The rationalization for why IMR works is that the release agent is marginally soluble in the polyol blend, but insoluble in the polymer. As reaction takes place, the release agent precipitates and exudes to the surface of the polymer, where it prevents the adhesion of the part to the tool. The evidence for this is sketchy, but it is known that:

1. If the release agent is not dissolved in the polyol blend, there is no beneficial effect on release.
2. The release agent *does* come out of solution during polymerization.

Release agents of this sort are not generally soluble in the typical glycol extended RIM system, meaning that condition 1 above is not satisfied.

The feasibility of IMR in amine-based systems gave yet another reason for completing the shift away from glycol-based systems. The relative processing ease with which fillers and short fiber reinforcements are incorporated is yet another, which is discussed in more detail in Section 4.2.4. Productivity increased in RIM plants between 10 and 40 % with the implementation of amine-based systems. It increased as much or more again with the implementation of IMR. What had been a method of meeting a government-mandated impact requirement (fascia that allowed the auto companies to meet MVSS-215) with a 3 minute cycle time, had become a way to make practically any vertical body part with cycles of 60–90 seconds at seriously competitive costs.

4.2.3 Recent Development in RRIM Resins

As noted in the Introduction, the chemistry based on amine chain extenders, which has almost completely replaced the standard chemistry of a decade ago, itself promises to be replaced. While for a number of years it has been recognized that urea linkages in a polymer are more thermally stable than urethane bonds (5), an appreciation for the role that catalysts play in the thermal stability of a polymer has developed more recently. A system that has urea bonds formed not only between the chain extender and the isocyanate, but also between the polyether and the isocyanate, then, has two advantages. The bonds themselves are more thermally stable, and they can be formed without catalyst. This latter property is an advantage because catalysts accelerate not only the forward (polymer formation) reactions, but also the reverse (polymer degradation) reactions.

Polyurea RIM systems have been under development at a number of companies for some years, most notably Texaco, Dow, and Mobay (6–8). In spite of the obvious material property advantages of a completely polyurea RIM system, even over the hybrid urethane–urea systems widely used today (see Section 4.6.1 and Table 4.13), processing limitations have slowed their implementation. These problems have arisen primarily from the very high reactivity of the systems (discussed in Section 4.5), and most of the development has centered on ways to make the systems slow enough to process (as opposed to the development of glycol-based urethane systems, where an increase in reactivity was needed). The basic materials used in a polyurea RIM system differ from the hybrid system only in that the polyether is now tipped not with ethylene oxide to give primary alcohol moiety, but with ammonia or other amine precursor. The same basic types of polyether are used, typically di- or trifunctional molecules of 2000–6000 molecular weight.

The process development work has been concentrated on three aspects, two of them material related. (The other is simply to use a mold temperature higher than the melting point of the oligomeric hard segment, to increase the time allowed for flow before the growing polymer gels.) In material terms, though, the standard approach of formulating around functionality and molecular weight has been supplemented by chemically altering the nature of the amine tip on the polyether to make it less reactive. The reactivity can be affected in two ways: sterically and electronically. Reduced reactivity can be accomplished by replacing one of the active hydrogen

111

atoms on the amine with a sufficiently bulky group, although many such groups are electron-donating groups, which tend to increase reactivity. Attachment of electron-withdrawing groups tends to have a greater effect and seems, at present, to have been the most beneficial to processing.

While equating thermodynamic data and kinetic data is theoretically inconsistent, some parallels can be drawn. The pK_a, a logarithmic measure of basicity, for methoxyethylamine, an analogue of a polyether tipped with ammonia, is about 10 (9). Substitution of increasingly bulky alkyl groups does little to change the basicity, as seen in the pK_a's of methylamine, dimethylamine, diisopropylamine, and di-isobutylamine, all around 10.5. Substitution with electron-withdrawing groups, though, has a great effect on basicity, as demonstrated by comparing the pK_a's of methylamine (methyl-substituted ammonia, 10.7), aniline (phenyl-substituted ammonia, 4.63), and glycine (carboxymethyl-substituted ammonia, 2.4). Basicity is not the same as nucleophilicity or reactivity with an isocyanate, but many things affect the two phenomena in similar ways. Thus, we can expect a simple alkyl- or oxyalkylamine to be orders of magnitude more reactive than a phenyl-substituted or carboxy- or cyano-substituted alkylamine. This electronic effect has been used with a good deal of success in the materials development that has helped bring polyurea RIM to the market.

Continuing development, using this type of concept as well as more conventional formulation techniques, promises to make processible polyurea RIM increasingly feasible. Whether it will completely displace the hybrid polyurethane-urea systems will depend on comparative costs and benefits. Production of amine-tipped polyethers, no matter how the amine is substituted, involves at least one extra step from the polyol and increases costs. Definite benefits, though, may more than offset the increased material costs; these include reduced demold and cycle times, higher thermal stability, and better surface appearance. The decade of the 1990s promises an explosion in the development of new materials of RIM and RRIM.

4.2.4 Reinforcements and Fillers Used for RRIM

As noted above (Section 4.2.2), the implementation of amine-extended systems allowed the relatively easy incorporation of fillers and reinforcements into RIM systems, commonly called RRIM. The word "relatively" is important. While short glass fibers have been added to the polyol blend in glycol-extended systems, few molders have been able to make this type of material work at the high production rates, and with the high reliability of amine-extended systems. Nonetheless, a great deal of equipment development and material development had to be done to bring the RRIM industry to its present state. Until fairly recently, though, that state was relatively settled. Various fillers and reinforcements had been used, but two materials had become standards: short glass fiber and small glass flake.

There are basically two reasons for adding a filler or reinforcement to a polymer: the first is to fill the system (i.e., to replace some of the more expensive polymer with less expensive "filler"); the second reason is for reinforcement (i.e., to add a material that actually improves the properties of the product). In some cases, both goals can be achieved. Fillers may be divided into categories: those with low aspect ratio and those with high aspect ratio. In RRIM, fillers with high aspect ratio do improve some properties at the expense of others. Those with low aspect ratio, though, generally increase only the hardness, the stiffness, and the coefficient of thermal expansion, while decreasing all other properties. Thus, where the properties of the RIM material more than suffice to fulfill the requirements of the application, calcium carbonate can be added to reduce the material cost. This has not been widely practiced, though.

Of the materials with high aspect ratio, glass fiber and glass flake are the two that have been used most commonly. Other material such as wollastonite have given some reinforcement, but at a significant sacrifice in impact strength. There is still a great deal that is not well understood about the interaction between the polymer and the reinforcement, but information from studies of various glass types does give some general guidelines. In this type of processing the glass is mixed into the polyol (or polyamine) blend. This slurry is then injected into the mold along with the isocyanate. Obviously the particle size of the glass must be small enough to fit through the mixhead orifices (< 2 mm). Additionally, the smaller the particle sizes, the higher the loadings of reinforcement, which can have some processing advantages. In terms of reinforcement properties, though, the longer the fiber, the better the reinforcement; thus there is a balance that give optimum properties and processibility. This would appear to be somewhere between 0.2 and 0.4 mm mean fiber length.

One of the troublesome aspects about glass fibers is found in the aspect ratio. Its length is much greater than its diameter. The fibers tend to orient with the flow of the material through the mold, giving that orientation to the physical properties, (Table 4.3) (10,11). In long narrow parts, long narrow reinforcement is not a problem. In large flat parts, this orientation, particularly with respect to shrinkage and the coefficient of linear thermal expansion (CLTE), can lead to part distortion. The solution to this is to use a platy reinforcement, glass flake. Table 4.3 shows the improvement in orientation between materials reinforced with glass fiber and with glass flake.

The only major problem with glass reinforcement, particularly glass flake, has been its effect on surface quality. An unreinforced part painted with conventional

TABLE 4.3   Properties in RRIM Systems Parallel (Par) and Perpendicular (Per) to Flow

Property	Par	Per	Par	Per
Unreinforced flexural modulus, MPa	240		520	
Reinforcement	20% glass fiber		20% glass flake	
Flexural modulus, MPa	960	490	1550	1340
Tensile strength, MPa			32	32
Elongation, %			20	20
Heat sag, mm, 10 cm OH[a] at 121 °C, 1 h	1.0	1.7	2.5	2.5
CLTE, $\times 10^6$/°C	45	135	55	63

[a] OH = overhang

113

TABLE 4.4 Comparisons of Distinction of Image for Paints and Fillers

Substrate[a]	Primer type	Paint type	Visual DOI (average)
Steel	1K Uniprime	1K Base/clear	88
Bayflex 110-35	1K Uniprime	1K Base/clear	65
Bayflex 110-35	2K Prime	1K Base/clear	75
Bayflex 110-35	2K Prime	2K Base/clear	95
Bayflex 110-80 20% glass flake	1K Uniprime	1K Base/clear	55
Bayflex 150 20% glass flake	1K Uniprime	1K Base/clear	60
Bayflex 150 25% milled fiber	1K Uniprime	1K Base/clear	65
Bayflex 150 DOI	1K Uniprime	1K Base/clear	88

[a] Bayflex is a trade name of Mobay Corporation.

paints typically looks far better than a part reinforced with either glass fiber or flake. There has been some work on addressing this problem with different paint systems. The effects of various two-component paint combinations are shown in Table 4.4. More recently, Mobay Corporation has commercialized a proprietary reinforcement reported to allow RRIM parts to match the surface quality of steel when painted with conventional paints. Surface measurements on parts made with this filler are also included in Table 4.4 (12). Developments in both the areas of reinforcements and painting technology also promise to increase the potential and usage of RRIM.

4.2.5 Resins Used in SRIM

Structural RIM makes use of the positive correlation of reinforcement length with better properties. In conventional RRIM, some properties are helped by the reinforce-

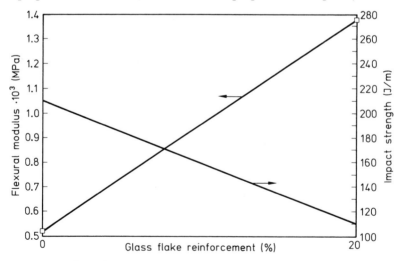

FIGURE 4.1 As flexural properties are increased by the addition of (conventional) short fiber reinforcement, impact strength is reduced.

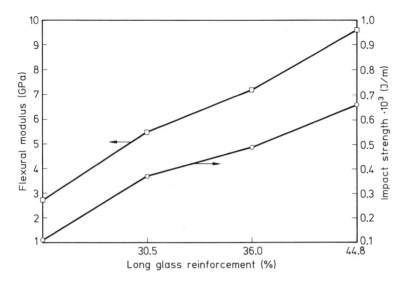

FIGURE 4.2   Long fiber reinforcement increases both flexural properties and impact strength.

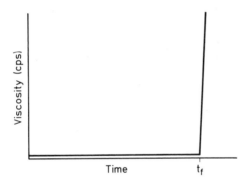

FIGURE 4.3   The ideal viscosity behavior of SRIM systems, with extended low viscosity and snap cure, is approximated by two-stage reactions schemes, such as isocyanurate and acrylamate formation, and urethane formation with the proper combination of catalysts.

ment while others suffer. Figure 4.1 shows the drop of impact strength with the increase in flexural modulus as glass fiber is added to the system. With much longer fibers this trend is reversed (Fig. 4.2). These much longer fibers, though, cannot be processed by injecting through the head but must instead be placed as a mat in the mold, after which the RIM resin is injected through the mat.

This is the most important constraint in SRIM resins: that they must flow through high levels of reinforcement fibers. This leads to two essential requirements. First, the component and mixed resin viscosities must be very low, and second, the viscosity of the mixed resin must remain low as it flows through the mold. Certainly advantageous is a resin having a viscosity that rises rapidly (i.e., the resin cures), as soon as the mold is full, as in the schematic ideal shown in Figure 4.3. Initial resin

115

developments focused on two-stage reactions to simulate this ideal. Ashland Chemical has developed systems based on ethylenically unsaturated esterols (13), which react with an isocyanate according to Eq. (4.3a).

$$2CH_2\!=\!CH\!-\!R\!-\!\underset{\underset{O}{\parallel}}{C}\!-\!O\!-\!R'\!-\!O\!-\!H \;+\; OCN\!-\!R''\!-\!NCO \tag{4.3a}$$

$$\downarrow$$

$$CH_2\!=\!CH\!-\!R\!-\!\underset{\underset{O}{\parallel}}{C}\!-\!O\!-\!R'\!-\!O\!-\!\underset{\underset{O}{\parallel}}{\underset{H}{\overset{|}{C}}}\!-\!\underset{H}{\overset{|}{N}}\!-\!R''\!-\!\underset{H}{\overset{|}{N}}\!-\!\underset{\underset{O}{\parallel}}{\overset{|}{C}}\!-\!O\!-\!R'\!-\!O\!-\!\underset{\underset{O}{\parallel}}{C}\!-\!R\!-\!CJ\!=\!CH_2 \tag{4.3b}$$

$$\downarrow$$

$$-\!\underset{H}{\overset{|}{N}}\!-\!\underset{\underset{O}{\parallel}}{C}\!-\!O\!-\!R'\!-\!O\!-\!\underset{\underset{O}{\parallel}}{C}\!-\!R\!-\!CH\!-\!CH_2\!-\!CH_2\!-\!CH_2\!-\!R\!-\!\underset{\underset{O}{\parallel}}{C}\!-\!O\!-\!R'\!-\!O\!-\!\underset{\underset{O}{\parallel}}{\underset{H}{\overset{|}{C}}}\!-\!\overset{|}{N}\!- \tag{4.3b}$$

The first reaction product (Eq. 4.3b) is an unsaturated diurethane oligomer. Through the proper selection of R and R', its viscosity is kept low. The heat from the first reaction and the hot mold, along with the free radical initiators in the system, kick off the second reaction, which is essentially a vinyl cross-linking reaction. Through this two-stage sequence, a low viscosity oligomer is formed up on mixing, flows through the mold, and cures rapidly to form what Ashland calls an acrylamate.

Another two-stage reaction approach being developed by Dow (14) for SRIM is based on more traditional isocyanate chemistry. Certain catalysts accelerate the urethane reaction under normal conditions, but rapidly accelerate the trimerization of isocyanates to isocyanurates at elevated temperature. When a hydroxy-functional material reacts with an isocyanate according to the following reaction:

$$R\!-\!(OH)_n \;+\; OCN\!-\!R'\!-\!NCO \tag{4.4a}$$

$$\downarrow$$

$$-\!O\!-\!R\!-\!O\!-\!\underset{\underset{O}{\parallel}}{\underset{H}{\overset{|}{C}}}\!-\!\underset{H}{\overset{|}{N}}\!-\!R'\!-\!\underset{H}{\overset{|}{N}}\!-\!\underset{\underset{O}{\parallel}}{C} \tag{4.4b}$$

a certain amount of heat is released. In the presence of trimerization catalysts, this heat combined with the heat of the mold can start trimerization.

$$3OCN\!-\!R'\!-\!NCO \tag{4.5a}$$

$$\downarrow$$

(4.5b)

If R is small enough and its concentration is low, the injected material will be essentially a little urethane oligomer dissolved in isocyanate, which will have a very low viscosity at mold temperature. The heat from the reaction to produce (4.4b) and the heat from the hot mold initiate the trimerization, thereby leading to a rapid cure.

The third approach has been a rediscovery of urethane chemistry. Dow also markets some "polycarbamate" (i.e., polyurethane) systems (14) for use in SRIM, although the viscosity of some of the polyol blends is somewhat higher than is typical of SRIM systems. Mobay Corporation has developed several systems based on a compatible mixture of a chain extender, an intermediate molecular weight polyol, and a portion of a long chain polyol (16). Through proper choice of the chemicals and their amounts, polyol viscosities are kept below 100 cP at room temperature. Flow studies (see Section 4.4.2) have shown the viscosity of the actual reaction mixture to be on the order of 10 cP at mold temperature. In spite of the traditional view of the urethane reaction as a single stage producing increasingly viscous materials, proper catalyst choice allows the approach of the ideal of a low viscosity mixture filling the mold and then curing very rapidly.

Obviously, elements of each of the approaches above can be combined. Reacting a blend of an esterol and a flexible polyether (or polyester) polyol with a diisocyanate produces a copolymer with urethane linkages, vinyl cross-linking, and a flexible polyether segment. Combination of the esterol with excess isocyanate and trimer catalysts leads to an acrylamate–isocyanurate copolymer. Practically all isocyanurate polymers contain some urethane linkages, making them copolymers of a sort. Most of the properties of SRIM composites depend more on the reinforcement level and arrangement than on the resin properties. Some neat, or unreinforced, resin properties are published or can be surmised from the published composite properties. Some of those properties for some SRIM resins representing each approach are shown in Table 4.5.

TABLE 4.5   Properties of Neat (Unreinforced) SRIM Resins

Property	Mobay urethane	Dow urethane	Dow isocyanurate	Ashland acrylamate	ICI urethane
Flexural modulus, MPa	3480				2140
Tensile strength, MPa	91				62
Izod impact strength, J/m	96				26.69
T_g, °C	120	147	167	~ 140	~ 150
Density, g/cm^3	1.21	~ 1.2	~ 1.2	~ 1.2	1.24
Resin viscosities, cP					
Isocyanate	50	100–200	30–60		55
"Polyol"	75	400–600	90–130		600

117

Some basic differences lead to some simple conclusions, which is about all that can be drawn from the limited data available. The more highly cross-linked systems are generally more thermally stable. They also tend to be more brittle in terms of the neat resin. While there is little indication of microphase separation in most SRIM resins, inclusion of a flexible chain material does increase the impact strength of the resin. Development of the resin materials will of course be motivated by the physical properties required in the applications. A great deal remains to be done, though, in the development of SRIM reinforcements (see next section) and processing (Sections 4.3 and 4.4).

4.2.6 Reinforcements Used in SRIM

There are a number of possibilities for reinforcements in structural RIM. The initial commercial applications of "RIM" polyurethanes reinforced with long strand glass were made from higher viscosity, rigid foam formulations poured onto a random continuous strand glass mat preplaced in an open mold. There are still a number of these applications in production in Europe. The continuous strand mat comes in a variety of grades differing in filament size, strand size, glass sizing, and binder formulations. This was the type of glass first used in the development of low viscosity RIM systems, which could be processed by injection into a closed mold with appreciable flow through the reinforcement.

The variations in glass type have arisen from the previous uses of the random continuous mat for reinforcement of other types of organic matrix composites. The filament size refers to the size of the individual fibers, several hundred of which may be bound together to form the strand. The strand size refers to the number of filaments in the strand. Glass sizing formulation is something of an art, with typical formulations having 10 or more ingredients. The main function of the sizing is to protect the filaments as they are processed, but there are also materials in the sizing that hold the filaments together in the strand (binders) and promote adhesion to the matrix resin (typically organofunctional silanes). The term "binder" can be somewhat confusing, since it can refer to both the material that holds the filaments together in the strand and the material that holds the strands together in the mat. The two materials are not necessarily the same, but both are basically glues.

Both binders and sizings for most commercially available continuous strand mat used in structural RIM were developed for other chemistries, particularly polyester-based resins. As discussed below, the matching of the sizing and binder formulations with the resin has a significant effect on the physical properties of the composite. Perhaps more important for the large-scale production of SRIM composites, though, is the economic effect of the continuous strand mat on the composite. Continuous strand mat comes on a roll as sheet goods. For any except a regularly shaped part of the proper size, there will be scrap cut off the edge of the mat to fit the part shape. This scrap ("offal") can be significant, particularly for parts with cut outs and holes, ranging from 5–50 %. The pricing on the continuous strand mat is presently in the range of $ 1.20–1.50 per pound.

TABLE 4.6 Comparison of SRIM Composite Properties from Continuous and Chopped Strand Mats

Glass/binder type	Glass (%)	Flexural modulus (GPa)	Flexural strength (MPa)	Tensile strength (MPa)	Notched Izod strength (J/m)
Acrylate, sized, continuous strand	40.2	8.5	250	165	720
Glass B/Binder B	43.6	10.0	253	157	950
Glass B/Binder D	47.9	7.9	204	141	972
Glass C/Binder D	47.3	8.2	223	157	905
Glass D/Binder D	47.3	8.3	224	160	1040
Glass E/Binder D	47.7	8.9	219	153	998
Glass F/Binder D	47.9	7.2	200	154	1180
Glass G/Binder D	48.0	8.0	227	158	1120
Glass C/Binder B	47.4	8.1	217	147	934
Glass D/Binder B	47.0	8.6	214	160	1040
Glass E/Binder B	48.3	9.1	232	172	1020
Glass F/Binder B	47.7	9.0	220	158	1230
Glass G/Binder B	46.5	7.6	215	159	1150

An alternative that has been demonstrated recently is chopped strand mat or chopped strand preformed glass. With glass costs in the range of $ 0.80 per pound and preforming scrap typically in the range of 1–6 %, even with technology developed for other processes, this represents a significant cost advantage over continuous strand mat. The disadvantage here is that the processes for preforming the chopped strand glass were developed earlier for composites of other types. While making preforms of a complex shape is somewhat easier with the chopped glass than with continuous strand mat, the process is labor intensive and operator sensitive. Nevertheless, if equivalent properties can be attained, the advantages — easier preforming, lower glass scrap, and glass costs at about half of the costs of the continuous strand mat — make the development of the chopped strand preforming process economically attractive.

Table 4.6 compares the properties that result from reinforcing a particular polyurethane resin with a continuous strand mat and a chopped strand mat. At approximately the same glass level, the properties of the composites are about the same, except for the tensile strength (where the continuous strand mat is slightly better) and the impact strength (where the chopped strand mat is somewhat better). The remaining rows of Table 4.6 demonstrate the effects of binder and sizing formulation on the properties of the composite. These data resulted from a screening experiment in which chopped strand mats with different sizings and binders were used to reinforce test plaques of a polyurethane SRIM resin. As indicated above, sizing and binder technology is somewhat complex, and information is not widely disseminated by the suppliers. The data in Table 4.6 show that the sizing and binder can make a significant difference in the composite properties, though. Once the critical properties

119

TABLE 4.7 Properties of a Structural RIM (Urethane) Composite with Random and Oriented Continuous Strand Reinforcement

Property	Glass configuration[a]			
	5 × R	1 × U 3 × R 1 × U	1 × U 4 × R 1 × U	2 × U 4 × R 2 × U
Total wt % glass	46	40	46	49
Specific gravity(theoretical)	1.60	1.54	1.60	1.63
Tensile strength, MPa	228	232/128	257/153	308/167
Tensile modulus, GPa	12.6	14.5/8.7	15.4/11.3	17.1/9.5
Flexural strength, MPa	285	280/169	331/218	301/201
Flexural modulus, GPa	11.1	14.5/5.8	16.9/6.9	18.2/6.2
Izod impact strength, notched, J/m	1360	1890/560	2230/693	2340/768

[a] R = random mat, $2\,oz./yd^2$; U = unidirectional mat, $8\,oz./yd^2$.

for an application have been determined, cooperation between glass and resin companies can be quite beneficial in optimizing the physical properties by balancing the sizing, binder, and resin. Fortunately, most SRIM resins are based at least in part on polyurethane chemistry; thus the sizing and binders optimized for one resin are likely to be close to the optimum for other SRIM resins.

Another type of glass reinforcement routinely used in structural RIM is oriented continuous strand mat or fabric. A mat can have any number of orientations, but the most common are the standard weaves and unidirectional and bidirectional stitched bonded fabrics. These mats, either used alone or in combination with some random glass, offer significantly increased properties in some of the directions. An indication of the properties attainable with this type of reinforcement is given in Table 4.7. With 46 w % random glass (glass configuration 5 × R), the flexural and tensile moduli are in the 11–12.5 GPa range, with the tensile strength around 225 MPa and the flexural strength around 280 MPa. At about the same level of glass, but with 20 % of the glass on the surface replaced by a unidirectional mat, the apparent "moduli" go up to the range of 16 GPa in the direction of the directional reinforcement. In the perpendicular direction the difference made by having the reinforcement on the surface is shown. The tensile modulus is governed to a great extent by the nature of the reinforcement throughout the composite; it drops only slightly with the orientation of a portion of the glass. The apparent flexural modulus, on the other hand, depends largely on the reinforcement on the composite surface, as shown by the large difference between the apparent modulus in the two directions.

Oriented continuous strand mat also comes as roll goods, so one must recognize that there will be some glass waste when it is used in an application. Consideration of this, though, along with the type of property differences shown in Table 4.7, demonstrates one of the potential advantages of making composite parts with this technology. If the property requirements of a part can be determined in the critical areas, the reinforcement can be engineered, or tailored, to those requirements. In a bumper beam application, for example, the flexural strength in the horizontal direction is more important than in the vertical. It is often more critical in the horizontal direction in a particular area of the beam cross section, as well. By placing unidirectional mat only in specific areas, glass cost and waste can be minimized while giving the part the necessary strength in the appropriate areas. Design of parts with this type of engineered reinforcement in mind promises to reduce the final part cost for SRIM composites for applications in which strength and modulus requirements vary considerably across the part.

Finally, there are the specialized reinforcements, such as carbon and polyaramid fibers. The theoretical modulus of a composite can be easily calculated using standard equations. Obviously, the use of carbon fibers, with a modulus exceeding 200 GPa, can give a higher composite modulus than can glass, with a modulus of 70 GPa. For certain specialized, low volume applications requiring very high properties, these reinforcements will make economic sense. For most of the high volume applications for which SRIM is being developed, though, the various forms of glass will be the reinforcement of choice for reasons of cost.

4.2.7 Additives

As with a discussion of formulation components of most materials, there must be a "miscellaneous" category for RRIM and SRIM. This category has really only one type of material in it, surfactants of one sort or another. Surfactants can play many different roles, depending on their chemical nature and that of the system. The various chemicals used as internal release agents in RIM are surfactants that change the interaction between the surface of the part and the surface of the mold. The interactions between the surfaces of a conventional polyurethane–urea RRIM part and a steel mold differ from the interactions of a polyurea RRIM or SRIM part with the same steel surface, necessitating the use of different surfactants to effect release from the mold for the different systems. Certain silanes with hydrophobic organic groups have been shown to improve the impact strength of RRIM parts. This is undoubtedly a function of changing the interaction between the surfaces of the glass reinforcement and the polymer matrix. Foam stabilizers (materials that change the surface tension of the polyol blend) are often used as nucleation and flow aids in RRIM. Foam destabilizers have been tried in SRIM as with other composites to release trapped air to help give a fully dense part.

To someone concerned with making parts from RRIM or SRIM, the issue of which particular additives are used in a formulation is usually not important, since the additive will be incorporated in the system and will not change until the system changes. When the system is changed, though, to accommodte part changes or for other reasons, the additives can become important. Many surfactants can have significant effects, both on processing and on part properties, at very low concentrations. The amount of residual material (i.e., that which is left in the machine after it is drained as much as possible) can be surprisingly large. The amount of flushing required to eliminate traces of surfactant can be correspondingly large. When systems are changed, processing and part properties can take some time to achieve expected levels because of the "hang up" of surfactants in the system, and this lag time should be considered whenever the system is changed.

4.3 RIM PROCESS TECHNOLOGY

4.3.1 Impingement Mixing

The heart of the RIM process is impingement mixing. For two streams of liquids to be mixed to make a solid polymer of high quality in a very short time, the reactivity of the liquids must be high, and most importantly, the mixing must be thorough. Figure 4.4 shows a schematic of one type of commonly used mixhead. In the recirculating position, the material circulates through the mixhead, so that when the head is opened, tempered (i.e., temperature-, pressure-, and nucleation-controlled) material will be used for the shot. In the open position, the control piston is moved back exposing nozzles, which spray the streams of reactive material at each other, much like garden hose nozzles. There has been a great deal of study of the mix phenomena, with the development of some fairly good models. Unfortunately, though, much of the practice of adjusting and optimizing the mix for any particular part is still an art.

Part of the reason for this is illustrated in Figure 4.5, which shows a trace of the pressure behind one of the nozzles throughout the shot cycle. In the recirculation stage (A), the nozzle pressures are adjusted through positioning of the nozzles. When the piston starts to move (B), there is a zero flow condition, even though the pumps continue to move material at a constant rate into any flexible hoses between the pump and the head. This causes a pressure spike immediately before the mixhead opens. When the piston exposes the nozzles to each other (C), the spike dissipates with the return of flexible lines to their prespike dimensions, and pressures return, with some time lag, to approximately their preset values (D), until the end of the shot (E), when there is another spike. (In mixheads of other types, the mechanics are different, but similar transients are observed.) In the mix chamber the pressure is essentially atmospheric until the shot begins. Then there is a gradient through the mix chamber, which can change as the part is filled. Pressure changes depend on the position of the chamber relative to the part, the materials and their reactivities, the part geometry, the nature of the reinforcement, and venting.

Thus, the process of impingement mixing in making a RIM part involves a number of changing parameters. Most of the theoretical treatments of impingement mixing have taken into consideration the nominal mix pressures and material flows. These are essentially the pressures and flows in A and D in Figure 4.5. Taken in combination with the viscosities, these parameters suggest that the Reynolds number of the streams of reactive material should be at least 150 and preferably more than 200 (17). In practice, this recommendation is a good starting point for a particular part, but adjusting the equipment to optimize the transients of the type observed in Figure 4.5, is still very much an art. To replace some of the art with science, an awareness of the variables mentioned and implicit in the description of Figure 4.5, as

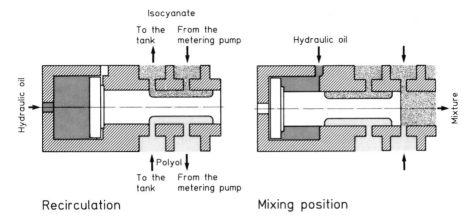

FIGURE 4.4   Schematic of a standard combined control/clean-out piston mixhead. (Diagram courtesy of Krauss-Maffei.)

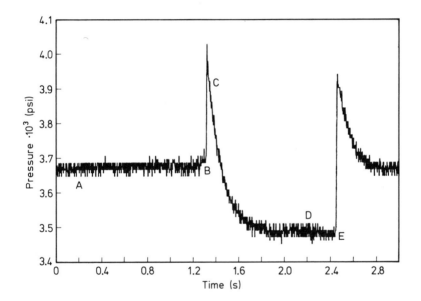

FIGURE 4.5   Trace of the pressure behind the polyol nozzle before, during, and after a shot from a mixhead of the type shown in Figure 4.4.

well as those described in earlier studies of the process, will be useful. The effect of the equipment-dependent variables will be discussed in more detail in the following subsection.

The effects of changes of the standard parameters on the mixing in the chamber of the mixhead have been studied by several methods, including visualization, calorimetry, and analysis of the physical properties of test plaques as a function of the parameters (18,19). In general, these studies have shown that the critical factor is the Reynolds number (Re) of the more viscous material stream,

123

$$\text{Re} \ = \ \frac{4Q\rho}{\pi d \eta}$$

where Q is the volume flow rate, ρ the density, d the nozzle diameter, and η the viscosity. For Re to be maintained above some minimum value (e.g., 200), the volume flow rate must be kept high enough and the diameter and viscosity low enough to satisfy the equation. The nozzle size is set in the machine setup, the flow rate can be adjusted within the range required by the part size and system reactivity, and the viscosity can be controlled to some extent by adjusting the material temperatures. There would be little need for discussion of mix quality if a good mix could be achieved by attention to these three points alone.

The purpose of intimately mixing the two streams is to bring the reactive materials close enough to be diffused together on a time scale comparable to the reaction time of the materials. Calculations based on some models have indicated that after mixing, the separate layers of materials are approximately 13 μm, a considerable distance when considering diffusion within the relatively short times allowed by the RIM reactivities. The other phenomenon that must be considered is the chemical compatibility at the interface between the layers of reactive materials. This was illuminated in one study in which the mixing of RIM chemicals was visualized on microscope slides rather than in a mixhead (20). When a drop of liquid MDI was allowed to contact a drop of an isocyanate-reactive material (such as a polyether polyol or polyamine and/or glycol or diamine chain extender), some tendency of the MDI to disperse in the other phase was observed. Observation of impingement mixed samples, either as they reacted or after reaction was complete, supported this formation of a dispersion at the interface of the two reactive streams. While the "dispersoid" or particle size in the dispersions tended to be smaller in the more reactive systems, this also generally parallels the compatibility of the systems. Long chain polyether polyol/butane diol/MDI systems gave particle sizes in the range of 3–5 μm, and amino polyether/aromatic diamine/MDI systems gave particles on the order of 0.1 μm. Obviously, the reactivity of the amine system is much higher than the glycol system; but just as clearly, diffusion across the much smaller distances is much faster.

A detailed description of the chemical physics in the mixing process is difficult because of the complex, changing nature of the system, but the key to understanding the process lies within that complexity. Several things determine the quality of the polymer resulting from the RIM process. Among these are the interplay between the mechanical mixing of the two streams, the reactivity of the chemicals, and the solubility of the chemical streams in each other as they are mixed and as they react. The importance of the differences in compatibility demonstrated by Wickert et al. (20) cannot be discounted. As mentioned earlier, part of the motivation for the switch from glycol extenders to aromatic diamines came from the better compatibility of the latter. Mix-related scrap categories were greatly reduced, if not eliminated, with this change in chemistry. The effect of compatibility on the mix in other systems is equally important. While achieving Reynolds numbers of 200 or greater is much easier with the low viscosities typical of structural RIM systems, the compatibility of the systems is still critical. This is especially true in filling parts large enough to require long gel times, where particularly incompatible systems can actually separate into their component parts.

4.3.2 Equipment Requirements

Much of the basic feasibility of the mixing is bound to the mechanical and chemical interplay just discussed; but with a given system, optimizing the mix quality depends primarily on the equipment. The basic requirements for RIM include:

1. Metering units that deliver the reactive materials at the flow rates and pressures necessary (as well as recirculate the material to keep them tempered)
2. A mixhead, where the mixing is actually accomplished
3. The interconnecting plumbing
4. A control system for switching from recirculation to injection.

Understanding the interplay between these components is the key to adjusting the mechanical system to the optimum for any chemical system.

There are two types of metering system: multiple-piston metering pumps of the Bosch or Rexroth type, and lance cylinder or syringe-type pumps. For RIM systems in which there is no filler injected with the reactive materials, the mechanical pumps have the advantage of lower cost and, for large SRIM parts, unlimited shot capacity. Lance cylinders, which are required when running filled systems, also have the advantage with low viscosity systems of better density control, provided the cylinders have the capacity for the complete shot. The cylinder size for very large shots (e.g., pickup truck beds) dictates a high cost for this type of metering unit, however. Basically, any metering unit must maintain the necessary accuracy within the range of delivery rate and at the pressures required by the part being made. With most systems the recommended index (molar ratio of isocyanate to isocyanate-reactive species times 100) needs to be kept within 5 points and sometimes tighter. All commercial systems available today are quite capable of maintaining ± 2 points in index when functioning properly. Within the range of all RIM and SRIM applications, pressures from 5 to 200 bar may be required. In an application where material is injected through a reinforcing mat, there may be a need for the pressure to rise as the shot progresses., Again, this is possible with all available commercial systems, although control systems may have to be altered to accommodate SRIM processing.

The commonly used mixhead types have been described in some detail elsewhere. There are potential advantages of each type in different applications. A brief description of the different mixhead types, though, along with the effects of other components on their operation, with reference to Figure 4.5, will be useful in optimizing machines for particular systems.

The pressure charted in Figure 4.5 was recorded by digitizing the signal from a transducer placed approximately 30 cm from the polyol nozzle of a mixhead of the type shown in Figure 4.4. Measurement at this point better reflects what is actually happening at the head than does measurement at the metering pump, which is separated from the head by about 3 m of flexible hose, As described above, there are

125

spikes in the pressure resulting from the zero flow conditions that occur as the control piston (which also serves as a clean-out piston) moves from the recirculation to mixing position and back. The length of these zero flow conditions depends on the control piston speed. Lag of the pressure in returning to its preset value is caused by the return of the flexible host to its preinjection size and by the return of any nucleation (finely dispersed gas) to its preinjection volume. The flow of material into the mix chamber is determined by the pressure drop across the nozzle.

Thus, in Figure 4.6, which shows the output of the flow meter for the same shot, the polyol flow starts at one rate (B) after the zero flow time (A) and diminishes until equilibrium is achieved between flow into the flexible hose from the pump and out through the head (C). The isocyanate side also sees a pressure spike for the same reasons as the polyol side; but because the density, pressure, and amount of dissolved or dispersed gas may be different, the response of the system is not necessarily the same. Any differences in flow response between the two sides results in a changing ratio. Although the flow differences apparent in Figure 4.6 are minor, great enough differences can create what is called a "lead/lag" situation—that is, an area of the part that is so rich in one component that polymer formation in that area is impossible. The key factors to control are the hydraulic pressure and the component pressures. Minimizing the length of the flexible hose can also be helpful.

These considerations also apply to any of the various modifications of this type of head because the recirculation–injection control mechanism is essentially the same. In one variation on this theme, recirculation is controlled by external valves. In such systems, the pressure spike can be adjusted by adjusting the time between the movement of the recirculation valves and the mixhead control piston; but because of the necessary zero flow condition, there will always be some spike.

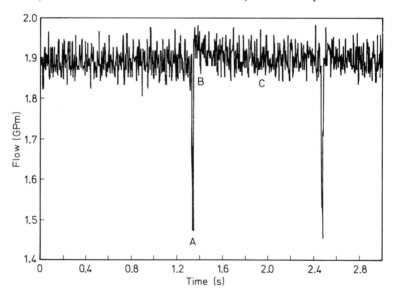

FIGURE 4.6   Trace of the flow of the polyol blend through the head for the same shot shown in Figure 4.5.

126

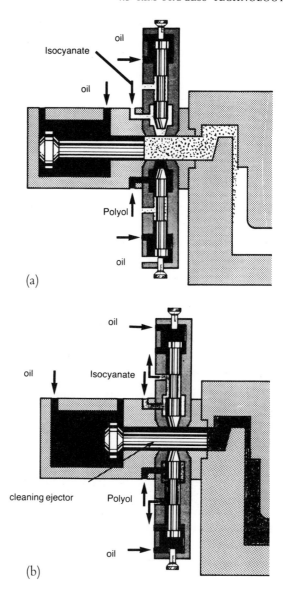

FIGURE 4.7 Schematic of a standard hydraulically actuated nozzle type mixhead. (Diagram courtesy of Mobay Machinery Group.) (a) Mixing position; (b) cleaning position.

The schematic of another type of control system is shown in Figure 4.7. The nozzles in this type of mixhead are opened by a hydraulic function separate from that which moves the clean-out piston. Recirculation pressures are set by external pressure control valves, and the injection pressures are set by adjustment of the nozzle set screw. Because the injection begins immediately when the nozzles shift, with no zero flow condition, it is possible to have a smooth transition between the recirculation and injection pressures. In practice, though, it is still quite possible to create a "lead/lag" situation if there is a mismatch between the pressures set by the recirculation valves and the nozzles.

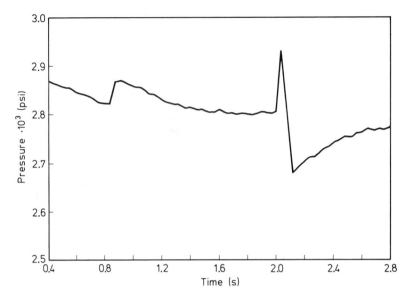

FIGURE 4.8   Trace of the polyol pressure before, during, and after a shot from a mixhead of the type shown in Figure 4.7.

Figure 4.8 shows a trace of the signal from a pressure transducer on the polyol supply of this type of head during an injection. The transition from recirculation pressure to the injection pressure is without a spike, but it is clear from the behavior of the pressure during the injection that adjustment of the recirculation pressure control valve and the nozzle is needed. Because there is no zero flow condition with this type of head, the length of the flexible hose to the head is less important, although keeping that length to a minimum has other advantages. Here the critical parameters to control are the hydraulic pressure and the proper nozzle and recirculation valve settings.

A variation on both the foregoing themes is to restrict the flow out of the mixing chamber, which works to transform the turbulent flow through the mixing chamber into laminar flow. Originally developed for open pour applications, these heads have found use in the closed-mold RIM applications, because they can improve the mix quality. Schematics of a combined control/clean-out piston version of this variation and of a hydraulic nozzle version appear in Figures 4.9 and 4.10, respectively. One possible advantage of this type of head is that it gives one more variable through which the mixing can be controlled. Obviously, the inherent disadvantage is that one more variable must be controlled.

A recent development in mixhead technology is the parallel stream mixhead, a schematic of which is shown in Figure 4.11. This technology is no longer appropriately called impingement mixing, since the two streams do not impinge against each other, but flow coaxially. Mixing is accomplished via the turbulence created at the interface of the concentric jets of material. Potential advantages of this type of head are the simplicity of operation, similar to combined control/clean-out piston heads

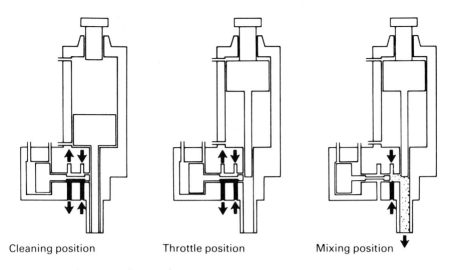

FIGURE 4.9 Schematic of a transfer or L-type combined control–cleanout mixhead. (Diagram courtesy of Krauss-Maffei.)

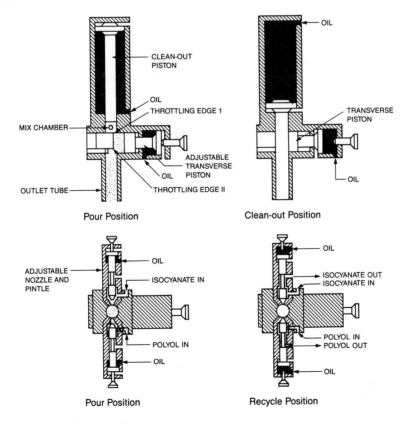

FIGURE 4.10 Schematic of an adjustable throttle, hydraulically actuated nozzle-type mixhead. (Diagram courtesy of Mobay Machinery Group.)

129

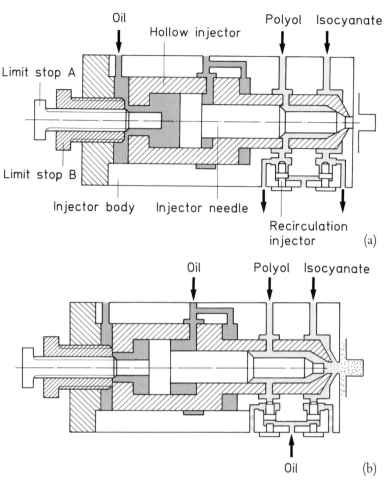

FIGURE 4.11 Schematic of a parallel stream mixhead. (Diagram courtesy of Mobay Machinery Group.) (a) Recirculation position; (b) pouring position.

except that since the piston needs to move only a short distance, only a small volume of hydraulic oil is required for shifting the head. While the annular nozzle arrangement would appear to be "lead/lag" prone, initial reports indicate that the parallel stream head has a somewhat wider processing window than comparable impingement heads, without such problems.

4.3.3 Process Parameters

The constraints inherent in impingement mixing, along with the reactivity of the chemical system, are the basic source for most process parameters in RIM. If a part weighing 3 kg is made from a system that runs at a weight ratio of 2:1 (polyol to isocyanate) with a maximum injection time of 1.5 seconds, the necessary flow of the polyol is calculated simply as 1.33 kg/s. Assuming a density of 1.25 g/cm³ and a polyol viscosity of 5000 cP, the nozzle diameter necessary to give Re > 200 is easily calculated to be 1.7 mm. Similarly, if one assumes a certain nozzle diameter, velocity, and density, one can calculate the viscosity necessary to give a Reynolds number indicative of satisfactory mix quality. With this viscosity calculated, examination of the temperature–viscosity relationship for the polyol will give the necessary temperature for the polyol to provide adequately mixed material.

The impingement pressures can also be calculated, but in practice, these calculations are almost never done and the process parameters are set mostly by trial and error. This generally works when using one basic type of system (e.g., a glass-reinforced polyurethane–urea), because the variables that determine the Reynolds number (e.g., the temperature–viscosity relationships) do not change significantly. When changing to other systems, though, the basic calculations can provide a better starting point for the trial-and-error process. Changing to higher nominal viscosity systems such as higher filler level polyurea systems can require higher material temperature. Similarly, changing to a low viscosity SRIM system can require lower initial impingement pressures for optimal mixing. Most system suppliers give basic processing parameters as starting points for mix optimization.

4.3.4 Mold Considerations

Probably the most important factor in determining the potential quality of a RIM part is the quality of the mold. While RIM tooling can be less expensive because of lower pressure requirements, attempts to cut costs on a part by skimping on tooling will almost always raise the ultimate cost of a quality part because of the added work of making the "lower cost" tooling work properly. This holds true for all types of RIM, but there are specific considerations that change depending on whether the part is RRIM or SRIM. Each will be covered, along with some basics common to both.

The material of choice for RIM tooling is high quality tool steel. While some parts are made from aluminum tooling, and some limited-run prototypes are made from epoxy tooling, the wear on tooling caused by glass reinforcement and cleaning dictates steel for high volume applications. Having replaceable aftermixer blocks and gates is also helpful, both because of wear and to allow changing aftermixer styles if required without removing a tool from production for too long a period. Relatively uniform control of the temperature of the tool is helpful to maintaining proper part dimension. While this is often done with temperature-controlled water systems for parts made from materials run at "conventional" RIM mold temperatures ($\sim 65\,°C$), oil heating systems offer higher temperature, if necessary, and extend tool life.

In RRIM, tools are typically edge gated. As a general rule, the cross section of the part should gradually increase with the flow through the part. That is, the sum of the channel sections through the aftermixer should be slightly higher than that of the runner leading into the aftermixer, and the gate section should be slightly larger than the aftermixer section. While sizing the gate to maintain a particular linear velocity can reduce or eliminate flow lines in the part, keeping the gate at or above 2 mm is recommended for both polyurethane–urea and polyurea systems. Keeping the transitions from the aftermixer to the gate and from the gate to the part as smooth as possible minimizes the chance for air entrapment in the part. For most RRIM tools, substantial vents at, or close to, the last points must be filled, as well as the minor venting around the parting line of the tool, which leads to flash around the edge of the part. Recently, shear edges were incorporated in a RRIM tool to allow the use of an afterpressure technique, which provides a better part surface and reduces the flash to a minimum (21, 22).

Most SRIM parts are gated in the center of the part. This helps prevent movement of the reinforcement, because the mat is pulled by the material flowing through it in all directions at the same time. Because of the low viscosities of most SRIM systems, substantial flash occurs around the parting line unless there is an additional provision for sealing the tool such as an O-ring gasket or alternative seal. Careful glass placement is necessary because of the differential flow rates of the material through the glass and through any low glass areas such as those that might occur around the edge of the part (see Sections 4.4.3 and 4.6.3). Shear edges have been used in SRIM tooling to help control the glass; but if there is too much clearance between the shears, the differential flow rates through the reinforcement and out the shear can cause excessive flash. As SRIM applications become more common, tool designs that make the technology more viable will become more widely known.

4.4 CURE KINETICS AND FLOW BEHAVIOR

4.4.1 Calorimetric Kinetic Characterization

The kinetics of the RIM process have been followed in a number of ways, each of which has given information on different aspects of the process. The formation of the final polymer can be thought of as occurring in two distinct steps: chemical reaction and separation into hard and soft phases. Each of these steps affects how the material flows through the mold. There are several ways to monitor each step, but the two main methods have been to measure the heat generated by the reacting material and to track the pressure at various points as the material fills a mold. The former gives an indication of the overall rate of reaction, while the latter is related to the viscosity of the reacting material, which is a complex function of the extent of reaction, phase separation, and gelation. For the practical problem of filling a mold, tracking the pressure is more directly significant, but knowledge of the extent of reaction makes the pressure data more meaningful.

While there is undoubtedly a thermal effect of phase separation, it is small compared with the heat of the urethane formation reactions. The heat of reaction has been monitored by measuring the temperature of the reacting material in a mold and, adiabatically, by measuring the temperature rise of mixed material in an insulated vessel (23,24). Of the two, the adiabatic temperature rise is the simpler in that it is not complicated by transfer of heat to or from the walls of the mold, although if the vessel is too large, heat-activated side reactions can blur the significance of the data. Relatively early, this method was used to determine basic rates of reactions in polyurethane systems, but unfortunately, polyurethane–urea and polyurea reactions are too fast for this method to yield actual reaction rate constants (25).

132

In SRIM, though, monitoring the temperature in the tool can give a very good indication of the extent of reaction, at least in the qualitative sense. With copolymer and isocyanurate systems, the lag between the mixing and the main exotherm gives a good indication of the allowable flow time, since the exotherm is closely correlated to the beginning of the "second step" of the reaction. With urethane systems there is no real second step, but there is a definite lag before the exotherm, which also gives a good indication of the flow time and the onset of gelation. In the development of new systems, this can provide useful information, but it also has practical use in optimizing catalyst levels for a part if there is a means to measure the temperature of the material in the tool.

4.4.2 Flow Characterization in RIM and RRIM

Although some methods for evaluating the viscosity of reacting materials have been reported and other proprietary methods exist, the most clearly meaningful method is to measure the pressure exerted by the reacting material on the mold walls as it flows through the tool. Interpretation of the data in terms of extent of reaction, gelation, and phase separation is not simple because of the interconnection of each of these with the viscosity. It is, nonetheless, possible to determine the actual viscosity of a system as it reacts by measuring the pressure at two points in a mold at a particular flow rate according to the equation

$$\eta = \frac{C_1 \Delta P}{v \Delta L}$$

where ΔP is the pressure drop measured between two points a distance ΔL from each other when the material is flowing past the points at a velocity v, C_1 is a constant that is dependent on the geometry of the mold, and η is the mean viscosity of the material between the two points.

This relationship has been used in two different ways. In a mold with two pressure transducers, the shape of the trace of pressure drop between the transducers gives an indication of how the viscosity rises. In a mold with multiple transducers, the pressure drop between the various points indicates how the viscosity varies as a function of distance from the gate. Figure 4.12a shows a schematic of a mold with several transducers spaced down the flow path of the mold. The traces of the pressure at each of six transducers appear in Figure 4.12b. From these traces and a knowledge of the flow pattern in the mold, it is possible to calculate the mean viscosity between the points in the mold. Since these points correspond to a time for which the material has been mixed (calculated from the distance and the flow rate), it is possible to assign a viscosity to the material as a function of time from mixing (Fig. 4.12c). It is then possible to study the effect of variables such as flow velocity, mold temperature, and formulation on the viscosity rise in the system. Such measurements have already proved to be of practical use in developing and optimizing systems, and in advancing our basic understanding of polymerization kinetics.

133

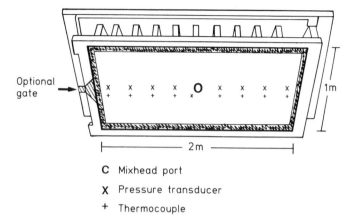

C Mixhead port

X Pressure transducer

+ Thermocouple

FIGURE 4.12a   Schematic of an instrumented plaque tool for characterizing the flow of RIM materials through a tool.

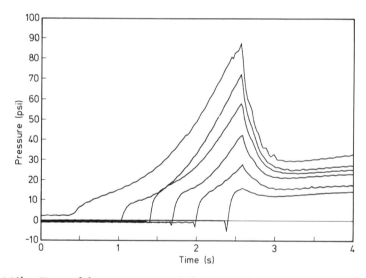

FIGURE 4.12b   Trace of the pressure at six different transducers in a tool such as that shown in (a).

4.4.3 Flow Characterization in SRIM

The flow of SRIM resins through a mold can be studied in the same way as RIM and RRIM resin flow is studied with a mold instrumented with pressure transducers, except that the equation is complicated by the resistance to resin flow due to the reinforcement. This can be handled by adding to the equation a quantity which factors in the permeability of the reinforcement:

134

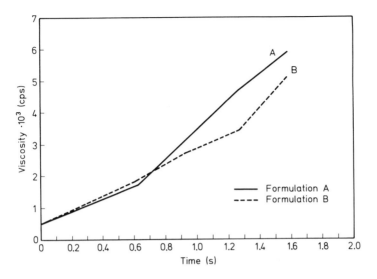

FIGURE 4.12c   Viscosities at different times from mixing calculated from the data in (*b*).

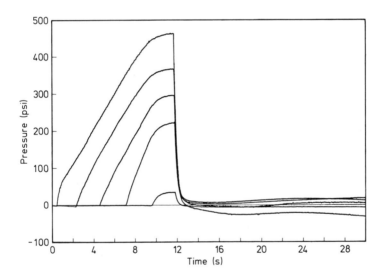

FIGURE 4.13   Pressure traces from five transducers in a plaque tool like that shown in Figure 4.12a for a shot of a polyurethane SRIM resin through random glass mat reinforcement.

$$\eta = \frac{C_1 C_2 \Delta P}{v \Delta L}$$

where all the variables have the same meaning as above and C_2 is a constant reflecting the permeability of the reinforcement. Figure 4.13 shows pressure traces for injection of an SRIM system through a reinforcing mat (26, 27) similar to those observed for

135

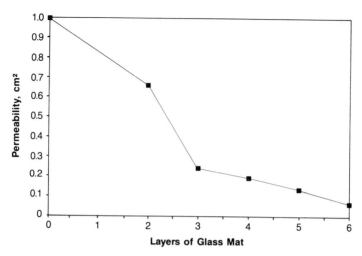

FIGURE 4.14   Permeability of random glass mat as a function of the number of layers of mat in the mold, calculated from pressure data as in Figure 4.13, using DIDP as the pumping fluid.

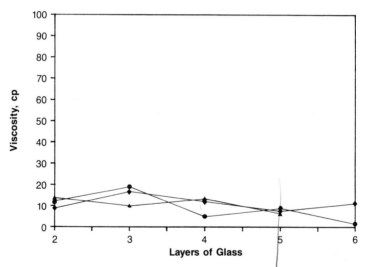

FIGURE 4.15   Viscosity of a polyurethane SRIM resin during a shot as a function of glass content in the mold and velocity (circles, diamonds, and triangles) calculated from permeabilities and during-shot pressure traces.

injection of a RIM resin into an empty mold except for the line shape, which indicates the different reactivity. The relatively straight lines as well as the relatively even spacing with distance indicate that there is essentially no viscosity rise as the material flows through the glass. To extract real viscosities from the data, it is necessary to know K. This can be accomplished by calibrating the "viscometer;" that is, by generating the same kind of data shown in Figure 4.13 with a fluid of known viscosity.

Figure 4.14 shows the results of such a calibration. As the glass level is increased, the permeability of the reinforcement decreases, to where it is 10% that of the open mold at 60 w % glass. This calibration factor can then be used to give the actual viscosity of the SRIM system as it flows through the mold.

Figure 4.15 shows the viscosity calculated from the pressure data using the calibration values from Figure 4.14. Note that the viscosity does not vary (within experimental error) either with glass content or with velocity, and that the viscosity is on the order of 10 cP, the range necessary for long flow through high glass levels.

The cure kinetics of an SRIM system can also be observed with this type of data. The small rise in pressure at around 20 seconds in Figure 4.13 corresponds to the gelation of the material and is consistent with the exotherm time for this system. The pressure rise observed with isocyanurate systems is more pronounced, probably because of the formation of by-product gases from side reactions.

4.4.4 Flow Modeling

A number of programs have been developed to calculate the flow of plastic materials through molds and predict flow patterns and pressures in the molds (28). While most of these programs were developed for thermoplastic injection molding, they have been used with considerable success with RIM and RRIM (at least for flow pattern prediction). These methods are based on iterative calculation of the flow, assuming the material flow to be analogous to wave propagation. Thus, the flow predictions are based only on what is occurring at the flow front. As long as all the material across the flow front has the same effective viscosity, such programs work equally well for thermoplastics with relatively high melt viscosities and with RIM materials. Pressure calculations depend on the viscosity of the material and its response to temperature and shear; therefore, until these are well defined for RIM materials, the pressure predictions from these programs will be less meaningful.

These programs can also be used for flow pattern predictions with SRIM if the permeability of the glass is factored in. This can be done with relative ease by considering the physical meaning of the permeability constant. This constant indicates the difficulty of resin flow through the reinforcement relative to that through the empty mold. If the constant is thought of as a relative thickness, it has the same effect on the relationship between viscosity, flow distance, and pressure drop. With sufficient knowledge of the permeability of various reinforcements, then, it is possible to set the geometry of a mold pattern calculation to give good predictions with existing programs.

Keep in mind, though, that these programs are meant to be used for calculating flow through elements that are about the same size as the mold thickness. Use of substantially larger elements can lead to spurious predictions because program smoothing functions are operating over too large an area. For large parts, this means considerable computer time to achieve reasonably useful predictions. The use of the predictions, though, can be quite beneficial if a troublesome air trap can be avoided, for instance, by correct tool design.

4.5 MATERIAL PROPERTIES

4.5.1 RIM and RRIM

RIM technology is rapidly changing, particularly in the area of material properties. Thus all current production of automotive fascia is in polyurethane–urea systems, but it is likely that polyurea systems will be in production by the early 1990s. The Pontiac Fiero vertical panels were made for most of the life of the model with polyurethane–ureas but were switched to polyurea during the last months of production. Future body panel applications will almost certainly be in polyurea. Because of the changing nature of the RIM materials picture, we try here to give some historical perspective and a snapshot of current materials.

Short glass fiber reinforcement was originally added to fascia systems to improve dimensional stability, both on the car and in the paint oven. Table 4.8 shows the material properties listed in a 1982 paper (10) on reinforcements for fascia systems. Table 4.9 lists the heat sag data from the same paper. Note that the sag test conditions were made more stringent [to 163 °C (325 °F) from 121 °C (250 °F)] to give larger, more statistically significant numbers. The properties viewed as critical were those associated with dimensional stability and impact strength.

Today there are a number of polyurea formulations available for fascia applications. Table 4.10 is from a 1989 paper on polyurea fascia systems (29). Many of the properties considered critical in 1982 remain the same today, but the polyureas give substantial improvement in heat sag. An addition to the list of items affecting dimensional stability is moisture growth, where the polyureas also show even better

TABLE 4.8 Reinforced Polyurethane–Urea Properties Affecting Dimensional Stability

Base flexural modulus (MPa)	Milled fiber (%)	Flexural modulus (MPa)		CLTE ($\times 10^6/°C$)		Shrinkage (%)	
		Par[a]	Per[b]	Par[a]	Per[b]	Par[a]	Per[b]
125	0	130		180			
	5	178	197	151	162		
	10	275	229	100	162		
	15	324	228	83	155		
175	0	172		184		1.41	
	0	245	257	139	157	1.23	1.28
	10	293	241	119	131	0.95	1.23
	15	392	248	74	137	0.61	1.16
240	0	276		157		1.21	
	5	361	301	130	144	0.97	1.05
	10	438	339	88	146	0.70	1.06
	15	644	386	67	135	0.53	0.99
	20	959	490	45	137	0.38	1.00

[a] Par = parallel
[b] Per = perpendicular

TABLE 4.9   Heat Sag Data for Polyurethane-Urea Systems

Base flexural modulus (MPa)	Milled fiber (%)	1 h at 121 °C, 100 mm OH[c]		1 h at 163 °C, 100 mm OH[c]	
		Par[a]	Per[b]	Par[a]	Per[b]
125	0	4.8		37.7	
	5	–	3.8	34.3	31.3
	10	2.0	3.7	15.7	28.5
	15	1.0	2.6	13.0	23.2
175	0	7.6		36.2	
	5	2.8	2.3	21.3	23.8
	10	3.5	2.8	17.8	27.3
	15	2.0	3.3	13.6	12.3
240	0	5.0		35.2	
	5	2.2	3.5	17.0	23.7
	10	1.7	3.0	10.5	16.3
	15	1.3	2.7	8.0	15.2
	20	1.0	1.7	4.3	9.8

[a] Par = parallel
[b] Per = perpendicular
[c] OH = Overhang

TABLE 4.10   Physical Property Comparison for Polyurethane–Urea and Polyurea Fascia Formulations

Property	Systems					
	25K Poly-urethane–Urea	25K Polyurea	35K Poly-urethane–Urea	35K Polyurea	50K Poly-urethane–Urea	50K Polyurea
Density, g/cm^3	0.99	0.93	1.01	1.00	1.00	1.00
Tensile strength, MPa	18.2	21.2	19.7	21.9	20.7	21.2
Elongation, %	230	230	210	230	200	190
Flexural modulus, MPa	160	167	234	208	331	393
Heat sag, mm; 1 h at 120 °C, 150 mm OH	25	11	16.5	10	15	6
Tear strength, kN/m	61	81	61	77	100	

performance. Another property that shows up in Table 4.10 is tear strength, which reflects increased concern with processing. Higher tear strength translates directly into less molding scrap in the fascia industry.

Perhaps a more meaningful comparison of the older technology with the new polyurea materials is offered by the data contained in Table 4.11, which shows the properties for a polyurethane–urea and a polyurea formulated for the same stiffness.

139

TABLE 4.11   Physical Properties of Polyurethane–Urea System and Lower Density Polyurea System Formulated for Same Stiffness

Property	Control: Bayflex 110-25 IMR	Polyurea 25K fascia system
Density, g/cm^3	1.0	0.93
Flexural modulus, MPa	160	156
Tensile strength, MPa	18.2	19.5
Elongation, %	230	220
Heat sag, mm; 1 h at 121 °C, 150 mm OH	25	15

Even at a reduced density, most of the polyurea properties are equivalent or better than the urethane–urea, and the heat sag is significantly better. This trend can be expected to follow for the reinforced systems—that is, though the material cost per pound is higher, less is required to meet a given set of requirements. This aspect, combined with the improved processing that polyurea gives, could mean that the industry is on the verge of a better product at the same or reduced piece cost.

The early 1980s saw the introduction of the first high volume model with polyurethane–urea body panels. Table 4.12 shows the physical properties of the material used in those panels. The material processed well and met the requirements set in 1983. There was room for improvement, though, particularly in the area of dimensional stability through the paint line. The late 1980s gave way to several commercially available polyurea body panel formulations. Table 4.13 lists the properties of different systems as reported in the literature (13,30,31). Each of these shows substantial improvement over the earlier material, particularly in thermal stability.

TABLE 4.12   Physical Properties of Bayflex 110-80 Polyurethane–Urea Body Panel Formulation

Property	Flow direction		
	Unfilled	Par[a]	Per[b]
Glass flake, %	0	20	
Density, g/cm^3	1.01	1.15	1.15
Flexural modulus, MPa			
RT	517	1550	1340
$-30°C$	965	2760	2230
$+65°C$	310	993	738
Modulus ratio $(-30°C/+65°C)$	3.1	2.8	3.0
Tensile strength, MPa	24	32	32
Elongation, %	110	20	20
Notched Izod strength, J/m	214	107	107
CLTE, $\times 10^6/°C$	55.8	63	144

[a] Par = parallel
[b] Per = perpendicular

TABLE 4.13   Physical Properties of Various Polyurea Body Panel Formulations[a]

Property	Bayflex 150	Spectrim HT	Spectrim HF	RIMline E-7070
Reinforcement	23% DOI	20% Glass flake	20% Glass flake	20% Glass flake
Density, g/cm^3	1.2	1.2	1.2	1.16
Tensile strength, MPa	31.0	26.2	27.6	21.4
Elongation, %	40	28	32	32
Tear strength, N/mm	105			
Flexural modulus, MPa	1740	1740	1630	1340
Heat sag, mm; 150 mm OH				
1 h at 121 °C	0.5	0.25	3.8	0.75
1 h at 162 °C	4.6	5.6	10.2	4.8
CLTE, × 10^6/°C	50	49	49	43
Shrinkage, %	0.60			

[a] Bayflex, Spectrim, and RIMline are trade names of Mobay Corporation, Dow Chemical Company, and ICI Americas, respectively.

TABLE 4.14   Processing Recommendations for Various Polyurea Body Panel Formulations

Property	Bayflex 150	Spectrim HT	Spectrim HF
Filled ratio	0.550	0.570	0.588
Index	105	103	103
Iso Temp., °C	48–51	38–41	43–46
Polyol Temp., °C	48–51	43–46	43–46
Mold Temp., °C	65	120	75
Cure time,s	20–30	20–30	10–20
Typical release/spray	~ 50	5–10	75–100

The various commercial offerings are distinctly different formulations, which is more apparent from the processing conditions recommended for each system as shown in Table 4.14. Both the Spectrim HT and Bayflex 150 materials are reported to withstand paint bake temperatures in excess of 175 °C. Further development along these lines promises on-line painting of RIM body panels.

4.5.2 SRIM

As with polyurea body panel formulations, there are a number of commercially available SRIM systems, as well as low density systems for use with long fiber reinforcements. Tables 4.6 and 4.7 give some properties of polyurethane composites with different types of glass; Table 4.15 presents the properties of various commercial systems with approximately the same level of glass (14,26,32,33). Between 34 and 40 w % random glass mat, the flexural modulus for the various systems varies from 7.2 to 9.2 GPa, and the flexural strength from 225 to 285 MPa. The tensile strength

TABLE 4.15   Comparative Physical Properties of Various SRIM Systems

Resin	Mobay (urethane)		Dow (urethane)	Dow (isocyanurate)	Ashland (acrylamate)	ICI (urethane)	
Glass, % random continuous	34	46	40	40	40	40	50
Flexural modulus, GPa	7.2	11.1	8.5	9.2	7.3	7.5	9.6
Flexural strength, MPa	225	285	279	285	231	–	–
Tensile strength, MPa	143	228	~ 160	~ 160	125	0.159	220
Notched izod, J/m	763	1360	693	530	790	750	96
HDT, °C				220	240	> 280	> 280

varies for the same samples from 125 to 160 MPa, and the impact strength ranges from 530 to 790 J/m.

At higher loadings (46–50 w % random glass), the flexural modulus varies from 9.7 to 11 GPa, with flexural strength around 280 MPa, and tensile strength around 220 MPa. The notched Izod strength does vary significantly, from 960 to 1360 J/m. The heat deflection temperature (HDT) also varies substantially, from 198 °C to 280 °C, but the difference between Mobay's urethane and Dow's isocyanurate is only about 20 °C for samples that have not been postcured. In fact, it is reported that while postcure brings a rise of + 70 °C in the glass transition temperature of the isocyanurate, this results in only a 20–30 °C rise in the HDT. Therefore, for parts under constant load, it would seem that the glass transition temperature would be the more accurate measure of part performance. In any case, comparison of Table 4.5 with Table 4.15 indicates that the glass is the main source of the basic properties for SRIM composites.

Other properties that will be important in some applications are the short- and long-term creep and fatigue behaviors. Many of the applications for which SRIM is under consideration involved extended subjection of the material to cyclic loading, sometimes under extreme environmental conditions. There is extensive testing planned and underway to determine the creep and fatigue behaviors of the various resin–glass combinations under a variety of conditions. Obviously, this testing must be completed before these materials may be used in critical load-bearing applications.

4.6 DEFECTS AND HOW TO AVOID THEM

4.6.1 RIM and RRIM Defects Arising from the Process

Avoidance of defects in RIM, as in most production processes, is primarily a matter of keeping process parameters within control limits. The key factor in this control is awareness of all the parameters from the start to the end of the process. While there is some chance of defects arising from contaminated material, as discussed below, even many of these relate back to the beginning of the process at the machine setup.

Machine flushing was mentioned earlier in connection with surfactants, but the problems arising from incomplete flushing are more general. Whenever there is a change from one chemical system to another, there will be some contamination of the second system with the first. This can be minimized with proper machine setup, but some effort to define the extent of cross-contamination should be made if material changes occur regularly. This can be done with a simple tracer experiment in which the system is drained, and a weighed amount of material containing the tracer is added and circulated with the hung-up material (i.e., that which remained in the machine after draining). Measurement of the concentration of the tracer in the mixture then quantifies the dilution and gives the amount of hang-up. There are a number of reactive dyes that make good tracer materials; or, if switching from one amine system to another, the nitrogen content of the materials can be used for this purpose. Depending on the arrangement of the system, circulation, and any potential dead legs, the hang-up can be surprisingly large.

Once the metering system has been set up, the process begins with arrival of material at the molding plant and finishes with shipping parts out the door. Awareness of all the controllable variables in between allows scrap to be minimized. Describing these variables, along with the possible defects that might arise from failure to control them, can be divided into eight basic categories. Raw materials are typically shipped with an analysis sheet listing the chemical parameters for the batch. Some molders reanalyze as a check; some depend on the supplier's analysis. In either case, the numbers should be checked for agreement with expected values before loading into storage tanks. The amount of material left in the system from the preceding batch should be recorded. Deviations in the temperature, blending, exposure to air, moisture, or heat in this premold material handling can lead to defects down the line. The primary defects caused by failure to control variables at this stage arise from poor green strength, which may be manifested by part tearing or excessive deformation upon demold. Poor release from the mold is another symptom that may be associated with improper premold material handling.

The next set of parameters consists of the proper tempering of the material and the associated variables. The tempering is really composed of two things: temperature and nucleation. Again, machine setup will affect this in that proper tempering requires sufficient recirculation. Improper raw material temperature can lead to poor green strength and associated defects, poor mixing, sinks in the part, or pregel. Nucleation, the dispersion of finely divided gas bubbles in the polyol blend, aids processing tremendously but must be controlled. Nucleation with nitrogen is recommended, and

143

the gas bubbles should be extremely small. The amount of gas dispersed, which will depend on the part and other processing parameters, should be held relatively constant once the proper level has been determined. Potential defects arising from inadequate control are air bubbles in the part, sink marks and flow lines, pregel, poor green strength, and improper part size due to changes in the shrinkage (which is affected by nucleation level to some extent).

Physical properties are directly dependent on the right ratio of isocyanate to isocyanate-reactive species in the mixed material. Indications of improper ratio control, other than physical properties, are poor green strength, poor release, and top coat paint discoloration.

Molding parts with a good surface depends on maintaining a clean mold. With IMR systems this is easier to do, but still critical. The tool should be stripped completely and rinsed with water. A heavy coat of mold release is then applied and buffed into the tool, after which a light spray is applied. Parts are then run on the tool either until there is sticking (in which case there is a light touch-up spray) or until there is a buildup of IMR (at which time the tool should be buffed again and lightly sprayed). Defects that might arise from improper tool maintenance are porosity, top coat blistering, poor release, and air bubbles.

The mold must also be maintained at the proper temperature across the mold. Failure to control this parameter can lead to poor green strength, sink marks, flow lines, pregel, poor release, and brittleness. Part dimension will also be affected, because the shrinkage of the material is correlated directly with the mold temperature.

The demolding time will also affect part shrinkage; more immediately, however, the green strength will be affected. Improper demolding can lead to part tearing and deformation. The release of the part from the mold will also be affected. Most machines are set automatically to open the mold at a preset time.

Postmold handling can also affect the final part appearance. Improper loading on postcure fixtures can lead to part distortion. As with any painting operation, proper washing of the part is necessary for good paint quality. With currently used IMR systems, the recommended procedure is to postcure the parts, then power wash. Dry-off temperatures out of the wash cycle must be kept below a certain threshold temperature. Improper postcure or washing procedures can lead to poor paint wetout or adhesion and other associated paint defects.

4.6.2 RIM and RRIM Defects Arising from the Materials

As stated in Section 4.6.1, material-related defects are typically related to contamination, which can have three sources. The first of these is material supply. Supply contamination can be as simple as having the wrong amount of reactive material or as gross as having foreign particles dispersed in the material. If incoming raw material control fails to catch supply contamination, it may appear as poor green strength or physical properties or, in the case of foreign particles, as poor surface on the parts or blistering or other defects out of the paint line.

The second possible source is the metering machine, as mentioned before. Contamination of one system with the previous material in the system can be avoided only by properly flushing the machine. For this to be accomplished with a minimum of effort and waste, the flushing procedure must be worked out quantitatively for a machine. Poor green strength or physical properties can result from contamination of the metering machine.

The third possible source lies in the machine itself: seal degradation or other foreign material. This type of defect will generally appear as a poor surface on the part or paint defect such as blistering or rash. Proper preventative maintenance on the machine will avoid defects of this sort. Another possible defect, which would appear to arise from the material but actually results from improper process control, shows up in much the same way (i.e., surface problems and paint blistering). Specifically, problems can arise from precipitated IMR agent, which is a type of "foreign matter" in the system. This is usually distinguished from a seal problem by the color of the foreign matter and by poor release in demolding.

Although setting up and controlling the process parameters to avoid defects is the best course, when defects occur, a troubleshooting guide can be helpful. Most material suppliers provide such a troubleshooting matrix, listing possible defects across the column headings and possible causes down the last column. Use of such a matrix with reference to the foregoing discussion may be helpful in identifying and eliminating the cause of a defect.

4.6.3 SRIM Defects

A number of the defects listed above can also occur in SRIM, particularly those associated with ratio control. Two other types of defect can occur, though, which are associated with the reinforcement. The first involves the flow and the second the surface appearance. As discussed in Section 4.4.3, the permeability of an unreinforced section is 1, while the permeability of any reinforced section is less than 1, meaning that the resin will flow preferentially through the unreinforced section. A common defect seen in prototyping parts is a void area, which results from the resin flowing down an unreinforced channel and back around to meet the flow through the glass. In moving out of the prototype stage, this type of defect must be eliminated. Improper placement of the reinforcement, however, can result in its recurrence.

The second type of defect that can be caused by the glass involves the surface of the part. Many SRIM applications are in nonappearance parts, but for those in which cosmetics are significant, glass handling can be important. Isocyanates react with water to produce carbon dioxide. If the glass is not relatively dry, this gas–producing reaction can create porosity on the surface of the part as well as reducing the overall density of the part. Even if the part is not painted, the porosity can cause other appearance problems. More porous parts tend to stick to the mold more than parts with a compact surface. After several parts have been run, the mold surface may begin to build up a residue that gives the following parts a rough texture. IMR systems recently developed help to prevent this buildup, but storing the glass in a dry place before molding will help minimize surface porosity even with non-IMR systems.

4.7 APPLICATIONS

Reinforcements were first added to RIM materials to improve the dimensional stability of automotive fascia. In North America, this is still by far the largest application of RRIM, with around 60 % of the cars produced having some RIM in the bumper or bumper cover area. The cost of more complex processing of RRIM (addition of glass, wear on the machines, etc.) necessary to achieve the improvements in appearance created a balance between RIM and RRIM in this industry. Until recently there was no offsetting factor in this equation, because the cost of the glass fiber was about the same as that of the resin. With the development of new, lower cost reinforcement types for use with RRIM, though, this balance is due to shift. Quite possibly, this shift will be complete. With demonstrations of polyurea materials used with DOI-enhancing fillers on the largest present fascia parts, the advantages of improved appearance and processing with little or no increase in material cost would seem to far outweigh the increased complexity of the operation. The same advantages may also be realized in the area of encapsulated windows, where glass fiber was originally added to give a better match between the dimensional stability of the RIM encapsulation and the window glass. In this area, particularly, the use of lower cost reinforcements and fillers promises substantial savings. The physical properties of most RIM systems far exceed what is truly required of the encapsulation, so that reductions in those properties inherent in the use of some low cost fillers should be quite acceptable. As this and other RIM applications grow, and as the fascia industry completes the shift to RRIM, the use of polyurethane, polyurethane–urea, and polyurea with various reinforcements can be expected to continue to grow.

The first RRIM body panels on production vehicles (the Oldsmobile Sport Omega and Porsche Carrera) demonstrated the feasibility of making body panels with the RIM process, as well as the potential savings from parts consolidation. The Pontiac Fiero demonstrated the viability of high volume production using this process, as well as the savings achievable in model changes and the quality that can be achieved in the molding process. While the cancellation of some programs for a variety of reasons in the mid–1980s was a blow to the use of RRIM body panels, the additional time allowed material developments, which ultimately will improve the appearance and quality of RRIM body panels. This, coupled with the development of machines capable of very high throughputs, is allowing progress in several body panel programs. Further developments in surface quality, thermal stability of the materials, and processibility promise substantial growth in this area in the next decade. Panels as small as lower side moldings for passenger cars and as large as rear quarter panels for pickup trucks are possible at high production rates.

The area of SRIM has the greatest potential for growth, but it also requires the most developmental work to realize this growth. With all the major automotive companies in North America actively involved in and supporting programs in this area, development will be rapid. Other than the basic materials development and characterization, preform production and placement, mold gating, venting and sealing, core placement, and process reproducibility have been identified as areas for concentrated developmental effort. Demonstrations of prototype automotive structural components are planned, and production has begun using long fiber reinforcements and low viscosity solid and foam systems on bumper beams and nonstructural parts such as spare tire covers and interior trim parts. Growth in the latter area is predicted to be rapid, supporting the production experience necessary to provide the confidence in material production required for applications such as floor pans and rail structures in automobiles.

The other area in which production experience will be gained in the short term is in non-automotive manufacturing, where production advantages may favor SRIM for some parts because of part size or complexity, or material advantages. Overall, with material and process developments already realized or underway, and with experience gained in intermediate-scale production on parts presently planned in SRIM, applications of this technology promise to grow markedly in the next decade.

REFERENCES[1]

1. W. E. Becker, Ed., *Reaction Injection Molding*, Van Nostrand Reinhold, New York, 1979.
2. R. E. Camargo, C. W. Macosko, M. Tirrel, and S. T. Wellinghoff, in J. E. Kresta, Ed., *Reaction Injection Molding*, American Chemical Society, Washington, DC 1985, p. 27.
3. For a more general discussion of structure–property relationships in polyurethane-based systems, see D. Dietrich, E. Grigat, and W. Hahn, in G. Oertel, Ed., Hanser, New York, 1985, p. 7.
4. R. P. Taylor and B. A Phillips, SAE International Congress and Exposition, Paper 810121, 1981.
5. R. J. G. Dominguez, D. M. Rice, and R. A. Grigsby, Jr., Polyurethanes World Congress 1987, Aachen, *Proceedings of the SPI/FSK*, 213.
6. R. P. Taylor and S. M. Abouzahr, SAE International Congress and Exposition, Paper 860282, 1986.
7. D. J. Primeaux, II, R. A. Grigsby, Jr., and M. Rice, SAE International Congress and Exposition, Paper 880353, 1988.
8. E. C. Martinez, W. J. Lepovitz, and M. C. Cornell, SAE International Congress and Exposition, Paper 870279, 1987.
9. *Handbook of Chemistry and Physics*, 64th ed, Chemical Rubber Company, Cleveland, 1983, p. D-164.
10. M. E. Cekorik, W. A. Ludwico, and R. P. Taylor, in *37th Annual Society of Plastics Engineers Proceedings*, 1982.
11. S. M. Abouzahr and R. P. Taylor, SAE International Congress and Exposition, Paper 850154, 1985.
12. F. Sanna, Jr., R. P. Taylor, and J. R. Kuzubski, SAE International Congress and Exposition, Paper 890694, 1989.
13. U.S. Patent 4,374,229.
14. D. Nelson, Polyurethanes World Congress 1987, Aachen, *Proceedings of the SPI/FSK*, 384.
15. D. Nelson and G. Ellerbe, SAE International Congress and Exposition, Paper 890197, 1989.

[1] *Editor's Note*: While the majority of RIM technology is based on polyurethane and polyurea resins, non–urethane RIM, utilizing Nylon 6, polydicyclopentadiene (DCP), expoxies etc. have also been developed. For a discussion on these non–urethane RIMs, refer to C. W. Macosko, *Fundamentals of Reaction Injection Molding*, Hanser Publishers, 1989.

16. U.S. Patent 4,279,576.
17. L. J. Lee, J. M. Ottino, W. J. Ranz, and C. W. Macosko, *Polym. Eng. Sci.*, 20, 868 (1980).
18. P. Kolodziej, C. W. Macosko, and W. E. Ranz, *Polym. Eng. Sci.*, 22, 388 (1982).
19. P. Kolodziej, W. P. Yang, C. W. Macosko, and S. T. Wellinghoff, *J. Polym. Sci., B, Polym. Phys.*, 24, 2359 (1986).
20. P. D. Wickert, W. E. Ranz, and C. W. Macosko, *Polymer*, 28, 1105 (1987).
21. H. Boden, L. Klier, and W. Schnieder, Polyurethanes World Congress 1987, Aachen, *Proceedings of the SPI/FSK*, 858.
22. W. Schoberth, SAE International Congress and Exposition, Paper 880357, 1988.
23. E. Broyer, C. W. Macosko, F. E. Critchfield, and L. F. Lawler, *Polym. Eng. Sci.*, 18, 382 (1978).
24. E. B. Richter and C. W. Macosko, *Polym. Eng. Sci.*, 18, 1012 (1978).
25. M. C. Pannone and C. W. Macosko, *J. Appl. Polym. Sci.*, 34, 2409 (1987).
26. G. H. Slocum, D. W. Schumacher, and M. F. Hurley, SAE International Congress and Exposition, Paper 880434, 1988.
27. G. H. Slocum, D. W. Schumacher, and M. F. Hurley, *Proceedings of the Fourth Annual ASM/ESD Conference on Advanced Composites*, 1988, 197.
28. H. Mueller, W. Mrotzek, and G. Menges, in E. Kresta, Ed., *Reaction Injection Molding*, American Chemical Society, Washington, DC, p. 237.
29. M. E. Cekoric and F. Sanna, SAE International Congress and Exposition, paper 890337, 1989.
30. N. N. Ghoussaini and R. H. Fowler, SAE International Congress and Exposition, Paper 860515, 1986.
31. J. Hemphill, SAE International Congress and Exposition, Paper 880361, 1988.
32. D. R. Brace, *Proceedings of the Second Annual ASM/ESD Conference on Advanced Composites*, 1986, 159.
33. J. A. Reitz and D. A. Bityk, *Proceedings of the 31st Annual SPI Technical/Marketing Conference*, 1988, 75.

5 · Resin Transfer Molding

Carl F. Johnson

Contents

Carl F. Johnson, Ford Motor Company, Dearborn, Michigan, USA

5.1 INTRODUCTION

In beginning this tutorial chapter on the relatively new process of rapid resin transfer molding, I briefly discuss the significance of the use of composites in the commercial sector of the economy. High performance composite materials are used frequently in applications in the military, space, and aerospace, but today their use in the commercial sector is only beginning. A large base of material, process, and design data has been generated for the classical laminated high performance composites typically used in the noncommercial applications. This design methodology, in combination with the classical aerospace fabrication techniques such as prepreg layup and filament winding, has been used successfully to fabricate high cost, high performance parts for such markets as sporting goods. The majority of other composite applications in the consumer sector are of a lower performance level and are manufactured by processes such as compression molding or spray-up as in the case of boats and recreational vehicles. There is significant opportunity for high performance composite materials to make further inroads into everyday applications such as automobiles.

Composite materials have demonstrated their ability to outperform steel in some demanding applications. They also can offer weight savings and unique performance characteristics achievable only by using the anisotropic mechanical properties of a composite. There is however a barrier that must be overcome if these materials are to become cost-effective substitutes for the materials used in today's design process. Manufacturing cost must be reduced by the use of rapid processes for the fabrication of a composite product. Development and use of low cost fabrication processes must be fundamental in the application of advanced composites to the consumer market. Design tools used for the labor-intensive aerospace fabrication methods do not accurately or completely allow the optimization of a part for rapid manufacture. The manufacturing process must drive the design rather than the reverse.

To utilize fully the benefits of advanced materials, the selected manufacturing process must permit the fabrication of large, complex, three-dimensional structures having anisotropic physical properties in selected areas. The ability to integrate multiple parts into single reliable and reproducible components may also be critical.

Resin transfer molding has the potential of becoming a dominant low cost process for the fabrication of large, integrated, high performance products for the consumer segment of the economy and ultimately for segments now dominated by the higher precision laminated fabrication techniques. This chapter details the current developing technology status of resin transfer molding and also discusses the materials, process equipment, and selection criteria for use of the process in composite fabrication.

5.2 PROCESS DEFINITIONS

Resin transfer molding (RTM) is used today to manufacture a wide variety of articles ranging from small armrests for buses to large water treatment plant components. The process in its most basic form is shown in Figure 5.1. A dry reinforcement material that has been cut and/or shaped into a preformed piece, generally called a preform,

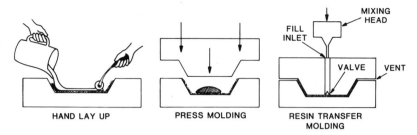

FILL INLET

MIXING HEAD

VALVE   VENT

HAND LAY UP   PRESS MOLDING   RESIN TRANSFER MOLDING

FIGURE 5.1   Schematic of the basic RTM molding process. (Diagram courtesy of Ford Motor Company.)

is placed in a prepared mold cavity. The preform must not extend beyond the desired seal or pinch-off area in the mold, to allow the mold to close and seal properly. Once the mold has been closed and clamped shut, resin is injected into the mold cavity, where it flows through the reinforcement preform, expelling the air in the cavity and wetting out or impregnating the reinforcement. When excess resin begins to flow from the vent areas of the mold, the resin flow is stopped and the molded component begins to cure. When cure is completed, which can take from several minutes to hours, it is removed from the mold and the process can begin again to form additional parts. The molded components may now require a postcure to further complete the resin reaction.

A very similar process is called structural reaction injection molding (SRIM). The schematic in Figure 5.2 shows the basic SRIM process (see also Chapter 4). Preform

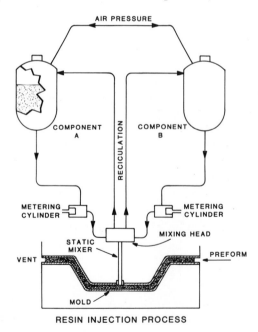

AIR PRESSURE

COMPONENT A   RECICULATION   COMPONENT B

METERING CYLINDER   METERING CYLINDER

STATIC MIXER   MIXING HEAD

VENT   PREFORM

MOLD

RESIN INJECTION PROCESS

FIGURE 5.2   SRIM molding process schematic. (Diagram courtesy of Ford Motor Company.)

and mold preparation are similar to the RTM process, with changes in mold release and reinforcement sizings made to optimize their chemical characteristics for the SRIM chemistry. Once the mold has been closed, the SRIM resin is rapidly introduced into the mold and reacts quickly to cure fully within a few seconds. This reaction is in progress as the resin flows through the reinforcement; therefore wetout and displacement of the air in the mold must occur rapidly. There is seldom much runoff of excess resin in an SRIM component. Cure is complete shortly after the resin reaches the extremities of the component, and the viscosity is generally too great to allow much resin to escape through vents. When cure is complete, the component is removed from the mold and the process is finished. No postcure is normally done.

Before looking at more sophisticated RTM and SRIM processes, let's examine in more detail the similarities and differences of the two processes.

We begin with resin chemistry because that is the area in which the processes exhibit the greatest difference. RTM resins are typically low viscosity liquids in the range of 100–1000 cP. Normally resin systems have two components and require a preinjection mixing ratio in the range of 100:1. The liquid reactants part A and part B can be mixed at low pressure using a static mixer. Figure 5.3 shows a typical RTM resin delivery system.

SRIM resins are similarly two–part, low viscosity liquids being in the viscosity range of 10–100 cP at room temperature. They are highly reactive in comparison to RTM resins and require very fast, high pressure impingement mixing to achieve thorough mixing before entering the mold. Mix ratios of typical systems are near 1:1, which is desirable for rapid impingement mixing. A schematic of the SRIM resin delivery system was shown in Figure 5.2.

Preforms for RTM and SRIM systems are similar in most respects. The rapidity of the resin reaction in the SRIM system, however, causes some additional concerns

FIGURE 5.3 Liquid controls RTM delivery system. (Photo courtesy of Liquid Control Corporation.)

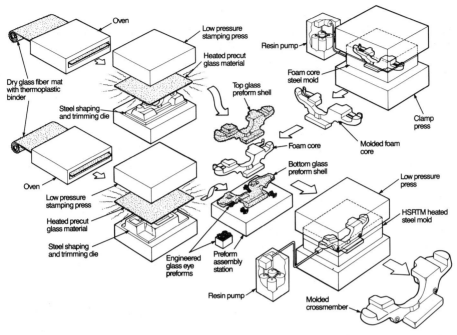

FIGURE 5.4   HSRTM process schematic (Diagram courtesy of Ford Motor Company.)

with respect to preform design. Once resin begins to enter an SRIM mold, its viscosity rapidly builds. After a few seconds of flow, the resin becomes much harder to pump through the reinforcement preform. Preforms for SRIM normally are lower in glass content or have additional directional glass, which acts as a resin flow channel. Thus, resin flow in SRIM is much more sensitive to preform construction than it is in the RTM process.

As previously mentioned, venting, or the planned release of trapped air from within the mold cavity as resin displaces it, is achieved differently in SRIM and RTM. The faster cure in the case of the SRIM system minimizes the need to design elaborate seals and overflows into a tool to manage the overflow of resin from the tool cavity. Management of excess resin and its related thin flash areas becomes a problem that must be solved when producing a large volume of components at short cycle times.

In numerous other respects, the two systems are similar. As RTM resin systems are developed to be faster reacting and SRIM systems are slowed down to produce a longer filling time capability, the two alternative systems are approaching a single process. Figure 5.4 shows such an idealized process schematically, as it could be applied to the manufacture of an automotive component. In this case, heated flat sheets of glass reinforcement are stamped into a three-dimensional preform shell and subsequently installed over a molded foam core. After the addition of some small, high performance attachment point preforms at certain high stress areas, the assembled preform is transferred to a heated steel tool for resin injection. Up to this point, if the details previously mentioned have been considered in preform design and venting, the

153

choice of RIM or RTM resins would be purely a matter of economics versus performance in the final part. The high speed resin transfer molding (HSRTM) system envisioned in Figure 5.4 typifies the direction industry is moving in developing a unified approach to liquid molding.

5.3 GEOMETRIC AND DESIGN CAPABILITIES

Now that we have defined the characteristics of liquid molding (RTM) as a process, let's examine this process in terms of part geometries. A long-time practitioner of RTM once told me he had never found a part that he could not make using RTM. Nevertheless, there are certainly many components that cannot be cost-effectively produced using the RTM process. A process that ultimately is chosen for an industrial application must permit the user to take full advantage of the unique attributes or capabilities it possesses. Otherwise, alternative technologies may offer a more cost-effective manufacturing process. I will briefly enlarge on each unique feature of the RTM process. For an in-depth look at design and process tradeoffs, several references are recommended (1–3).

Size capability is among the major benefits of RTM. Large area designs are feasible because pressures needed to mold components can be maintained at low levels. In some applications where cycle time is not critical, pressures can be maintained below $0.703 \, kg/cm^2$ (10 psi). In the case of the Lotus VARI ® (vacuum-assisted resin injection) process, pressures in the tool remain below atmospheric pressure throughout the molding cycle. With very low pressure for injection, only minimal mold clamping pressures are required. Alternative composite technologies such as compression molding can require pressures up to $140.6 \, kg/cm^2$ (2000 psi) over the part surface to achieve good physical properties and appearance.

In addition to large area parts, RTM is well suited to manufacture of parts that have deep draws and minimum draft on the sides of the part. Such a component is shown in Figure 5.5. Alternative fabrication technologies such as thermoset compression molding or thermoplastic compression molding are limited in the depth to which they can form by the pressures required for consolidation. Pressure on a deep section is related to the axial pressure over the sine of the included angle; hence as the draft angle decreases (i.e., the wall of a part becomes more vertical), the pressure also decreases. This is not completely the case, because for a very short time during molding, compression-molded materials act as an isostatic liquid. This time is very short however, and the depth of draw over which good consolidation can be achieved by indirect pressure tends to be small [on the order of a maximum of 30.5 cm (12 in.)]. In the case of RTM, low viscosity liquid is present everywhere in the mold before cure and can be readily pressurized to improve wetout or to minimize voids.

Although RTM is well suited to molding large objects, part complexity may restrict the use of the process. Unlike processes in which the resin flows into the complex areas of the mold carrying the reinforcement with it, in RTM the reinforcement must be in place before the resin is introduced. While preplacing the reinforcement provides superior physical properties and eliminates flow-induced property

FIGURE 5.5   Electrical transformer case, demonstrating a deep draw with minimum draft. (Photo courtesy of Morrison Molded Fiber Glass.)

variations, it also makes the use of features such as ribs and bosses very difficult. Any small detail or molded-in feature (e.g., holes and grooves) may be better achieved in other processes. There are of course alternatives to the use of some design features. Ribs can conveniently be replaced with closed, cored sections where stiffness is required. RTM allows the use of foam cores for added rigidity in a structure, as well as for providing increasingly complex three-dimensional structures to be molded in one piece. The low pressures used in this process allow foams in the 0.064–0.096 g/cm^3 (4–6 lb/ft^3) density range to be molded in place without significant deformation. Some alternative processes also allow foam cores to be used but they are generally higher in density to avoid crushing; thus the structure will be heavier.

One-piece molded RTM components can in some cases provide performance in critical applications that is unattainable using a multipiece assembly of components fabricated by other processes. The single-piece Ford Escort front structure shown in Figure 5.6 integrates 45 individual steel stampings into a single RTM molded component. The resulting component reduces the weight of the structure 30 % over the normal steel structure, while promising improved durability and crash performance. The one-piece nature of the front energy management rail (seen in Fig. 5.7) is essential to the structure's performance.

Attachment areas between the structure and the loads it must carry are often a difficult design challenge in composite materials. In designing an RTM structure, attachments must be included in the preform construction, and they must have sufficient continuity with the bulk of the structure to ensure that stresses are distributed properly. An example of a preformed attachment is shown in Figure 5.8. This braided glass preform subassembly retains the lower suspension arm of a truck crossmember. Stresses are distributed into the structure through the flange areas of the preformed attachment eye.

155

FIGURE 5.6   Escort composite front structure. (Photo courtesy Ford Motor Company.)

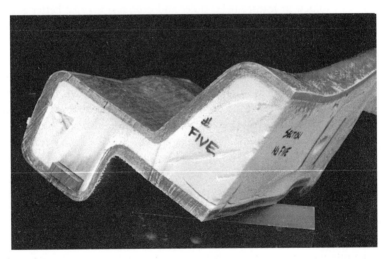

FIGURE 5.7   Cross section of a composite component showing the relationship of the core and laminate. (Photo courtesy of Ford Motor Company.)

Other possibilities for attachments of RTM structures include adhesive bonding and molded-in metallic inserts. RTM provides design flexibility with respect to attachments, a feature that can and should be exploited in a good design. Compression molding, for example, requires the same attention to attachments but presents the additional complications of placement of attachment materials precisely in a hot tool and control of fiber movement during molding. The presence of flow and knit lines in compression-molded materials tends to make attachments less reliable.

156

FIGURE 5.8   Braided suspension attachment eye preform. (Photo courtesy of Ford Motor Company.)

Finite element methods were mentioned briefly early in the chapter, but now let's look at what is available and what is likely to be developed in the near future for design with the RTM process. In most computer-assisted component designs done thus far for RTM manufacturing, a classical NASTRAN or similar finite element analysis is used (4). As a subprogram to the finite element analysis, some type of laminate analysis is also used to develop the element properties for the main line NASTRAN analysis. Because of the flexibility available in designing fiber orientations in a preform, the design in RTM can be very efficient with respect to material use. Only the reinforcements carrying load need to be included. Presently available laminate analysis can place reinforcements in the x and y planes only, but as more sophisticated preform technology is developing, such as three-dimensional knitting and weaving (5), work is in progress on more sophisticated analytical techniques (6).

Finite element and finite difference techniques are also under development (7, 8) for the prediction of resin flow through the RTM preform during manufacture. Ultimately, prediction of flow may be as critical as prediction of physical properties when a unified liquid molding design code is finally available. There is currently significant work in progress with respect to optimized preform designs and computer-assisted process selection.

5.4 PROCESS MATERIALS

There are numerous resin chemistry choices for a resin transfer molded component. Any resin must have the ability to fill the mold completely, wet out the reinforcement, and cure or solidify to a solid state with the desired physical properties. RTM offers such a broad variety of capabilities that it is difficult to further define resin attributes.

157

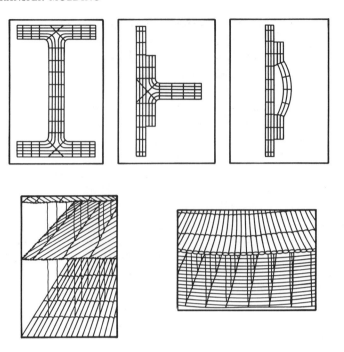

FIGURE 5.9   Example of the application of aerospace cut-and-sew preforming techniques. (Diagrams courtesy of Xercon. Inc.)

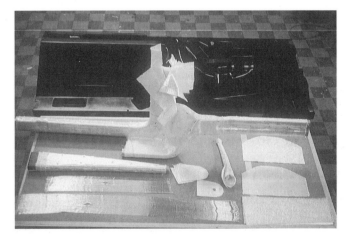

FIGURE 5.10   A preform for a composite front structure made using automotive cut-and-sew techniques. (Photo courtesy of Ford Motor Company.)

In high performance applications viscous epoxy resins can be molded using slow cycles. Where cycle time is critical, low viscosity vinyl ester, acrylamate, or RIM resins based on urethane chemistry can be injected very rapidly into the mold. Even the use of some thermoplastic resin systems such as polycarbonate or nylon may be feasible in the near future.

Reinforcement material selection is likewise seemingly limitless. Glass (both E- and S-2), aramid, and carbon fibers are commonly used. Wood fiber and other synthetic fibers such as polyester are possibilities, depending on requirements. Metal can be an excellent choice for local reinforcement in a structure. Metals should be noncorrosive or properly protected against moisture or other environments prior to inclusion in the preform.

Any reinforcement must be capable of being preformed into the desired configuration. The economics required of the preforming process will depend on the application. Aerospace and low volume applications can utilize cut-and-sew preforming as shown in Figure 5.9. In a cut-and-sew preform, areas of material, typically fabric, are defined based on the requirements determined in the finite element analysis. The general size and shape of each area is cut from a conformable material and fit to the part mold or a part model. Once fit has been achieved by cutting, trimming, and sewing, a final template is constructed and the actual reinforcement is cut and sewn on the preform. Figure 5.10 shows the assembly of a cut-and-sew preform for an automotive prototype. This process lends itself to easy translation from finite element model to componentry, but it is slow and labor intensive.

Higher volume preforming calls for a faster process, such as the stamping process shown in Figure 5.11. Materials used for stamping are typically continuous strand random glass mats in which the reinforcement is in a swirled configuration. Several manufacturers produce materials for use in stamped preforms with a variety of physical properties available. Both thermoplastic and thermoset binder systems are available to retain the formed shape after stamping. Binder proportion typically ranges from 4 to 8 % by weight and can vary through the thickness of the preform if desired. Figures 5.12 and 5.13 give an idea of the range of properties available for tensile strength and tensile modulus. The use of conformable mats is limited to applications requiring low tensile strength and stiffness. It is apparent from these

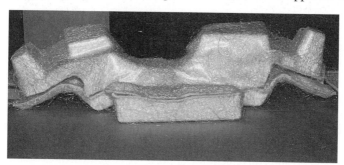

FIGURE 5.11   A preform stamped from multiple layers of continuous strand glass mat. (Photo courtesy of Ford Motor Company.)

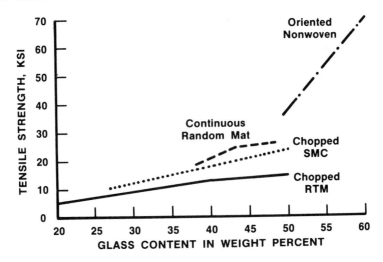

FIGURE 5.12   Tensile strength of typical random materials.

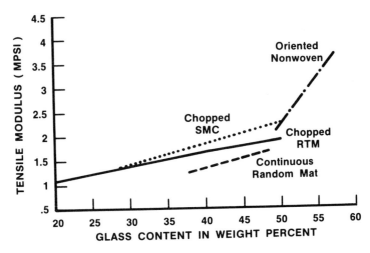

FIGURE 5.13   Tensile modulus of typical random materials.

graphs that glass content in the final part will be limited to approximately 50 % by weight; this is where the curves flatten out, and only marginal benefits are gained in overall physical properties. If the data are presented statistically as shown in Figure 5.14, it is observed that the B-10 strength level actually decreases between 40 and 50 % glass content. This decrease can be attributed to damage of the glass reinforcement, which occurs during the preforming, and subsequent compression in the molding tool, of higher glass levels.

Although rapid and convenient for preforming large structures, stamping of fully random materials is limited in its level of performance. To be more competitive with alternate materials such as steel and aluminum with respect to weight and cost,

160

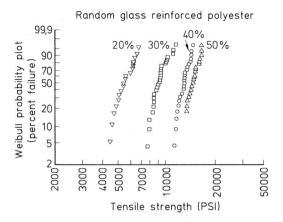

FIGURE 5.14   Probability of failure for some selected random glass RTM materials.

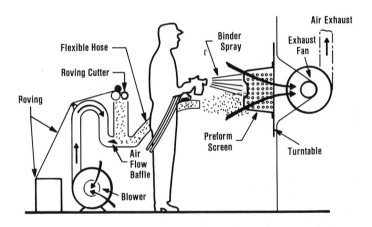

FIGURE 5.15   Spray-up preform fabrication schematic. (Diagram courtesy of PPG Industries.)

composite structures will normally require the use of either oriented glass or some additional reinforcement type such as carbon or Kevlar. It is possible to include limited amounts of conformable oriented materials with the random material, subsequently forming the entire sandwich of materials by stamping.

Spray-up of reinforcement presents an additional option for the formation of preforms. Figure 5.15 presents a schematic of the random spray-up technique. A perforated screen is made which conforms to the shape of the desired preform. Vacuum is applied to the rear of the screen, and normally the screen is simultaneously rotated about the x axis. A chopper system cuts reinforcement into short lengths and a binder spray is also applied. Air is blown through a large chopping manifold to

161

disperse and randomize the fibers and direct them to the preform screen. The vacuum behind the screen holds the fibers securely on the screen in the location at which they first contact it. Once the desired thickness of reinforcement has been achieved, the chopping system is turned off and the preform and screen are transferred to a baking oven, where the binder cures and rigidizes the preform. Once stabilized, the preform is cooled, removed from the screen, and trimmed if required. Spray-up systems have demonstrated a capability to manufacture large preforms at rates as fast as one per minute (9).

Future preform systems will likely use a combination of highly automated technologies. Computerized braiding and filament winding combined with the three-dimensional screen of spray-up are being investigated for use in fully integrated design and preform manufacturing systems.

Foam cores are an additional feature of the RTM preform. The ability to design and fabricate fully three-dimensional components with no joints or seams is a significant advantage of RTM, since in some other processes the use of foam cores either is not possible or is more limited. Foam cores used in RTM must be capable of withstanding molding pressure exerted by the resin without excess deformation. If the foam core deforms, excess resin will be pumped into the mold cavity, resulting in higher cost and weight. Cores must also be stable at molding temperature and ultimately at service temperature. Unstable foam systems will expand in a heated mold, causing undesirable compaction of the preform and flow restrictions. Cores that are not stable in service can expand and distort the finished component. Thermal expansion difference between the laminate and the core must also be considered in some cases, particularly when laminate skins are thin and the foam volume is large.

In many applications the foam serves only as a tooling and fabrication aid. If the core is structural (i.e., is relied on to supply a portion of the rigidity of the structure), a core with adequate shear and interfacial strength characteristics must be selected. The long-term environmental and fatigue degradation of the laminate–core interface is a critical consideration in such a design.

5.5 PROCESS EQUIPMENT

5.5.1 Tooling

The process of resin transfer molding has a number of advantages with respect to the tooling used to fabricate components. The process can be carried out at low pressures and in some cases pressures below atmospheric pressure. Low pressures reduce the cost and complexity of tooling required. Low cost epoxy tooling can be used and is in fact used for the majority of today's low volume RTM production. Although tools can be very simple and clamping can be accomplished by a simple system of perimeter clamps, the tool must be closed and sealed during the filling process to achieve complete and reproducible resin flow. A typical two-piece epoxy tool is shown in Figure 5.16. Epoxy tools are normally used where volumes of production are low; an upper production limit of several thousand parts is typical. As the complexity of

FIGURE 5.16   A typical two-piece epoxy tool. (Photo courtesy of Ford Motor Company.)

a component increases, the number of parts that can be produced will decrease because of tool wear and damage occurring during the preform loading process. Epoxy tools are not normally multipiece tools but can sometimes include removable inserts in one or both tool halves. The epoxy tool will have an inner surface of high quality epoxy resin and will be backed up with multiple layers of glass cloth and epoxy resin. The total tool surface thickness may approach 2.54 cm (1 in.). Plywood egg crate is added to the back surface of the epoxy tool to provide rigidity during molding. To gain full benefit from the egg crate, the tool must be clamped uniformly in a press rather than perimeter clamped. During tool construction, heating or cooling lines can be attached to the rear of the tool surface, if desired; but the low thermal conductivity of the tool surface will reduce their efficiency, increasing thermal response time.

Sealing of the mold is required to contain the resin during molding. Two sealing and preform containment methods are normally used in RTM molding. For less complex parts a "pinch-off" technique will effectively retain the preform and seal the mold. More complex parts may require an O-ring design in which the preform is fully contained in a tool cavity. These techniques are shown in Figure 5.17.

Mold releases must be carefully selected when using epoxy tools. Internal mold releases (i.e., releases that are contained in the resin system) cannot be used if they require a heated tool surface to be effective. Coatings of wax, silicone, or polyvinyl alcohol (PVA) are normally used with epoxy tools and may require application after

163

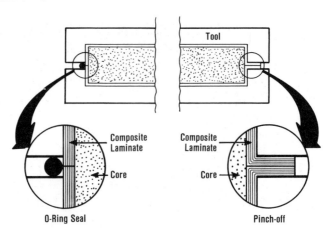

FIGURE 5.17   Two methods of tool sealing: pinch-off versus O-ring sealing.

every molding. If resin adheres to the tool surface, the inner surface of the tool can be pulled off rather easily.

Venting of the epoxy tool is required to allow air to escape during injection. Vents may be gaps in the tool seal or tubes inserted through the tool surface. In the case of very low volumes, holes can be drilled through the tool surface at critical points, such as high points where air would naturally be trapped during resin fill. Vents will require cleaning between moldings, so they should be designed for easy maintenance.

The accuracy and surface finish of an epoxy tool will depend strongly on the quality of the pattern used to make the initial tool casting. Parts having acceptable surface finish can be made using these tools, and with hand finishing, parts can be made to automotive Class "A" specifications. Epoxy tools tend to degrade quickly though, and read-through of the backup egg crate structure will occur after only a few moldings, necessitating increased hand finishing of surface critical parts. Where high quality surfaces are required, higher quality tools should be used.

For better surface quality, improved durability, faster thermal response, and higher molding pressure, nickel shell tools present an alternative to epoxy tools. Although more expensive, they may provide benefits that quickly outweigh the added expense. Figure 5.18 shows a two-sided nickel shell tool for molding an RTM component. Tools may also have one side or half made from nickel and a second side made from epoxy. The one-sided tool is easier to manufacture because a proper match between tool halves can be achieved easily. Two-sided nickel shell tools are prone to warpage during manufacture and must be restrained in a holding fixture or ground to shape after manufacture. Two-sided nickel shell tools normally cost three to four times as much as single-sided tools.

High quality surface finish can be achieved using nickel shell tools, and because the tool can be readily heated to elevated temperatures [149 °C (300 °F) typical], a wider variety of low profile resins and mold releases can be used. Cycle times can also be reduced using heat-activated catalyst packages, and there is no danger of damaging the tool with excessive exotherms during cure, as in the case of epoxy tools. Cure can

FIGURE 5.18   A typical two-piece nickel shell tool. (Photo courtesy of Ford Motor Company.)

even be controlled from a hot surface to a cooler surface to give improved surface finish. Typical volumes for nickel shell tooling range from several thousand upward to 20,000–40,000 parts annually, which is significantly better than epoxy tools.

Venting and sealing of these tools is accomplished in a manner similar to the epoxy tools but vents are more difficult to modify. It is often wise to prove out the filling and venting system for the mold by making one epoxy tool before making nickel shell tools.

Cast tooling presents an additional option for medium volume RTM tooling. Solid zinc alloy castings as shown in Figure 5.19 or thin shells of cast aluminum or steel can be effectively used for RTM molding. These alternatives give advantages similar to the nickel shell tools but eliminate some of the distortion problems associated with two-sided nickel shell tools. Modification of the sealing and venting

FIGURE 5.19   A typical two-piece zinc alloy tool. (Photo courtesy of Ford Motor Company.)

165

are again more difficult than for the epoxy tools, but easier than for nickel shell tooling. Care must be taken to design tooling that will not be too heavy for available molding equipment. Some presses used for RTM molding are not capable of handling a large cast zinc alloy tool. High quality surfaces are possible with all the cast tool materials, but zinc and aluminum tend to be more porous and may require additional mold release. Normally cast RTM tools are of two-piece construction with an upper and a lower mold half. Multiple-piece tools are feasible using castings if subsequent machining is done to provide sealing surfaces between tool sections. Cast tools can compete at volumes similar to those of nickel shell tools and may have a longer life because of the absence of residual stresses, which tend to degrade the surface quality of nickel shell tools over time.

At high volumes or where complex multipiece tools are required, a machined tool steel mold should be used. Figure 5.20 shows an example of a four-piece automated tool used to mold an automobile component. In addition to the four main tool sections, there are location pins and ejectors to further automate the RTM process and reduce cycle time. Steel tools are high in cost but offer life times of 500,000–1,000,000 parts. The complexity of steel tool design dictates that gating, venting, and sealing of the die be determined prior to tool construction.

There are design tools available (10) for determining the flow of resin through a preform and selecting gate and vent locations. A complex mold such as the one pictured in Figure 5.19 also requires the use of a heat transfer design program to optimize the thermal management system. In an automated RTM tool, resin inlets and vents utilize self-cleaning hydraulically actuated valves such as those described in connection with resin injection (Section 5.5.4). A steel tool provides the optimum in cycle time, surface finish, part-to-part repeatability, tool durability, and automation; but because of its high cost, volumes must be in the 10,000–50,000 range to justify the added expense.

FIGURE 5.20   A production-capable automotive HSRTM tool. (Photo courtesy of Ford Motor Company.)

5.5.2 Presses

Resin transfer molding is a low pressure process that achieves maximum economic viability when used to fabricate large area, highly integrated, composite structures. Parts molded often include deep sections and molded-in foam cores or other inserts. Press equipment used in the process will reflect these characteristics of the typical molded components—that is, large platen area, low clamping tonnage, and often large daylight openings to facilitate loading of a complex preform.

Required press tonnage can be readily calculated as a function of the pressure required to completely fill the mold and the projected area of the component. The experimental crossmember shown in Figure 5.21 required a press capacity of 85 tons to maintain mold closure during injection. If the tool is designed to be of a self-locking design, this value can be substantially reduced.

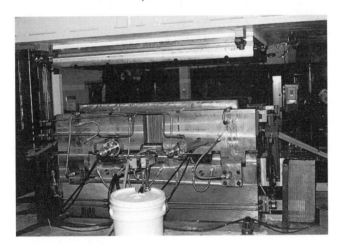

FIGURE 5.21   A production HSRTM tool in the molding press. (Photo courtesy of Ford Motor Company.)

There are several available options for low cost, low pressure presses for RTM molding. An airbag press such as that shown in Figure 5.22 uses a series of pressurized rubber bags or canvas hoses to apply clamping pressure to a mold. This is a good alternative for low volume or prototype fabrication but is limited both with respect to daylight and by its inability to lift the upper mold half during demolding. Adding a mold shuttle as shown in Figure 5.23 or using a secondary low pressure cylinder to manipulate the upper mold half overcomes this limitation but adds cost to the press system. Another alternative for press systems is the low pressure hydraulic press. An example is shown in Figure 5.24. The clamping and platen movement of this press is totally hydraulic, resulting in improved control and faster cycles, but at a higher cost. A shuttle system can also be added to this type of press to improve productivity.

Conventional high pressure presses can, of course, be used for RTM molding, but the added expense will reduce the advantages of RTM as a candidate fabrication process. When using high pressure presses with low cost or lightweight tooling, care

FIGURE 5.22   A light–duty, low cost airbag press. (Photo courtesy of Snow Inc.)

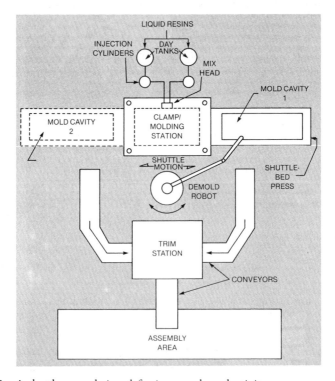

FIGURE 5.23   A shuttle press designed for improved productivity.

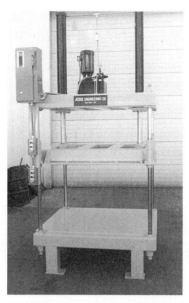

FIGURE 5.24   A light-duty hydraulic press. (Photo courtesy of Jesse Engineering.)

must be taken to ensure that the press is controllable at the low pressures required to avoid overcompression of the tool.

Preform loading of a complex part adds some complication to press selection. Often if several preform pieces must be placed in the mold before closing, productivity will be improved sufficiently to justify the additional cost of a shuttle system utilizing two lower molds and a single upper mold. If the preform to be loaded tends to be slightly oversize (this is typical for structural components, where reinforcement must reach the edges of the component), additional clamping pressure may be required to close the mold. It should also be remembered that additional press daylight may be beneficial in loading the preform into the mold where shuttles are not used.

5.5.3 Preform Equipment

Several preforming techniques have been previously mentioned, all requiring equipment significantly different in type and complexity. The development of an economic preforming process for the component to be molded is a key factor in the overall economic viability of an RTM alternative. Selection of the appropriate preform technique, reinforcement material, and preform equipment can make the difference between a successful RTM application and an economic failure.

Cut-and-sew preforming requires a minimum of equipment but is least efficient with respect to productivity. The process can consist of hand-held cutters and a cutting table, or it can be automated by adding steel rule-cutting dies or possibly a robotically controlled cutting table. Preform assembly can be done by hand, which is typical for cut and sew, or it could be automated. Automation of the cut-and-sew process is being slowly implemented for prepreg layup for aerospace applications, but

169

this preform technique is likely to be very expensive and primarily suited to low volumes of complex, very high preformance components.

The spray-up preform process as shown in Figure 5.25 yields near net shape preforms with higher productivity levels possible. The equipment is large and consumes considerable floor space but is low in cost. Some manufacturers have used such equipment to make very large preforms such as entire automobile floor pans. A cycle time of one preform per minute is achievable using this technique.

Stamped preforms can be produced at very fast rates. Equipment used for preform stamping is pictured in Figure 5.26. Although the process requires more trimming equipment than the spray-up method, continuous reinforcement can be used with this technique, increasing the performance level of components made.

FIGURE 5.25 Preform spray-up of a transformer housing. (Photo courtesy of Morrison Molded Fiber Glass.)

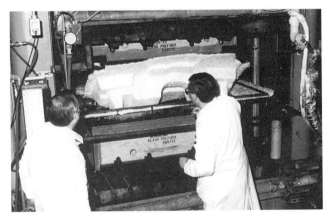

FIGURE 5.26 Preform stamping equipment with preform. (Photo courtesy of Owens Corning Fiberglass.)

FIGURE 5.27   Large multiaxial ring braider. (Photo courtesy of Drexel University.)

Preforming processes such as braiding and filament winding may be appropriate for preforming simple geodesic shapes. Braiders such as that pictured in Figure 5.27 can provide an interlocked triaxial preform that exhibits high values of tensile strength and stiffness in the finished composite. Braiding applies all three orientations of reinforcement simultaneously and normally produces a preform structure faster and with less capital cost than filament winding. Filament winders normally apply reinforcement along a single well-defined path, resulting in slower preforming rates but increased placement accuracy and an associated increase in performance. Equipment complexity and cost are higher for filament winding relative to braiding. In either technique, care must be exercised to keep reinforcement content below the upper volume percentage limit required to achieve the cycle needed time for resin impregnation. Both processes are capable of placing reinforcement in a preform at levels that are too dense to impregnate with resin.

5.5.4 Resin Injection Equipment

Liquid molding resin handling equipment varies in complexity and cost with cycle time and resin reactivity. The most basic RTM setup uses a one-component resin system and a simple pressure pot for resin injection. A pressure pot system and a sample resin chemistry are presented in Figure 5.28. The pressure pot system offers a low cost option for resin handling where volumes are low. The system is limited to low reactivity systems, and where long pot life is required, a heated tool is needed to start the resin cure. Resins that cure at room temperature can be used with a pressure pot, but the system must be thoroughly cleaned after each shot to prevent resin cure in the system. A note of caution: Inexpensive systems often contain zinc

171

FIGURE 5.28   Pressure pot system resin injection system. (Photo courtesy of Ford Motor Company.)

or brass components, which can react with components in the resin system, causing acceleration or retardation of the cure; therefore, to be safe, use stainless steel or plastic if there is concern with respect to a reaction.

More rapid cycles, a wider selection of resin systems, and less frequent clean-up can be achieved using a two-component pumping system similar to that shown in Figure 5.29. In this type of system an A-side resin composition is mixed with a B-side resin composition in a static mixer just before entering the mold. Once combined, the resin reacts rapidly to begin the cure process, with a 1–5 minute cure time being typical. Resins can easily be made highly reactive, and care must be taken to provide adequate time to flush out the mixing portion of the system before the resin cures

FIGURE 5.29   Air-powered piston pumping system. (Diagram courtesy of Grayco Inc.)

in the static mixer. The mixer must be flushed clean after each shot. Care must also be taken to keep the resin system exotherm low if epoxy tools are being used, to avoid degradation of the tool surfaces. Most production RTM employs a two–pump system with a static mixer. Several advanced, high volume, liquid molding companies are presently developing equipment for resin handling based on modified RIM equipment. A schematic of this type of equipment is shown in Figure 5.30. Once again A and B resin components are contained in separate holding tanks. Positive displacement pumps (either piston or gear) circulate the reactants through the system. When an injection is requested, the circulating resins are diverted to the mixhead and through the mix head into the mold. After mold fill, the resins are again allowed to circulate in their respective A and B side loops, and the mixing head is automatically flushed with solvent and dried with compressed air. This type of equipment allows the use

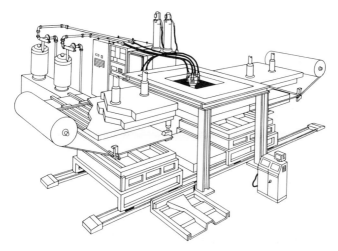

FIGURE 5.30   RIM–based molding system. (Diagram courtesy of Mobay Inc.)

of resin systems having higher reactivities to achieve rapid cycle times. The equipment is highly automated and is capable of producing high volumes of components.

Injection of the resin into the mold is normally accomplished using either a hand–held gun (Fig. 5.31) or a fixed injection nozzle (Fig. 5.32). The hand–held gun with an integral static mixture will normally be used with the pressure pot or two–component piston pump system. The automated fixed nozzle is used for the RIM–type molding system. In addition to accommodating resin inlet to the mold, both injection systems provide for solvent flush and air dry after a molding shot. Fixed inlet nozzles must be liquid cooled when used with heated tooling.

Resin and solvent disposal are becoming increasingly difficult and expensive. Any system should minimize amounts of waste generated, and care should be taken to properly store and separate waste materials. Disposal and handling must be planned before any molding program is started.

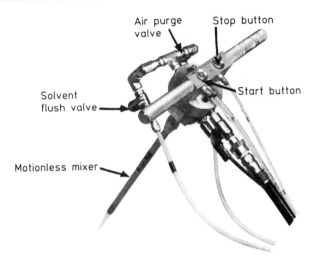

FIGURE 5.31   Hand-held resin injection gun. (Photo courtesy of Liquid Controls, Inc.)

FIGURE 5.32   Automated resin injection nozzle. (Photo courtesy of Morrel, Inc.)

5.6 PROTOTYPING WITH RTM

Resin transfer molding is an excellent process choice for making prototype components. Unlike processes such as compression molding and injection molding, which require tools and equipment similar to those used in the actual process to accurately simulate the physical properties achievable in the production level component, RTM allows accurate prototypes to be molded at low costs. It should be noted that in some cases, RTM can also be used to prototype components designed for these other processes, such as compression molding. When simulating other processes such as SMC, the RTM component will typically have properties that exceed the production level product.

When prototyping with RTM, generally less reactive resins are used, allowing long fill times and easy manipulation of the vents. Tooling is normally low cost epoxy, but it could be made from any impervious material that would contain the resin such as plaster or wood. Prototype preforms can be made by cut-and-sew methods, and any foam cores used may be machined to shape. Sizes can range from small components to very large, complex three dimensional structures.

174

TABLE 5.1   Physical Properties of Some Typical RTM Materials[a]

Material designation	Glass (%)	Tensile strength (ksi)	Tensile modulus (Mpsi)
R–35	37.7	18.0	1.24
R–40	43.1	24.5	1.48
R–50	48.2	26.2	1.66
C-10/R-40 (0-Deg)	49.3	35.5	2.08
C-20/R-30 (0-Deg)	50.6	37.7	2.68
C-40/R-20 (0-Deg)	57.3	70.0	3.67
C-10/R-40 (90-Deg)	49.3	19.3	1.56
C-20/R-30 (90-Deg)	50.6	15.7	1.46
C-40/R-20 (90-Deg)	57.3	10.0	1.38

[a] Resin: Dow 411-C50 vinyl ester, no filler

In prototyping with RTM, a wide selection of realistic resins and reinforcements for commercial applications is currently available from a number of manufacturers. The table of physical properties shown below (Table 5.1) indicates the flexibility of design possible.

Other processes used for prototyping such as hand layup and wet molding give only a single finished surface; dimensions in the thickness direction are uncontrolled. Resin transfer molding provides two finished surfaces, controlled thickness, and it requires no auxiliary vacuum or autoclave equipment.

5.7 PRODUCTIVITY

Cycle times for resin transfer molding can be adjusted as required to meet productivity targets. If a long cycle time is acceptable, equipment complexity and cost can be significantly reduced. For example, if several large composite structures per day were to be molded, a low cost system could be easily configured. A one-part resin system having a cure time of 1 or 2 hours could be premixed and a pressure pot system used to inject the resin into the tool. For a large structure, an epoxy tool would provide a moderate number of moldings at low cost.

Tool damage normally determines tool life. The number of parts molded will depend on the care with which the preform is loaded into the tool. With low production rates, no special resin inlet or vent provisions are required. After closing, the tool resin can be injected and allowed to flow from vent tubes until there is no evidence of bubbles in the vent stream. Resin flow can be started and stopped, the tool can be tilted, or vibration and rotation can be used to further enhance wetout. For very large moldings, windows have even been built into a tool to allow observation of the completeness of wetout. With slow cycle times there is sufficient time to remove and clean the pumping equipment while the component is curing. In general, if the resin gel time t_{gel} is the variable, cure time will be 4–5 $\times$ t_{gel}. In the case of our 1 hour gel time, the component will be ready for demolding in 4–5 hours.

Cycle times for resin transfer molding can be reduced from hours to literally seconds by changes in resin chemistry and associated changes in equipment complexity. The distinction between RTM and RIM resin systems in continuing to diminish as the area of liquid molding develops. Cycle times of 1–2 minutes are achievable for moderate-sized components. Tooling for these rapid cycles must be heated metal, and resin handling equipment must be two-component, self-cleaning units.

Process automation is rapidly developing for both low productivity and high productivity molding. Shuttle presses similar to the press previously shown Fig. 5.22 allow a single upper mold half to use two lower mold halves alternately, permitting preform loading, demolding, and mold preparation to take place while the next component is curing. Computer-controlled resin delivery systems have been developed by a number of manufacturers, and some are in daily use, molding at annual rates in excess of 1 million components. Automation is also being developed for preforming as previously discussed.

Preforming automation is an area to be thoroughly studied relative to productivity requirements. The cost associated with operations such as preform assembly can be very high. If volumes are low, hand assembly may be preferable. If high volumes are required, the preform and resulting physical properties must be optimized for automation. Processes such as stamping and trimming of preforms are being automated in systems such as that shown in Figure 5.33. It is unlikely that automation of a

FIGURE 5.33   Automated preforming system for stampable materials. (Photo courtesy of Snow, Inc.)

cut-and-sew preforming approach, similar to aerospace prepreg layup, can be cost effective in the commercial sector.

Automation of the deflashing, trimming, and machining of RTM molded components is no different from that used on components molded using other processes. The good dimensional accuracy associated with RTM molded components aids in ensuring the repeatability of location required in automated systems.

Two additional issues that can affect productivity are postcure requirements of the components and scrap management. Some resin systems may require a postbaking or postcure to achieve their optimum physical properties. When such a resin is used, a postcure line will be required as well as postcure fixtures to maintain critical dimensions during the bake cycle. Without a fixture, components often distort during postcure because of resin shrinkage. Parts with very complex geometry may require extensive fixturing. Obvious alternatives are to allow the part to cure completely in the tool or to use resin systems that do not require a postcure.

Scrap management can significantly affect the overall economics of the RTM process. Preforms are often large and expensive to produce. Inspection and repair of a preform are necessary when large components are produced. After molding, disposal of a large molding composed of mixed materials is complex and expensive. Increased process control is an effective way to ensure that fewer molded components are rejected because of molding variations.

5.8 SUMMARY

Large complex components that facilitate large-scale part integration are well suited to manufacture by the liquid molding process. At low volumes the process can be used for a wide variety of part geometries, resin systems, and reinforcements. At higher volumes larger scale part integration, a fast resin system, and low cost preforms, are required for the process to be competitive economically.

RTM provides a wide variety of design options because of the ability of the process to utilize a variety of reinforcements, foam cores, and resin systems, resulting in a broad range of physical properties. Production volume can range from prototype level to high volume production, with physical properties changing very little as process equipment is scaled up.

REFERENCES

1. Richard D. Pistole, "Compression Molding and Stamping," in *Engineered Materials Handbook*, Vol. 2, ASM International, Metals Park, OH, 1988, pp. 325–343.
2. C. F. Johnson, "Resin Transfer Molding and Structural Reaction Injection Molding," in *Engineered Materials Handbook*, Vol. 2, ASM International, Metals Park, OH, 1988, pp. 344–357.
3. H. J. Borstell, "Hand Layup, Spray-Up, and Prepreg Molding," in *Engineered Materials Handbook*, Vol. 2, ASM International, Metals Park, OH, 1988, pp. 338–343.
4. C. F. Johnson, N. G. Chavka, R. A. Jeryan, C. J. Morris, and D. A. Babbington, "Design

and Fabrication of a HSRTM Crossmember Module," in *Proceedings of the Third Advanced Composites Conference*, Detroit, ASM International, September 1987, pp. 197–218.

5. F. K. Ko and C. M. Pastori, "Structure and Properties of an Integrated 3-D Fabric for Structural Composites," Special Technical Testing Publication 864, American Society for Testing and Materials, Philadelphia, 1985, pp. 428–439.

6. *Proceedings of the Second Textile Structural Composites Symposium*, February 1987, Fibrous Materials Research Laboratory, Drexel University, Philadelphia, PA 19104.

7. John P. Coulter, "Resin Impregnation During the Manufacturing of Composite Materials," Center for Composite Materials Report CCM88-7, University of Delaware, 1988.

8. S. Pilitsis and A. Beris, "Calculations of Steady State Viscoelastic Flow in an Undulating Tube," Center for Composite Materials Report CCM88-23, University of Delaware, 1988.

9. R. S. Morrison, "Resin Transfer Molding of Fiber Glass Preform Reinforced Polyester Resin," 36th Annual Technical Conference, February 1981, 15-D.

10. G. Q. Martin and J. S. Son, "Fluid Mechanics of Mold Filling for Fiber Reinforced Plastics," *Proceedings of the First Advanced Composites Conference*, Detroit, ASM International, November 1986, pp. 149–159.

6 · Continuous Fiber Molding Processes
A. Filament Winding

Howard S. Kliger
Brian A. Wilson

Contents

Howard S. Kliger, Edison, New Jersey, USA
Brian A. Wilson, Folsom, California, USA

6A.1 INTRODUCTION

Filament winding is usually thought of as a process by which a filamentary yarn or tow is first wetted by a resin and then uniformly and regularly wound about a rotating mandrel. The finished pattern is cured and the mandrel removed. The result could be something as simple as a piece of pipe or as complex as an aircraft fuselage or an automobile frame. Typical materials include fiberglass, carbon, or aramid fiber coupled to polyester, vinyl ester, or epoxy resin.

The principal advantages of filament winding over other composite material fabrication methods are its low material and labor costs and its reproducibility due to the robotic motions. The greatest disadvantages are the tooling limitations for removable mandrels and the inability to wind on negatively curved (concave) surfaces.

Mass production for filament winding is conceptually somewhat different from mass production for other manufacturing modes. While compression or injection molding may see rates of thousands or more per day, a filament-wound mass production part rate is more properly characterized as hundreds per machine station per day. Winding speeds are limited to about 61–122 m (200–400 ft) per minute. This is a practical limitation of the fibers' ability to thoroughly wet resin to all its filament surfaces. Volumetric or mass laydown rates are then set by the number of wetted yarns in the winding bundle. For instance, a machine running 25 ends of 225 yield glass in epoxy resin could lay down about 409 kg (900 lb) per hour.

The most visible application areas for mass-produced filament-wound parts are pipe, pressure vessels, missile launch tubes and motor cases, automotive leaf springs, fuse holders, and bearings. The largest material volume is filament-wound pipe, where approximately 63 million kg (140 million lb) of fiberglass and 27 million kg (60 million lb) of resin were consumed in 1987 in the United States.

6A.2 MATERIALS

There are a number of material options available for filament winding. Typical fibers are fiberglass, carbon, and aramid. The most commonly used resins include thermoset polyesters, vinyl esters, epoxies, and phenolics. The material combination can be classified as either a wet system (i.e., the fiber is wetted with resin just before winding on the mandrel) or a prepreg system (i.e., the resin has been applied to the fiber in an earlier operation and a "staged" yarn is delivered at the winding station).

The predominant system for mass-produced filament-wound parts is fiberglass with polyester or vinyl ester resin, delivered in the wet mode. This system, with a raw material cost in the range of $ 2.27–$ 4.54 per kilogram ($ 1–$ 2/lb) competes effectively against specialty metallic systems such as stainless steel, aluminum, or coated tubing. Aside from the economic issue, fiber choice is dictated by requirements such as strength and stiffness, electrical and thermal conductivity, chemical resistance, and damage tolerance. Resin choice is usually keyed to thermal and chemical resistance, and elongation.

181

6A.2.1 Fibers for Filament Winding

Fiberglass for filament winding is available as either single-end or multistrand roving. The single-end roving is one strand of glass filament collected into a discrete bundle during the spinning operation. Because of this, there is less of a tendency to form catenaries than with gathered strands. Single-end rovings are available in yields of 47–747 m/kg (113–1800 yd/lb). The fiber is normally sized with a silane coupling agent.

Most glass fiber used for filament winding is either E-glass or S-2 glass. The S-901 form of S-glass, widely used in the 1960s and 1970s for high performance filament-wound structures, is no longer available.

Fiberglass is packaged on either internal or external payoff spools. For the majority of filament winding operations, it is desirable to have the spool fed from the outside, to maintain tension through the machine. Where any kind of tensioner system is involved, particularly the wind-back type (which is discussed later), the spools must be outside fed. However, for filament winders that generate tension by taking the fiber through the payoff system of the machine, it is economical to use the interior payoff spools.

Aramid fiber, and most notably du Pont's Kevlar, was widely accepted as a filament winding fiber beginning with the strategic missile motor case industry. Its principal asset is its high strength-to-weight ratio, which is considerably improved over the fiberglass that previously had been used for motor cases.

Aramid is available in yields of 124–9540 m/kg (298–23,000 yd/lb), depending on roving type and fiber grade. For filament winding, Kevlar is available as 1307 and 1626 m/kg (3150 and 3920 yd/lb) single-strand roving and 261 and 406 m/kg (630 and 980 yd/lb) multistrand rovings, all on outer pull packages.

The fibrilative nature of aramid fiber gives it the property of being able to withstand abrasive wear. This property suggests aramid's use as an external layer for structures that receive considerable wear and abrasion. As a companion property to its ability to take abuse, aramid itself will abrade materials that come into contact with it. This is a major consideration for the pulleys and the various components of the payoff system in a filament winder. A chromed steel surface or a highly polished ceramic surface would not be as effective as a 55–80 rms mat finish. Under a microscope, the latter appears like a pebbled surface over which the aramid travels undamaged.

Carbon fiber for filament winding is generally available as 3000, 6000, 12,000, and 50,000 filament tows. This corresponds to yields of 996, 498, 249, and 62 m/kg (2400, 1200, 600, and 150 yd/lb). Unlike glass and aramid, carbon is not made as multistrand rovings. Nor is it available in center-pull packages. Carbon fiber is very different from fiberglass and aramid. Being brittle, it has a tendency to abrade and break; thus more care must be taken in handling it. To minimize breakage of the filaments, the number of turns and twists in the process must be kept to an absolute minimum.

From the standpoint of initial material damage, wet winding is better than prepreg tow winding, since the additional unrolling and rerolling that occur during prepregging is eliminated. Whether a wet winding process or a prepreg fiber is used, the key factor is still to minimize the number of turns to which the carbon fiber is subjected. Also, chrome plate on steel or high gloss ceramic should be used for payoff contact points.

Also, unlike the other fibers, carbon is electrically conductive. This means that dry fiber fly must be controlled to avoid electrical shorts in nearby equipment. This control is normally achieved by enclosing the creel with negative pressure, vacuuming the eyelet boards, and impregnating the fiber at the earliest possible point in the delivery stream.

Fiberglass for filament winding is available from Owens Corning Fiberglass Corporation, PPG Industries, Certainteed Corporation, Manville Sales Corporation, Nitto Boseki Ltd, and Pilkington Brothers Ltd. Aramid is available from du Pont (Kevlar) and Enka (Twaron). Carbon fiber is manufactured by a host of companies, most notably Toray, Toho Beslon, Hercules, Amoco, BASF, Akzo, RK and Courtauld's.

6A.2.2 Resins for Filament Winding

Filament winding can utilize resin in three distinct physical forms. The predominant one is as a liquid, where the fiber is wetted as it passes through a resin bath. Another form is as a prepreg tow, where the fiber is impregnated in an early step, staged to a tack-free consistency, and rewound on a bobbin. A third form utilizes thermoplastic resins, which may be in the form of a dry ribbon, a powdered coating, or a commingled fiber.

Wet thermoset filament winding requires a resin to have a viscosity in the range of 1000–3000 cP. Resin components can be chosen so that the combination of pot life, winding temperature, viscosity, gel time, and cure time can be optimized. Wet resins include epoxy, vinyl ester, polyester, phenolic, polyimide, and bismaleimide. With each resin there are a number of recognized curative and accelerator combinations to give acceptable end-use performance. Epoxies have the widest range of properties of the resins used in filament winding. They are the predominant resin used in the aerospace market. However, as end-use temperature requirements have increased, more interest has developed in more difficult-to-process systems such as phenolics and polyimides.

Preimpregnated tow, typically carbon with epoxy resin, has the advantages of accurate resin content control and a higher level of QC and traceability as compared with wet systems. Also, the resin systems may be identical to conventional prepreg tape forms, and this can be particularly attractive when aerospace material qualifications are needed. The disadvantage is higher cost (i.e., higher than the wet systems).

The use of thermoplastics in filament winding is just beginning. Although thermoplastic matrices have been available for many years, only recently have material forms emerged that are useful for filament winding. These include reinforcing yarn covered with heavy concentrations of finely dispersed thermoplastic powders, commingled yarns made up of reinforcement fiber and melt-spun multifilament thermoplastic filaments, and resin-impregnated or encapsulated reinforcing yarns. The latter are usually stiff and "boardy" in appearance. For all the thermoplastic systems, the delivery system must be heated in a way that serves to soften and consolidate the winding yarn onto the mandrel.

A concerted research effort is underway to develop winding techniques for thermoplastics. This undoubtedly will result in the development of one or more acceptable techniques for thermoplastic filament winding as a production process.

6A.3 THE FILAMENT WINDING PROCESS

The filament winding process can generally be classified as either helical or polar winding. (Figures 6A.1 and 6A.2). In the first case, fiber is fed from a horizontally translating delivery head to a rotating mandrel, while in the second case, a delivery

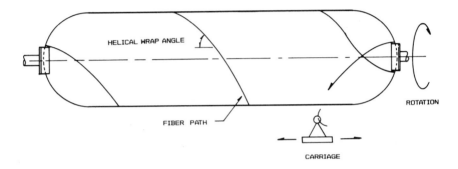

FIGURE 6A.1   Helical winding.

FIGURE 6A.2   Polar winding.

unit races around a slowly indexing mandrel. All filament winding processes have a number of required subsystems. These include fiber delivery (resin bath and band alignment), mandrel, mechanical and/or electronic control, curing, and mandrel preparation and removal. All these are discussed in subsequent sections.

6A.3.1 Helical Winding

In helical winding, the fiber band is fed from a translating carriage onto a rotating mandrel. The wind angle is specified by the ratio of the two relative motions. As the translating carriage reaches the end of the mandrel, it will slow down, dwell, and reverse direction, continuing to wind fiber onto the mandrel with a negative angle. As the motion and reversals continue, a diamond pattern will develop on the mandrel. A conventional winding pattern on a cylinder will close on itself in the final traverse, yielding a structure with a uniform thickness and constant $\pm \beta$ angle everywhere. If the winding mandrel is tapered or irregular, the angles and thicknesses will vary accordingly.

Helical winding is the predominant method used today. It is particularly well suited for long slender geometries such as pressure pipe and launch tubes, where winding angles of 20–90° (hoop) are needed. Most pipe is wound at 54.7°, which is derived from netting theory and assumes a 2:1 (hoop-to-longitudinal) stress field in a cylindrical, capped pressure vessel.

There are a number of practical limitations to helical winding. These include machine bed size, mandrel and wound part weight, and clearance of the turning diameter. Most filament winding machines have beds ranging in length from 2.4 to 7.3 m (8–24 ft). Considering necessary clearances at the head and tail stocks, wound structures can be from less than half a meter up to 7 m (23 ft) in length, and up to 3.7 m (12 ft) in diameter. Special machines can accommodate even larger structures.

The largest helically wound structure is understood to be the windmill blade wound by Hamilton Standard on a special winder designed by Engineering Technology Corporation; it is 46 by 4.6 m (150 × 15 ft). Another large and unusual helically wound structure is the Glasshopper rail car. This part contains 8175 kg (18,000 lb) of fiberglass and polyester resin, and was wound on a machine designed by McClean Anderson. Its dimensions are 15 × 4.5 × 3 m (50 × 15 × 10 ft).

In helically winding large structures, special consideration must be given to mandrel design. Mandrel weight causes wear on bearings. Mandrel deflection is a critical and often controlling issue, particularly for sand and plaster units. Inertia considerations often limit winding speed. Also, very low angles (0–10°) usually are avoided in helically wound parts with large length-to-diameter ratios. These low angles are difficult to capture during dwells at the ends of the mandrel, and the accelerations associated with these motions can lead to unacceptable vibrations in the delivery head. However, in large diameter structures such as rocket motor cases with length-to-diameter ratios of 2:1 or greater, low angle helicals are essential to absorb the longitudinal stresses in the structure. Low winding speeds and precise geodesic winding paths are dictated here.

Hoop winding is a special case of helical winding in which the angle approaches 90°. This type of winding provides reinforcement only in the circumferential direction. Hoop winds are often used in conjunction with other angles not only to provide the hoop reinforcement, but also to compact the wet winding at intermediate fabrication points. Winding with constant tension, hoop winding provides more normally directed force to the previously wound plies than does any other angle. Conversely, a 0° wind has no normal force component. Note also that when winding thick structures, winding tension is often varied from inner to outer ply.

Examples of helically wound structure include pressure pipe, osmotic tubes, cylindrical pressure vessels, tactical missile launch tubes, auxiliary fuel tanks for aircraft and helicopters, engine nacelles and cowls, main rotor driveshafts, strategic and tactical missile motor cases, torque tubes, truck driveshafts, electric motor driveshafts, and core sample holders.

6A.3.2 Polar Winding

In polar winding, the fiber delivery system rotates in a single plane or racetrack— either horizontally or vertically—while the mandrel is incrementally rotated within the plane. The result is a dual layer of fiber at $\pm \beta$ with all crossovers occurring at the mandrel ends or pole pieces. The mandrel is usually vertical and supported at the base. Consequently, very large mandrels can be supported without sag. For the few horizontal polar winders in existence, the mandrel is supported as a cantilever and its weight as well as the final wound part weight are limited by deflection criteria.

Contrary to helical winding, polar winding favors very low wind angles. Since the fiber is wound in a plane intersecting the mandrel ends, the angle must be less than about 20°. Typically, polar angles are 5–15°. A special case of this is the winding of a spherical pressure vessel. The initial layer is tangent to pole pieces or bosses. The second layer is one bandwidth away from the boss, the third layer is two bandwidths away, and so on. The final layer is an equator or hoop band.

The first stage motor of the Polaris missile was perhaps the largest polar-wound structure. It was about 0.9 m (3 ft) in diameter and 3 m (10 ft) long. Most polar-wound parts have L/D ratios of 1:2.

The principal advantage of polar winding is that it is a simple and rapid winding technique for short, stubby geometries with L/D of less than 2, where balanced fiber placement is required. The disadvantage is that since it winds in a plane and not on a geodesic, the dome geometry is not always weight efficient and the stress distribution is less than optimum. Another disadvantage is that it is difficult to wind cases with large differences in boss diameters. Finally, a polar winder does not conveniently wind hoops, and a second winding head is usually required.

In polar-wound structures that require combinations of polar and hoop wraps, it is common practice to intersperse the patterns. This tends to more evenly load the hoop fiber as well as to provide a compaction for each set of polar wraps.

Examples of polar-wound structures include virtually all third-stage missile motors, apogee kick motors, inertial upper stage motors of space shuttle payloads, and even certain spherical petroleum storage tanks.

6A.3.3 Special Winding Modes

A number of winding systems in use today that generally fall under helical winding are so special and unusual that they demand separate identities. One system that has been used successfully for pipe and automotive driveshafts is a 360° fiber delivery system. Here a ring surrounds the mandrel and simultaneously delivers hundreds of filament ends to the translating and rotating mandrel. (In some cases, the ring translates while the mandrel only rotates). The result is a single ply at a $+\beta$ orientation. Tying down the filament ends and reversing direction results in a full $-\beta$ ply (Figs. 6A.3 and 6A.4).

Another pipe manufacturing mode utilizes a continuously translating nonrotating mandrel that travels through a series of rings, each of which has a number of fiber spools mounted to it. These in turn rotate around the ring and pay out fiber onto the mandrel at an angle set by the rotation to translation ratio. This system closely resembles braiding.

A third variation couples an articulated robot to a rotating or stationary mandrel. Though winding speed is slower than the conventional modes, the opportunity exists to wind much more complex shapes.

Hoop winding has been mentioned previously and is a special case of helical winding. Certain types of low pressure pipe are wound with a pure hoop layer of fiberglass over a chopped glass strand spray-up. Another example of hoop winding is the Corvette leaf spring. Here, the glass strands are rapidly wound into a trilobal channel which becomes the mold for the compression-cured leaves.

FIGURE 6A.3   Full circumference delivery system. (Courtesy of McClean Anderson, Milwaukee, WI.)

FIGURE 6A.4   360° delivery ring. (Courtesy of McClean Anderson, Milwaukee, WI.)

Tumble and racetrack winding are two other special cases. In tumble winding the mandrel appears to tumble or rotate over itself as fiber is fed from a stationary eyelet. Tanks and spheres can be made from this configuration. In a racetrack mode, an indexing mandrel is inserted into the plane of the track, and the eyelet pays out fiber as it translates around the track.

6A.4 EQUIPMENT

6A.4.1 Winders

6A.4.1.1 Common Factors

Filament winding equipment is as varied as the parts produced from it. There is no clear meaning of the term "universal winder," though it normally connotes a gear-and-sprocket or computer-controlled helical machine with 2–4 degrees of freedom, capable of winding spheres, cylinders, cones, box beams, and combinations thereof. Special machines, which in many cases are simplifications of the universal types, are usually required for mass production. Most filament winding machines have two degrees of freedom: mandrel rotation and delivery head translation. As machine control becomes more complex, additional motions of the delivery head can be regulated. These include a vertical and horizontal axis as well as three head rotations (Fig. 6A.5).

Machines for pipe winding are typically two-axis machines. As the product increases in sophistication, more axes or degrees of freedom are added. The require-

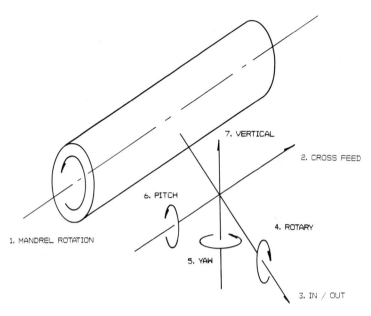

FIGURE 6A.5   Degrees of freedom.

ment to have the payoff head deliver the fiber band normal to the surface of the mandrel has increased the emphasis on sophistication of the delivery system. Many four-axis machines are now being upgraded to five- and six-axis capability. Microprocessor control is normal with these systems. Some delivery systems are now being provided with overhead delivery. The three axes of end effector motion can be replaced with a wrist unit for which the computer software now exists from Cincinnati Milacron.

The principal advantage of filament winding is the ability to accurately place a fiber band on a rotating mandrel. In a basic gear-driven machine, this implies direct coupling between the machine headstock (angle of rotation) and the chain drive on the traversing eyelet. Control of this system is by mechanical or numerical means. Mechanical control is the predominant method. Here a single motor drives both the winding mandrel and a sprocket gear linked to the drive chain and delivery head. The principal disadvantage of this system is the inability to rapidly change winding angles. Reprogramming requires the assembly of a new gear, idler, and link combination, which can take several hours. This has been partially overcome with the development of dual gear systems, which allows the winding of two preset helical angles and a $90°$ hoop using a single switch.

The limitation of one or two wind angles is not a serious constraint for many mass-produced structures such as pipe or pressure bottles where the machinery is dedicated to a single product and winding pattern.

Numerically controlled winders also have been developed with punched tape operation. The tapes control hydraulic servo drives, and each axis has its own hydraulic motor. Another control method has replaced the punched tape with an

189

optical tracking device. Here the servo is controlled by an electric eye following a black–white interface on a rotating disc or cylinder.

Within the past 10–15 years, numerical control with microprocessor-controlled servo motors has been developed and refined. The two principal advantages of a microcomputer-controlled machine are the ability to rapidly change and store winding patterns and the ability to more accurately place the fiber on the mandrel through eyelet manipulations. Also, the ability to program in a nonlinear mode is extremely desirable for complex geometries. Some microprocessor-controlled machines have as many as 6 degrees of freedom coupled with path smoothing options, acceleration controls, and independent yarn tension control.

Recent advances in the microprocessor control have included off-line program development, where most of the patterns and parameters can be determined in advance of actual prototype winding. Off-line programming is typically used with a graphics terminal that can simulate the mandrel shape, the fiber band path, and the path of the delivery point in space. The same program can graphically display the closing error and, given the coefficient of friction between the band and the existing layers, can define the permissible offset from the optimum geodesic path.

6A.4.1.2 Helical Machines

In a very basic helical filament winding machine, the drive motor rotates the mandrel. The translating fiber payout system is controlled by a chain, sprocket, and idler assembly, which is driven and timed by the main drive motor. The basic system then has 2 degrees of freedom. The third degree of freedom is the motion of the payoff head toward and away from the wind axis of the machine. This permits placement of the payoff head adjacent to the surface of the mandrel, regardless of mandrel diameter. It also permits the delivery head to follow a change of mandrel diameter —for instance, in winding a conical surface or when a dome is used. For dome winding the fourth degree of freedom is required. This is a delivery arm rotation. The fifth axis is the yaw axis of the end effector, which permits delivery normal to the mandrel surface under all geometric conditions.

A two-axis machine may be nothing more than a modified lathe with the bidirectional screw shaft used to control crosshead movement. This type of system has been used to wind cotton fiber oil filter cartridges and metallic filter tubes from flattened stainless steel wire. On a larger scale it is used for winding fiberglass pipe for water distribution, sewage, and oil field applications.

All filament winding processes require that the crosshead motion reverse itself at the end of the traverse. This can be done in three ways depending on the type of structure. With tubular structures, as wound on two-axis machines, turnaround is normally accomplished by winding extra length on each end of the tube, with the result that the instantaneous reversal of the crosshead builds up a disarray of fibers beyond the required length of the tube. After cure, this section can be cut off.

The primary method of turnaround is to use a dome on the end of the mandrel. This dome can be part of the structure (i.e., in cylindrical pressure vessels or solid rocket motor cases), or it can be a false dome that is incorporated in the tooling and is later cut off. Domed structures are normally wound with three- and four-axis machines. However, they can be processed on the simple two-axis machine if a delay feature is included in the traversing program for the crosshead and payoff system. This is accomplished by using a continuous chain, which is set around two sprocket gears, one at each end of the travel. The crosshead is then attached to the chain, and the system, in effect, slows down at the end of the traverse as the attachment to the chain is going around the sprocket. Changing the size of the sprocket relates to changing the size of the cylinder and dome being wound, hence to the period of motion around the gear. The third method of turnaround is to use protruding pins in the end of a cylindrical mandrel; the fiber is wound across the flat end of the cylindrical mandrel and back onto the winding surface at the opposite side. The pins in this situation prevent the fiber from moving and slipping off the abrupt end of the mandrel.

A special case of using pins in the winding process occurs when holes are required in the end of a filament-wound structure (e.g., motor case or launch tube). This method, known as "wound in place holes," was developed and patented by the Brunswick Corporation. Protruding pins are placed in the end of the mandrel, and the machine is programmed with complex software to loop the fiber band around and between the pins. The pins have a special flattened head that retains the fiber in place. After curing, the pins are pulled, and the flash is trimmed from the surrounding area, leaving a system of holes that do not have cut fibers and provide full strength of the wound helical system in a tension mode. There is, however, a shadow effect on the backside of the pin that tends to reduce the compressive strength in this area.

The majority of helical machines (see, e.g., Fig. 6A.6) operate on the principle of using the geodesic path for laying down the fiber band. The geodesic path is defined as the shortest distance between two points on a curved surface. While the use of the

FIGURE 6A.6   Helical winder. (Courtesy of McClean Anderson, Milwaukee, WI.)

geodesic path is certainly the most efficient method of winding, frequently the combination of part geometry and the design fiber orientation does not permit the fiber to be placed precisely on the geodesic path. Under such conditions the friction factor for resistance against slippage of the fiber band on either the mandrel or the wound surface must be taken into account. This friction factor can be input into a machine control program to determine the maximum amount of deviation from the geodesic path before friction is overcome and slippage takes place. This will define the available fiber orientation from a manufacturing standpoint.

While the majority of helical machines are horizontally mounted and use a horizontal position of the wind axis and mandrel, there is nothing to prevent a helical winder from being used in a vertical position. The vertical machine has several advantages over the horizontal unit: the winder takes up significantly less floor space, and the vertical machine rotates its mandrel on a pedestal bearing, where the load is uniformly distributed on the bearing. In the horizontal mode, the load from the wind axis and mandrel is taken out through side load on the main bearings in the headstock and tailstock of the machine. Additionally, in the horizontal machine, deflection problems of the structure, mandrel, and wind axes may be created by the weight of this entire system exerting a beam-type loading on the wind axis. Since the diameter, and hence structural stiffness, of the wind axis is frequently controlled by the boss openings of the pressure vessel, this deflection can assume serious proportions. A disadvantage of a helical machine in a vertical position is that the top of the structure being wound is so high off the floor that a small elevator is needed to transport production workers to the top for working access. In addition, the vertical design makes a floor-mounted creel desirable.

6A.4.1.3 Polar Machines

A polar machine (Fig. 6A.7) is the ultimate in simplicity, comparable to the helical pipe winders. Only 2 degrees of freedom are used. These are the mandrel rotation and the rotation of a large vertical arm around the mandrel.

The polar winding process lays the fiber band immediately adjacent to the band laid on the previous rotation. Subsequent bands are laid next to each other rather than forming a net or diamond-shaped structure as in helical winding. Thus the crossover points of the fiber bands are all in the boss area at the top and bottom of the mandrel. (In the helical process, the crossover points are in the cylindrical area.) The band is laid in a perfectly straight plane tangential to the bosses at the two ends of the mandrel. While the process can accommodate bosses of different sizes, the optimum condition would be to use bosses of identical size. While there are occasional exceptions to the rule, polar winding is usually an optimum process with structures that have L/D ratios of less than 1.5:1. This is why the majority of applications for polar winding are in upper stage motor cases for the solid rocket industry, where the typical structure is short and stubby.

A disadvantage of the polar winding process is that the creel of fiber spools cannot be mounted in a fixed position relative to the winding machine structure. To prevent

FIGURE 6A.7 · Polar winder. (Courtesy of Aerojet General, Sacramento, CA.)

twisting of the band as the rotating arm describes its path around the mandrel longitudinally, the fiber spool creel must be mounted on a board that rotates with the arm. As a result of this, the bandwidth of polar winding is usually limited to about 25 mm (1 in.), corresponding to eight spools. An additional consideration in the use of polar winding is that the machine does not adapt well to the winding of circumferential or hoop fibers. A clever adaptation of the polar machine enables a hoop layer to be wound by positioning the payoff arm in the horizontal position, then moving the axis of the arm slowly in a vertical position.

Polar winders, because of their simplicity, have been in operation for many years. The market achieved saturation in the early 1970s, and very few polar winding machines have been manufactured and supplied to the industry in the past 15 years. An exception to this is the small laboratory polar winder known as a tumble winder. The tumble winder has a fiber payoff system normally consisting of one or two spools, maintained in a fixed position. The mandrel is mounted on a spindle inside a large rotating arm, which turns the mandrel end over end as well as rotating it on its axis, one bandwidth for every revolution. Since these machines only use one or two spools, the bandwidth is between 2.5 and 5 mm (0.1 and 0.2 in.).

6A.4.1.4 Pin Winders

Wrist and pin winders are chosen for operations that do not use mandrels in the classical sense and have motions that are repetitive but not necessarily rotational. The mandrel is a series of pins and channels around which a highly articulated arm delivers wetted fiber. Arm and pin motion are coupled with gears and cams in the case of a mechanical pin winder, or through electrical signals to dc drive motors in the case of

193

a robotic system. This type of machine was used to integrally wind the spokes, handle, and hub of a prototype steering wheel.

6A.4.1.5 Multiple-Spindle Machines

Multiple-spindle machines are an effective means of increasing productivity. An eight-spindle machine is shown in Figure 6A.8. The extra spindles are geared to a main drive spindle, which in turn is coupled to the carriage unit. This slaving arrangement allows for multiple, identical parts to be wound simultaneously. Fewer, larger diameter parts can be wound using only selected spindles. Normally, only 2 degrees of freedom are used in a multiple winding mode. A three-axis capability is possible if a tandem carriage is added to the vertical axis drive. Disadvantages of the multiple spindle system include longer setup times and the difficulties associated with restarts and repairs in the midst of a cycle.

FIGURE 6A.8   Multiple-spindle winder. (Courtesy of McClean Anderson, Milwaukee, WI.)

6A.4.1.6 Hoop Winders

Hoop winding has already been discussed in connection with helical and polar winding. In a polar machine, the hoop winding is at times accomplished by moving the mandrel to a separate and unique hoop winder, whereas with the helical machine it is simple to program the machine in the hoop winding mode in addition to the helical mode. Hoop windings are used for two separate purposes. The first is to withstand the circumferential stress in a cylindrical member, since that stress is always twice the stress in the longitudinal direction for a given pressure vessel. Additionally, hoop layers are used for compacting the helical or polar layers as they are wound on the structure. Thus, the hoop windings are nominally applied after layers of helical or polar winding.

6A.4.1.7 360° Delivery Systems

One disadvantage of conventional filament winding is that only a single band of fibers can be laid down in each pass. Many time-consuming passes are required for full mandrel coverage. To speed up the fiber delivery, a number of special machines have been developed which are referred to as 360° fiber delivery systems. One such concept was shown in Figures 6A.3 and 6A.4. Instead of building up a $\pm \beta$ layer pattern with a single band of fibers wound with a reciprocating eyelet and a rotating mandrel, many bands of fibers are simultaneously wound on the mandrel using a 360° delivery ring.

In one configuration, the mandrel is stationary and the ring translates and rotates, laying down a full layer of $+ \beta$. Reversing translating direction but continuing with the same rotational direction will place another layer at $- \beta$. A variation on this concept is to have the delivery ring stationary with the mandrel translating and rotating as the fiber layers are applied.

Another variation on the 360° delivery system uses a continuous disposable paper mandrel. This is currently being used in the manufacture of high pressure pipe. Here, a spiral-wrapped paper mandrel is formed at one end of a production line. As the mandrel translates down the line, it passes through a series of stationary fiber delivery rings. Each ring has a secondary ring, which has mounted on it spools of fiber. This second ring then rotates clockwise or counterclockwise, producing a single layer of fiber. The rotational speed and the number of spools on each ring can be changed to affect the wind angle and layer thickness.

6A.4.1.8 Machine Manufacturers

A number of companies manufacture turnkey filament winding machines. Most of them are also capable of manufacturing specialty machines for mass production or research. The most notable companies are listed below:

Automation Dynamics
2961 Junipero Avenue
Signal Hill, California 90806
USA

Cincinnati Milacron
Cincinnati, Ohio 45209
USA

Dura-Wound Inc.
P.O. Box 23
Washugal, Washington 98671
USA

Engineering Technology Inc.
145 West 2950 South
Salt Lake City, Utah 84115
USA

Goldsworthy Engineering Inc.
23930 Madison Street
Torrance, California 90506
USA

McClean Anderson Inc.
10600 West Glenbrook Court
Milwaukee, Wisconsin 53224, USA

Retek Inc.
7125 Saltsberg Road
Pittsburgh, Pennsylvania 15235
USA

Venus Products Inc.
1862 Ives Avenue
Kent, Washington 98032
USA

Josef Baer Maschinenfabrik
Postfach 1140
7987 Weingarten/Württ.
West Germany

Bolentz and Schafer
3560 Biedenkopf/Eckelshausen
West Germany

Pultrex Limited
Brunel Road
Clacton on Sea
Essex CO15 4LT
Great Britain

6A.4.2 Mandrels

The mandrel is the geometric basis for the final part. As such, it must support the uncured composite during winding and through cure without deforming beyond acceptable limits. Mandrels can be classified as permanent, removable, and reusable.

A permanent mandrel will become an integral part of the final structure. In a scuba tank, for example, fiber is wound over a thin metal wall that essentially acts as a gas barrier. Removable mandrels must be separable from the cured part without damage to the part. Reusable mandrels must be removable in a way that maintains the integrity of the part and the mandrel.

When the filament-wound part is open ended, and the opening is the largest diameter of the part, mandrel design and material considerations are straightforward. For instance, pipe and driveshaft mandrels are straight steel cylinders, often with chromed surfaces and very slight tapers [0.17 mm/m (0.002 in./ft)] to facilitate extraction. Aluminum has also been used, but it is more prone to damage.

When the end opening is smaller than the midsection, as would be the case for a rocket motor with an integral exit nozzle, a collapsible mandrel is needed.

6A.4.2.1 Metal Mandrels

Two of the primary types of mandrel structure require the use of metal. One is a permanent shell, while the other is reusable. The permanent use of a metal mandrel is normally for high pressure (i.e., gas pressure) vessels. For this application, the metal mandrel acts as a leakage barrier for gas inside the vessel. Most composite structures are not impervious to the passage of gas, particularly for helium applications. Hence the internal metal structure is a necessity. The structure also has the advantage of being used as a winding mandrel that does not have to be removed after cure of the part. Aluminum, stainless steel, Inconel, and titanium have all been used for this application.

Thin metal liners are typically formed in two halves, approximately 1.0–1.3 mm (0.04–0.05 in.) thick, and then welded together. This type of construction permits the use of threaded steel bosses, which are bonded into the dome ends of the aluminum liner halves before the whole structure is welded closed.

The other type of metal mandrel, known as the net metal mandrel, is removed from the composite structure after cure. It is reusable without any practical limits on the number of times of reuse. Net metal mandrels have been used in the solid rocket motor industry. These mandrels, which are assembled over a skeleton structure, consist of aluminum stave or ski type sections with ogive pieces being used in the dome areas. The aluminum segments are pinned or bolted together onto the skeletal support structure. These net metal mandrels provide the precise geometry for the internal dimensions of the solid rocket motor case and its insulation. After cure, the skeletal support structure and the aluminum sheet metal shapes are disassembled and removed from the motor case through the largest boss opening.

A third type of metal mandrel, the straight or slightly tapered tubular steel tool, is used in the filament winding of fiberglass pipe and launch tubes.

6A.4.2.2 Expandable Mandrels

All mandrels expand during the cure process by normal coefficient of thermal expansion action. However, certain mandrels, notably those made out of rubber, can be artificially inflated from the inside to provide either shape or pressure to the curing composite part. In one configuration for petroleum storage tanks, thick-walled rubber, circular balloons are air inflated to a preset diameter, gel coated with a chopper gun, and polar wound with glass and polyester resin. After an ambient cure, the pressure is released; then the balloon collapses and is withdrawn through the tank top opening. In a second configuration, a thin rubber bladder is formed over a hard tool. When winding is complete, the unit is placed in a clamshell mold and the tool–rubber interface is pressurized to force the windings outward against the clamshell during cure. The result is a dimensionally controlled and smooth outer surface. Note that it is difficult to accurately control inside dimensions using rubber tooling.

6A.4.2.3 Single-Use Mandrels

The oldest of single-use devices is the plaster mandrel. This structure is created by assembling a skeleton iron core with iron cover plates, wrapping the core with burlap to increase the adhesion of the plaster, and then trowelling approximately 10–15 cm (4–6 in.) of plaster in place on the external surface. This plaster is trimmed to shape while it is still soft, using a template and a rotating carriage. The whole structure is then cured in an oven to harden the plaster. During the construction of the mandrel, a steel cable is embedded in the middle of the plaster layer to assist in the removal process. After cure, the surface is sealed with an epoxy coating, the insulation layer is laid up by hand, and the motor case structure is filament wound over the outside. After the composite structure has been cured, the iron support is disassembled and removed through the largest of the two boss openings. The burlap is then removed, and the steel cable is pulled out. This serves to break up most of the plaster.

A major disadvantage of this type of mandrel is the presence of significant quantities of water in the plaster in its cured condition. This moisture boils off at resin curing temperatures and degrades the matrix by a plasticizing process. Gel coats are not totally effective in preventing moisture escape during resin cure.

A second type of single-use mandrel is the soluble structure. Originally this type of mandrel was manufactured by casting a salt paste into a mold and heating the mold to drive off the water, leaving a cast salt block of the required shape to be used as the mandrel. After cure of the composite structure, the mandrel was removed by dissolving it in water. The salt mandrel was very susceptible to damage; also, sections of salt tended to break away from the structure. This type of mandrel was then changed to one using ultrafine sand with polyvinyl alcohol (PVA) as a binder. The sand is cast in a suitable mold, with installation of lightening tubes to reduce the weight of the subsequent structure. The sand mandrel is cured at approximately 93 °C (200 °F). The resulting mandrel has a very hard surface that can be machined using carbide or diamond tools to introduce features that could not be created in the mold. Tight tolerance structures also require machining of the mandrel. When the composite structure is cured, the PVA binder is softened by allowing water to percolate through the sand. PVA is soluble in water; ultimately this results in the conversion of the sand and PVA mixture to a slurry, which flows out of the motor case into a receiving tank or pit.

Two disadvantages are inherent to this system. First, it is the heaviest of all mandrel materials, and the lifting capabilities in plants where this type of mandrel is used must be high. The second disadvantage is that it cannot be used at temperatures about 150 °C (300 °F); consequently, alternate methods must be found for high temperature curing resins. The PVA forms a varnish above at this temperature, and after the conversion, the material is no longer completely soluble in water. A recent solution to this problem is the use of sodium silicate or "water glass" as the binder. This binder is more expensive than the PVA, but it is not affected by temperatures of up to 343 °C (650 °F). This type of mandrel has been used with polyimide resins.

6A.4.3 Tension Control Systems (Figure 6A.9)

Controlling fiber tension in filament winding is an integral part of the process and has been key to optimum performance of the resulting composite structure. Yarn tension levels of between 0.45 and 1.8 kg (1–4 lb) have typically been applied to helical and polar windings. Hoop windings generally use a higher tension [2.7–3.6 kg (6–8 lb)] to provide a compaction layer for the structure. Although servo-based tensioners (Fig. 6A.10a) are common today, early tensioning technology relied on a standard spring-loaded, rotary motion take-up. The tension was provided by the calibrated setting of an internal spring, and the take-up of slack fiber to maintain the required tension was controlled through the rotary motion of two spring-loaded arms and the action of a small brake. This type of tensioner is shown in Figure 6A.10b. The minimal take-up capacity of these units requires that they be mounted on the crosshead of the filament winding machine, and in larger machines where 24–36 fiber spools are used, a very heavy crosshead structure results. High drive torques are

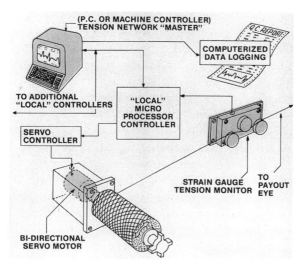

FIGURE 6A.9   Tension control system. (Courtesy of Electroid Co., Springfield, NJ.)

FIGURE 6A.10a   Servo-based tensioner. (Courtesy of Electroid Co., Springfield, NJ.)

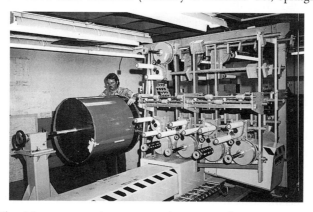

FIGURE 6A.10b   Non-servo tensioner system. (Courtesy of Aerojet General, Sacramento, CA.)

needed for braking and accelerating, and the use of 100 horsepower motors is not uncommon.

In the late 1970s electronic tensioners were conceived. This type of tensioner usually senses fiber tension by means of a strain gage activated load cell mounted in the fiber payoff system. Through use of either digital or analog conversion, this tension level is maintained to a very close tolerance by controlling the speed and rotation direction of each individual spool drive motor. These tensioners can be individually calibrated and controlled from the console of the filament winding machine, and digital readout of the tension of each fiber can be provided during the winding process. The take-up capability of these tensioners permits them to be used in a floor-mounted mode. The weight of the crosshead system is thus reduced considerably, and a much smaller drive motor is needed. This particular type of tensioner can have the capability of providing a printed record of the tension in each fiber bundle for quality assurance purposes.

Electronic tensioners are being offered by several manufacturers. Since the cost of this type of tensioner is typically three times that of the mechanical tensioner, the units are normally being used only for the more sophisticated type of winding machine, mainly in prototype development laboratories.

Along with the electronic tensioners has come the development of magnetic particle brake tensioners. Operating on the principle of regulating magnetic force in a clutch-type system by changing the electrical energy supplied to the magnetic field, this unit looks much like a mechanical tensioner. However, the substitution of the magnetic particle brake for the original spring-loaded version permits a much larger take-up capacity, and much closer control of the actual tension level. These units cost less than the electronic tensioners and are more likely to be found on production winders.

6A.4.4 Resin Bath and Fiber Delivery Systems

There are three common types of resin bath in use today: dip bath, roller, and metered orifice (Fig. 6A.11). There are a number of variations of each. The most common user alterations to filament winding machines are to the eyelet delivery and impregnating bath equipment.

Dip baths, the simplest devices, often have no moving parts. Typically they consist of an inlet comb, a resin reservoir area with submerged bars, and an exit comb with a wiper blade. Multiple yarns run through the first comb and under the bars, picking up resin, then through the second comb and wiper and onto the mandrel. Yarn wetout can be improved, particularly at higher running speeds, by having in the system a rotating roller that tends to spread out the fiber bundle. Also, a metered eyelet board at the exit comb can aid in controlling resin content delivered to the mandrel. Figure 6A.12 shows three types of fiber delivery combs.

The dip bath is best suited for glass and aramid fibers, which are considerably more damage tolerant than carbon. The roller drum impregnation system tends to be more gentle with the fiber and allows more controlled impregnation at lower tension levels. In this device, only a small portion of a rotating drum contacts the resin bath.

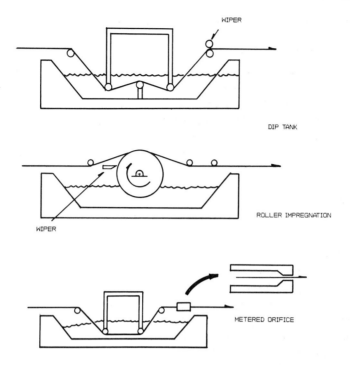

FIGURE 6A.11   Resin impregnation.

The drum picks up resin, which is then metered with a doctor blade. The fiber passes through the metered resin film on the topside of the rotating drum. Fiber peelers (i.e., broken filaments that adhere to the rotating drum and tend to grow with time) are a disadvantage here. If allowed to continue, these peelers will affect the fiber/resin ratio and will increasingly damage the passing yarns. At some point, the machine must be stopped to clean the drum.

A third system of resin application is through metered orifices. Though there are a variety of designs for these systems, the basic concept is to introduce fiber and resin at the large end of a converging channel and meter the resin to the fiber as it passes through the channel and a very accurately machined hole. The system can attain the highest level of resin control, but there are major drawbacks: the orifices must be changed for each fiber–resin combination or ratio, and the orifices are not capable of passing knots or splices.

In another highly controlled impregnating device, the mix ratio for epoxy–resin systems is set by a dc servo motor driving two pairs of precision gear pumps. The motor speed is proportional to fiber delivery speed. The resin components are mixed in a dynamic head and dispensed onto a heated impregnating drum with a calibrating roller. The fibers make multiple passes over the drum, effectively picking up all the dispensed resin.

The overall fiber delivery system must be designed for minimal fiber damage and optimal fiber wetting and placement on the mandrel. Fiber damage can be minimized

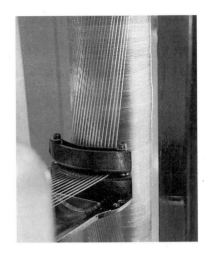

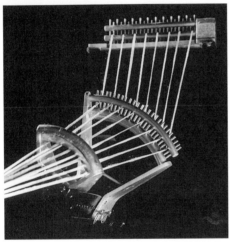

FIGURE 6A.12a–c   Fiber delivery combs.
(Courtesy of ABB Composites,
Irvine, CA.)

by using large, smooth bend radii on eyelets and guide rings. Ceramics and chromed metal surfaces can be used to minimize wear on the components. (See Section 6A.2.1 for comment on aramid fiber).

Fiber coverage in a winding pattern is based on an assumed bandwidth. Variation in the bandwidth can result in open patterns or undesirable fiber buildups. The bandwidth is partially controlled by the selection of the final delivery ring. Combs, O-rings, and D-rings are common. Selection is often based on observations during the prototype winding stage.

6A.4.5 Curing Systems

In filament winding, the requirements for long pot life of the resin and fast cure of the part are contradictory. In general, resins that can maintain acceptable winding viscosities for long periods of time—from a few hours to days—will require either longer cure times at lower temperatures or much higher initiation temperatures, depending on the accelerator and catalyst. Higher curing temperatures may result in thermally induced microcracking in the finished laminate. Other factors that will influence microcrack formation include the relative coefficient of thermal expansion between mandrel, fiber, and resin, the heating and cooling rates, resin shrinkage due to curing and cross-linking, and fiber directionality.

With the exception of microwave energy, the curing process for filament-wound parts is initiated at either the inner or outer surface, depending on the location of the heat source. Both methods have advantages. When curing initiates at the inner surface, the parts tend toward higher fiber content because the resin can be bled out. Also, void content is reduced, since there is less tendency to trap air pockets. With external surface curing first, higher resin contents are possible, and drippage can be eliminated.

In filament winding, it is usually the curing system that is already in place. The resin system must be chosen to fit those limits.

6A.4.5.1 Ovens

Gas-fired or electric air-circulating ovens are the predominant curing mode. They are inexpensive, and they can be very large. Any supplemental curing pressure must be applied with shrink tape or vacuum bag. In many cases, for pipe and other circular forms, the part is rotated during the cure to minimize sag and drippage. Energy costs associated with oven cure are higher than with many other methods because the heated mass consists of the part, the surrounding air, and all associated hardware, including mandrel and support stand. Also, ovens are not desirable from a facilities standpoint because they take up considerable floor space. This is particularly objectionable in high bay manufacturing areas, where much filament winding takes place.

In addition to the standard batch process ovens used for curing composites, continuous high production volume automated systems use a belt carriage with sufficient serpentine path to provide as much curing time in the high temperature zone as is required for the resin system.

6A.4.5.2 Hot Oil

The hot oil system is typically used with a very fast curing resin system, normally with the ability to cure in less than 15 minutes. The use of hot oil ensures a very rapid heat-up of the mandrel and eliminates the need for the curing oven. Passages are typically provided throughout the mandrel, and this permits the mandrel to be heated early in the process. This has a distinct advantage in relation to mandrel removal. Composites frequently shrink during the cure process and in doing so, attach themselves firmly to their winding mandrel, which ultimately heats up and expands, then contracts upon cool-down, with the part still firmly attached. In the hot oil system, the mandrel heats up first. It expands to its hot condition, at which point the composite heats up and cures, and contracts at the expanded condition. Then when the hot oil flow ceases and the mandrel is cooled down, it shrinks away from the cured composite and permits relatively easy removal from the structure. Hot oil systems are typically used in the range of 150–204 °C (300–400 °F).

6A.4.5.3 Lamps

Heat lamps, used in conjunction with reflective surfaces and a rotating mandrel, can also provide cure temperatures on the order of 171 °C (350 °F). Heat lamps are often thought of as being portable or backup units. Because the source is highly directed, care must be taken to provide curing of all sections of the part.

Infrared lamps are frequently used as a means of providing heat to curing resin systems in a composite structure. However, these lamps are rarely used for the full production process. They do, however, achieve considerable use in the advancing of the resin system from the A to the B stage. This is normally done using banks of lamps, shaped to conform to the structure receiving the heat energy. The structure frequently rotates while heating takes place. The ultimate change of the resin system to the B stage, in effect, freezes the resin into the structure, but still leaves the capability for full flow and cross-linking in the final curing process. B-Staging prevents flow and drainage of the resin system in a cylindrical structure and also makes the part easy to handle, since the B-staging produces a tack-free external surface condition.

Another lamp curing method utilizes a capacitive discharge, pulsed xenon lamp. To promote the curing process using this type of lamp, a light-sensitive catalyst is used in the resin system. The major problem with this lamp is the tremendous amount of noise produced as the capacitor discharges. A continuous illumination, high intensity lamp known as the Macbeth lamp is also used with light-sensitive catalysts.

One disadvantage of the light curing system is that a skin frequently forms over the outside of the resin, and this skin precludes further heat transfer into the system under cure.

6A.4.5.4 Steam

A number of pipe manufacturers use hot steam for resin curing. The metal mandrel ends have adaptors to allow the passage of steam and water. After the tube has been wound, hot steam is circulated through the hollow mandrel. When cure is complete, a cold water flush cools the mandrel for handling and provides enough shrinkage to facilitate mandrel extraction.

6A.4.5.5 Autoclave

When aerospace quality laminates are needed or more sophisticated epoxy, bismaleimide, or polyimide resins are used, it may be necessary to cure the filament-wound parts in an autoclave with vacuum assist. Though not normally associated with mass production techniques, autoclave curing can provide pressures of 1.4–2.1 MPa (200–300 psi) with vacuum augmentation at temperatures as high as 371 °C (700 °F). The principal disadvantage of autoclaves is the long cycle time, coupled with limited size and availability.

The autoclave is the primary curing means for aircraft components. In the production of missile structure, some parts are wound from preimpregnated tows, which, depending on the resin system, may require the autoclave process to thoroughly cure and eliminate voids. Condensation-type polyimides require the vacuum bag to capture off gassing products. Because pressure is applied to the outside uncured surface of the composite, the outer fibers are compacted, and some initial tension is released. Care must be taken to avoid folds and other material motions before set.

6A.4.5.6 Microwave

It has been shown in several development programs that microwave curing can be a significant advantage with fiberglass and aramid fiber composites. Microwave energy is absorbed rapidly by both resin and fiber, and results have shown that cures can be effected in minutes for systems that previously took many hours. The energy level required for this type of system is high; hence the process is costly. A major drawback, however, is the inability to use microwave curing methods with conductive fiber such as carbon. For this reason, the majority of composites fabricators have abandoned consideration of microwave energy as a means of curing composites. Hercules is the only major fabricator to use this technique as an assist in temperature ramp-up.

6A.4.5.7 Other Curing Methods

Electron beam, laser, radio frequency (RF) energy, ultrasonics, and induction curing methods have all been studied with filament winding systems, with varying degrees of success. Ultrasonic energy has proved to be unreliable because the energy is imparted to the resin system in a very rapid and nonuniform manner. This has resulted in uncontrolled exotherms. nonuniform curing, and actual burning and charring of the composite structure. The Grumman Corporation has devoted significant research and development efforts to the RF curing method and has demonstrated a preliminary feasibility.

Laser methods and electron beam heating have been studied and abandoned, primarily because they do not work with carbon fiber systems. Both these forms of energy are unable to penetrate the outer layer of carbon fiber in a composite structure.

Laser-directed energy, however, may be an advantage in the consolidation process for thermoplastics. ICI has demonstrated the use of a laser beam for melting the zone surrounding the laydown point of prepreg fiber in a filament-wound polyetheretherketone (PEEK) system. This method should be applicable to all forms of thermoplastic matrices for filament winding. Since only the top layer is being melted, the inability of the laser beam to penetrate through the carbon fiber layer is not important.

Because carbon fiber is conductive, induction heating may have the most promising future for use with this fiber system in a variety of matrix resins. Induction heating is a process that has been used for more than 50 years in a variety of production processes in various industries. However, its application to the curing of resin systems has not been studied until recently. Aerojet has performed several demonstrations which suggest that 2—3 hour cure times can be reduced to only a few minutes.

6A.5 SELECT APPLICATION AREAS

6A.5.1 Aerospace/Defense

The primary aerospace applications for the filament winding process are motor cases, launch tubes, pressure vessels, and survivable fuel tanks. Filament-wound composite motor cases have been used in the United States since the middle 1960s, beginning with such programs as Minuteman, Polaris, and Poseidon. All stages of both Polaris and Poseidon were filament-wound using S901 fiberglass, as was the third stage of the Minuteman II and III. Following these early successes, the Sprint anti–ICBM missile motor case was filament-wound, also from fiberglass.

In the early 1970s, the Navy continued its emphasis on composite motor cases by developing a filament-wound C4 missile for use on the Trident I submarine. Kevlar 49 was the material of choice. The largest ICBM, developed in 1975–1980, was originally known at the MX missile. Now known as the Peacekeeper, this 2.4 m (92 in.) diameter motor case is filament wound using Kevlar. The Pershing field artillery weapon, developed in the late 1970s for the army, is also filament-wound, using Kevlar 49.

The first use of carbon fiber in missile motor cases did not occur until 1980. The LoAD, later known as the Sentry missile, was an intercept system developed for the Army. It used a hybrid of Kevlar and carbon fiber in its high strength, tapered case. This missile was canceled under terms of the SALT Agreement and never went into production.

In 1982 the small ICBM or Midgetman system became the first to use an all-carbon-fiber, filament-wound motor case. The change to carbon fiber was made feasible by the development of a high strength, high elongation, intermediate modulus carbon fiber.

U.S. Army and Marine Corps ground-to-ground and ground-to-air missiles have, with very few exceptions, utilized filament-wound composite launch tubes. The Redeye, Dragon, and Tow systems used launch tubes that were filament wound using E-glass in the late 1960s and 1970s. The follow-on to the Redeye program became the Stinger ground-to-air missile, which was initially developed in the late 1970s with Kevlar and is still in production.

The Viper antitank weapon was developed in the early 1980s and used a small filament-wound motor case for its propulsion system. It also had a filament-wound E-glass two-piece telescoping launch tube. This system became the first man-rated, shoulder-fired, antitank weapon system, and it was to be manufactured in a highly automated production facility. The program was canceled before going into production in favor of the AT4 antitank system, which is being fabricated for the Army. The only portion of the AT4 that is filament wound, however, is the launch tube, which has a diameter of 203 mm (8 in.).

The multiple launch rocket system (MLRS) is a ground-to-ground, field artillery missile system in use by the U.S. Army. The motor case for the missile is metallic, but the 279 mm (11 in.) diameter launch tube is filament wound using E-glass. After a successful development program, the system went into full production in 1983. The system is unique in that it uses helical ribs, cocured to the internal surface of the launch tube, to impart a stabilizing spin to the missile. These ribs increase the complexity of the mandrel and make the tube difficult to remove.

The aerospace pressure vessel industry uses sophisticated filament winding techniques to produce optimum design, lightweight pressure vessels for containing gases at pressures of 21–42 MPa (3000–6000 psi). These vessels can be either cylindrical or spherical and are used for flotation devices, pneumatic seat ejection mechanisms for aircraft, pressurization systems for pressure-fed liquid propellant rocket engines, and environmental and breathing systems for space vehicles and spacecraft. All these pressure vessels utilize thin metallic liners that serve both as leakage barriers to contain the high pressure gases and as winding mandrels for the filament wrapping process. The majority of the liners are aluminum, but titanium and Inconel have also been used. The overwrapping materials have been predominantly fiberglass and Kevlar because of their high strength and elongation, damage and abrasion resistance, and fatigue endurance.

The most notable use of filament-overwrapped pressure vessels is in the space shuttle vehicle. Seventeen pressure vessels are used throughout the space shuttle in its environmental, pneumatic actuation, and pressurization systems. These vessels are all spherical in design and use a Kevlar overwrap as the reinforcing fiber. There are three basic diameters: 0.45, 0.66, and 1.0 m (18, 26, and 40 in.). The majority of the liners for these pressure vessels are made from Inconel 618. The oxygen storage unit uses Ti-6Al-4V. The fatigue cycling capabilities of these spheres make them ideal for this application, since the high pressure gas storage systems of the space shuttle vehicle are constantly used and refilled.

The major fire incidents that occurred on the Navy aircraft carriers *Forrestal* in 1968 and *Enterprise* in 1974 emphasized the dangers of having flammable fuels stored in uninsulated and unprotected tanks under combat and combat-ready conditions. New requirements were written for survivability coverings for jet fuel tanks. The Navy emphasis in this area has been for the carrier-borne F-18 fighter, and for the army in CH53 and CH47 helicopters. The survivability layers that are added to the tank consist of fire-retardant foam contained within a honeycomb structure, carbon fiber cloth, and a filament-wound hybrid layer of fiberglass and carbon fiber.

6A.5.2 Industrial

The commercial and industrial markets for filament-wound products are varied. Some of the product lines include commercial pressure vessels, high pressure piping for oil field use, and corrosion-proof underground petroleum storage tanks.

The majority of commercial pressure vessels fall into the general classification of breathing apparatus, which includes backpack vessels for firemen, medical oxygen packs for both hospital and home use, mountain climbing and spelunker backpack units, and scuba diving tanks.

Filament-wound underground fluid storage tanks have been available since the late 1960s. The major advantage of these tanks over their metal counterparts is that they do not corrode and leak the contents to the surrounding environment. In recent years, state governments (notably California) have instituted regulations to prevent the pollution of groundwater. This has caused a renewal of interest in storing underground fluids in essentially leak-proof tankage. Initially, the tanks that were designed for this market were of single-wall, fiberglass, filament-wound construction. Now, double-walled tanks are being fabricated. These tanks have a ribbed construction between the two walls to maintain a separation distance, and the between-wall cavity is used to sense any leakage that may take place, either outward or inward.

Filament-wound high pressure pipe is used extensively in oil field installations, both for surface transport and down-hole injection. Though most pipe is rated between 2.7 and 13.8 MPa (400–2000 psi), there are products available that have continuous-use pressure capabilities at 41 MPa (6000 psi). The joints (elbows, tees, etc.) are also filament wound. All pipe uses E-glass as the reinforcing fiber, primarily because of cost. There has been some interest, as evidenced by prototype tests, in carbon fiber reinforced pipe, particularly for containment of extremely corrosive fluids where the only alternatives would be precious metal pipes such as zirconium.

6A.5.3 Light Truck Driveshaft

In the mid 1970s, interest developed in a one-piece composite driveshaft to replace a two-piece steel shaft used in light trucks. The steel configuration with the two short sections had a sufficiently high natural rotational frequency to detune the drivetrain vibrations. Analysis showed that a composite material, if sufficiently stiff and light, could do the same with a single-piece construction. The challenge was to develop a material combination and fabrication method that would be cost competitive with the steel assembly. Limited production runs of a filament-wound composite driveshaft occurred during the mid-1980s, and the shafts were used on Ford Econoline and Astrostar models.

Requirements included a shaft yoke-to-yoke centerline distance of about 1.9 m (75 in.) with a 101 mm (4 in.) diameter. Also, it must support a torque load in excess of 3400 N·m (30,000 in.·lb) and have natural rotational frequency in excess of 6500 rpm. Maximum operational temperature was between 99 and 121 °C (210–250 °F).

Two distinctly different designs developed. The first, made by Hercules, combined a near-axial wrap of carbon fiber with a high angle ($\pm$ 45 °) wrap of glass in vinyl ester resin. Steel end sleeves were wound in place and adhesively bonded and mechanically pinned into the composite shafts, and steel yokes were welded to the sleeves. The second shaft, made by Ciba Geigy, combined the two fibers into a single wind angle of about 10–20 °. The resin was epoxy. Steel end sleeves were first welded

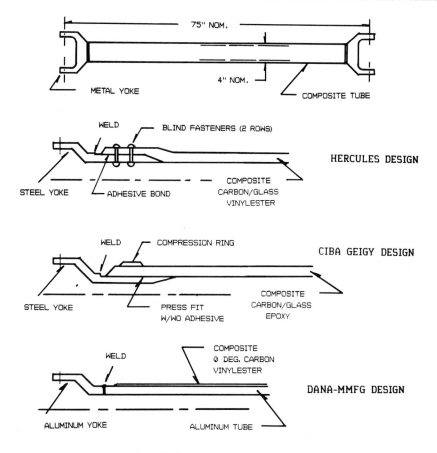

FIGURE 6A.13   Composite driveshaft concepts.

to the yokes and then attached to the composite shaft using an outer steel compression ring. In both cases, filament winding was the manufacturing mode. A third design, currently being used for a General Motors van, though utilizing hoop filament winding, is basically a pultrusion of $0°$ carbon fibers with a hoop-wound glass cinching ply over a permanent aluminum core. Aluminum yokes are beam welded to the aluminum core. This shaft is manufactured by Morrison Molded Fiber Glass Division of Shell (MMFG) and Dana. All three versions are shown schematically in Figure 6A.13.

The different layering configurations were the result of nonunique optimization studies. Stiffness and torque requirements can be satisfied with discrete combinations of $0°$ carbon (for stiffness) and $±45°$ glass (for torque) or mixing the two fibers at an intermediate angle, about $15°$, for both stiffness and torque. Optimization was directed at minimizing the weight percent of the more costly carbon fiber. In both cases, prototype shafts were first built using conventional filament winding machines. The mass-produced shafts were built with specially designed, dedicated winding machines.

The Hercules shaft used about 1.8 kg (4 lb) of E-glass fiber in an inner $\pm$ 45° layer, followed by 0.45 kg (1 lb) of conventional 234 GPa (34 Msi) carbon fiber at 0°. The total composite shaft (without metal ends) weighed approximately 3.6 kg (8 lb) and had a thickness of about 5 mm (0.2 in.). Fiber content was about 70 % by weight.

The dedicated winding machine was built by Goldsworthy Engineering and appears to follow a fabrication concept developed for continuous pipe winding. Long, straight steel mandrels pass sequentially through a series of staggered rings on which spools of carbon or glass rotate and deposit resin-wetted fiber onto the mandrel. Sets of end sleeves, treated with an adhesive, were located on the bare mandrel in the preparation stage. Winding was continuous over the mandrel such that the sleeves were wound in place. Layer thickness was determined by the number of spools and ring stations, and fiber angle by the relative mandrel translation and spool rotations. The mandrels were rotated during oven cure, and the outer surface was unfinished, leaving a resin-rippled appearance. Using a puller, the mandrel was extracted from the cured tube, which was then cut into individual shafts. The end sleeves were bonded in place, and the adhesive bond was sufficient to take the torque load. However, as a precaution, at each end, 16 pop rivets, located on two staggered bolt circles, were fastened through holes drilled in the composite tube and metal sleeve.

The Ciba Geigy shafts also utilized a special filament winding approach to deposit approximately the same amount of carbon and glass fiber on the same diameter mandrel. The winding consisted of a stationary creel with a 360° fiber delivery ring and a conventional headstock–tailstock configuration for rotating the mandrel. Carbon and glass ends were distributed around the delivery ring, and as the mandrel rotated, the delivery ring translated from one end to the other and back again, placing first a $+\beta$ degree ply and then a $-\beta$ ply on the mandrel. Repeating the process developed the proper thickness. Some of the shafts were cured in this configuration, while others were hoop overwrapped with a narrow fabric or tape of polyester, nylon, or glass. The tape provided additional compaction and gave the appearance of a narrow spiral down the length of the tube.

The end attachments of the Ciba tube were considerably different. The attachment consisted of a tapered inner steel sleeve, to which the yoke was welded, and an outer steel sleeve. Assembly was by press fitting the inner and outer sleeves over the composite tube (i.d. and/or o.d. machining may have been required). Though Ciba claimed that the frictional effect under the compressive load was sufficient to absorb all torque, it is believed that an adhesive was used in some of the production shafts to increase load-bearing capability and also to provide lubrication during the press fit.

6 · Continuous Fiber Molding Processes
B. Pultrusion

Clint Smith
Jerry Stone

Contents

© Morrison Molded Fiber Glass Company

Clint Smith and Jerry Stone, Morrison Molded Fiber Glass Company, Bristol, Virginia, USA

6B.1 INTRODUCTION

Pultrusion is a continuous manufacturing process used to produce high fiber content reinforced plastic structural shapes. Pultrusion is distinct from filament winding in that filament winding places the primary reinforcement in the circumferential (hoop) direction while pultrusion has the primary reinforcement in the longitudinal direction. Accordingly, while good mechanical properties can be achieved in pultrusion in the transverse (crosswise) direction by using special reinforcements, the primary strength will occur in the longitudinal direction. In some specialized pultrusion applications, such as oil field sucker rods, all the reinforcement is in the longitudinal direction.

Since the first patents in 1951, pultrusion has developed into a widely used method of manufacturing straight sections of fiber-reinforced plastic shapes having constant cross sections. It is possible to produce curved sections by using an auxiliary molding process with B-staged pultruded parts. It is estimated that there are more than 650 pultrusion machines worldwide producing $1016 \times 10^5 \, kg$ (100,000 tons) annually.

Pultrusion was initially used to produce simple, small diameter solid rods for electrical applications such as transformer spacer sticks and for consumer applications such as fishing rods. In 1956, Universal Molded Products Corporation, which is now Morrison Molded Fiber Glass (MMFG), began producing structural shapes and became the early leader in pultruded product development. The advent of structural shapes in reinforced plastic was a milestone in pultrusion history and led to the myriad of applications for pultruded products seen today. By the early 1970s pultrusion began to appear in various codes and standards. The American National Standards Institute (ANSI) developed standards for pultruded side rails for fiberglass step and extension ladders. The American Society for Testing and Materials (ASTM) established a special section (ASTM Section D 20-18.02) to develop specifications particularly related to the pultrusion industry.

6B.2 PROCESS DESCRIPTION

In the pultrusion process fiber reinforcements that have been impregnated with a solution containing resin and other additives are pulled through a heated die. This is to be compared with the extrusion process in which heated material is pushed through a die. The reinforcements entering the die are generally saturated with a liquid resin solution but are solid when exiting the die. The pultrusion process can produce solid, open-sided, and hollow shapes, which can be cut to length and packaged for shipment at the machine.

The pultrusion process is continuous, manufacturing 1.5–60 m/h depending on the shape. The machine may be utilized 24 hours a day, 7 days a week with the only scheduled stoppage required to perform routine cleaning, perhaps once every 2 weeks. The operation of a pultrusion machine places severe constraints on the raw material and resin mixture combinations that can be used, as discussed in detail later. The continuous nature of the process also creates opportunities and constraints for the

213

quality control system to be utilized in a pultrusion operation. This will also be discussed later.

A pultrusion machine can be conceptually divided into five basic functional areas for its operation with each area performing a special task. The machine schematic as shown in Figure 6B.1 represents a particular type of machine that is widely used in the pultrusion industry. The description of the pultrusion process will be centered around this particular type of machine which, except for the pulling mechanism and possible additional heat initiation area, is basically the same as other machine designs. The five basic functional areas are listed and described next.

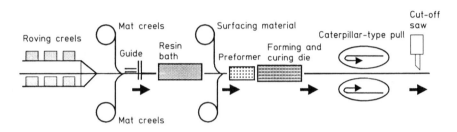

FIGURE 6B.1   Pultrusion machine.

6B.2.1 Fiber Reinforcement Feed and Placement Area

The correct position of the reinforcements within a shape is determined in the design of the composite by production engineers. Pultrusion yields high strength laminates, but the maximum strength for the shape is attained only when all reinforcements are properly placed in the composite. The process tooling must be engineered to guide the reinforcements into the designed positions within the shape.

1. Longitudinal reinforcement racks (sometimes called roving creels) are the storage areas for the longitudinal reinforcements. Longitudinal reinforcements generally are packaged in a cylindrical form with the reinforcement exiting from either the center or the outside circumference of the package. These packages are sometimes known as doffs when the reinforcements are fiberglass. These racks have either shelves or horizontal rods (axles).

2. Mat racks (sometimes called mat creels) are the storage areas generally for reinforcements used for transverse properties in pultrusions. These reinforcements in sheet form (mats, stitched fabrics, knitted fabrics) are packaged on a core to be placed on a horizontal rod. This allows the reinforcement package to rotate just as the "outside pull" longitudinal reinforcement rotates.

3. Guide plates are utilized to position the longitudinal and transverse reinforcements in their designed positions in the shape being pultruded. Different reinforcement forms can require different guide plate designs, to allow the use of several different concepts.

6B.2.2 Fiber Reinforcement Impregnation Area

Saturation of the reinforcements with the resin mixture is as important as correct fiber placement. Failure to achieve the full impregnation of the reinforcements can produce a pultruded shape with lower mechanical properties than intended.

Most pultrusion processes operate by passing the reinforcements through a solution containing the polymer and other additives such as filler, UV inhibitor and catalyst for proper impregnation. A good pultrusion practice is to incorporate into the resin bath various structures, sometimes known as breaker bars, to force the reinforcement materials to change directions while passing through the resin bath. These changes of direction serve to slightly spread the longitudinal reinforcements, offering more access to the interior filaments of the bundles for proper impregnation by the resin mix solution. Resin baths are occasionally heated to provide a better impregnation of the reinforcements by the resin matrix, but this heating may present problems with certain low temperature catalysts.

Some pultruders use other means of impregnating reinforcements, such as using resin mix injectors as a closed resin bath or moving the fiber impregnating area to the front of the die and injecting the resin directly into the die.

After the fiber has exited the resin bath, a layer of surfacing material is often added on the top and bottom surfaces of the impregnated reinforcements. Adding surfacing material is generally considered a good pultrusion practice, the reason for this is discussed later.

6B.2.3 Preforming Area

Preformers are guides to gently and gradually bend the impregnated reinforcements to form the shape being pultruded. They vary from simple guides at the die entrance to complex guides that are 0.6–1.2 m (2–4 ft) long. Some reasons for using preformers are: (*a*) the impregnated reinforcements exiting the resin bath are essentially flat, (*b*) bending them into the proper shape configuration will prevent excess stress from occurring during the cure cycle, and (*c*) by removing excess resin from the reinforcements, one ensures that the maximum fiber volume within the shape is pultruded.

6B.2.4 Curing Area

One of the more challenging steps in the pultrusion process is the continuous polymerization process that takes place in the die. The shape is cured within the die which typically can range in length from 30 to 155 cm (12–60 in.): these lengths are not absolute constraints. Dies may be electrically heated or heated with hot oil. Pultruders will incorporate one to four heating zones in the die. The choice of the number of heating zones is dictated by several factors, such as the type of resin being utilized, the desired line speed, and the length of the die. Many very successful commercial manufacturers of pultruded products use only one zone, while other manufacturers use multiple zones. There does not appear to be a rule of thumb for the choice of the number of heating zones.

Sometimes overlooked in good pultrusion practices is the need for consistent location of the controlling thermocouple. This probe must be consistently maintained relative to the location of the heat source, whether the heat source is electrical or hot oil. Many pultruders find that controlling the peak temperature within the die is extremely desirable; however, to control the peak exotherm temperature implies a consistent process setup with consistent raw materials and consistent mixing of the resin solution.

6B.2.5 Pulling Area

The pultrusion process takes its name from the mechanical pulling action utilized to engage the process. Extensive engineering has gone into the various types of pullers, to make them dependable and to keep maintenance costs minimal. The speed must be adjustable because different resins, catalysts, pigments, and shapes require different pulling speeds.

Two basic types of pullers are being used in the pultrusion industry: the reciprocating pulling system (Fig. 6B.2) and the caterpiller system (Fig. 6B.3). Both pulling systems are widely used.

The reciprocating pulling system has two pull blocks that alternately move forward (in the pultrusion direction) and backward. These puller blocks have mechanically operated clamps that open and close to maintain a steady pulling force on the part. When pull block 1 moves forward with the clamp closed, it is pulling the part. While pull block 1 is pulling, the clamp on pull block 2 is open and the block returns to its starting position by travelling backwards. As pull block 1 completes its forward travel, the clamp on pull block 2 closes, the clamp on pull block 1 opens, and pull block 2 is pulling the part. As pull block 2 pulls the part, pull block 1 returns to its starting position, completing the cycle, and is ready to start the next cycle, which begins when pull block 2 completes its forward travel.

The caterpiller-type pulling system has a set of lower tracks and a set of upper tracks. A set of chains having a clamping system for attaching the pull blocks travels on each of these tracks. The space between the upper and lower tracks is variable, permitting the correct pressure to be transferred to the part being pulled through the machine. The caterpiller-type pulling system remains stationary relative to the part movement with potentially less mechanical problems.

Both pulling systems have a means to vary the pulling speed and clamping pressure; however, the caterpiller system typically has a higher clamping force, which could cause problems on complex, thin-walled shapes.

All pultrusion machines have some mechanism for cutting the part to length after is has passed through the pull blocks. In some operations the part is simply cut with a hand-held circular saw; in more complicated saw systems the part is maintained at the proper length during the manufacturing operation. A diamond-tipped blade is recommended for the saw to prevent premature blade failure.

The "block diagram" used in this chapter will be essentially the same for all pultrusion machines. However, at first glance there might appear to be substantial changes due to the nature of the part being pulled. For instance, to pull a tubular shape

216

FIGURE 6B.2   Reciprocating pulling system.

FIGURE 6B.3   Caterpillar pulling system.

it is necessary to form the hole in the center by using a mandrel that is inserted from some position behind the die to the estimated cure point within the die. The mandrel should be as short as possible because it is attached at only one point behind the die and the cantilever loading may cause symmetry problems in the subsequent wall thickness. Terminating the length of the mandrel at the cure point of the die will prevent excess drag of the molded part against the mandrel. The pull force required will increase with the additional surface area imparted by the inside of a hollow section.

The next noticeable change in the pultrusion process from the standard block diagram approach would occur if the manufacturer wishes to install a core in the center of the part during the pultrusion process. Some cores can easily be added in

217

the center, replacing the mandrel. Materials that may be used as cores include thermoplastics, wood, structural foam, and metal. The cores can add rigidity to a thin-walled plastic structure and can also impart an increased thermal barrier. The temperatures within the die as the part cures can easily reach 135–190 °C (300–400 °F), which places limits on the types of core material that can be used. A core with a low melting point is obviously unsatisfactory for the pultrusion application.

Not mentioned above is the method of transferring the premixed resin solution to the resin bath. Many different techniques have been used, including operations as simple as ladling the material from a mix tank to the resin bath to a sophisticated automated pumping system. The temperature of the resin mix exiting the pumping system is an important processing parameter. In some situations it may be desirable to heat the resin mixture to improve wetout; in other situations it may be better to maintain the resin mixture at ambient temperature or lower because the mixture contains highly heat-sensitive catalysts.

A pultrusion machine used for all glass longitudinally reinforced shapes often has a radiofrequency (RF) heater inserted between the resin bath and the die. This unit serves to preheat the resin to initiate inside/out curing of the composite. An RF unit cannot be used in the pultrusion of shapes having graphite or other conductive reinforcements because the microwave energy will flow through the reinforcements and discharge into the guidance system or other metallic parts of the pultrusion machine, causing a short circuit and shutting down the RF heater.

6B.3 RAW MATERIALS

As with any process that uses fiberglass/filament reinforcement, the nature of the process forces constraints on the properties of the raw materials. Conceptually, different reinforcement schemes, whether the part is composed of all longitudinal reinforcement or of a combination of longitudinal and continuous strand mat, mean that different viscosities will be required from the resin mixture solution. This in turn requires good consistency of the viscosity of the incoming resin from the resin suppliers. The nature of the pultrusion process is to enter one end of the die as a reinforcement saturated with resin mix and to exit from the other end as a solid part in a short period of time. This places severe requirements for the cure process, which in turn constrains the choice of resin and catalyst raw materials.

6B.3.1 Resins

Some prior knowledge of the basic resin differences is assumed in this chapter. If questions arise on the generic differences among the resins, the reader should contact the resin suppliers for more detailed information on properties. This section includes only thermoset resin matrices. There is considerable interest in the pultrusion industry today in the possible processing of thermoplastics, but this is not currently a common commercially available product. The traditional thermosets, which have been more completely developed in pultrusion, are polyesters, vinyl esters, and epoxies, with the pultrusion of phenolic resin matrix potentially available in the future.

While many properties of the resin could be tested in accordance with SPI procedures, the viscosity and the cure have the primary interest for pultruders. As the reinforcement combinations are altered, the viscosity of the resin mix solution must also be altered to accommodate the reinforcement scheme. All longitudinal roving reinforcement will require a lower resin mix viscosity than a part composed of continuous strand mat and roving. The resin mixture must be thin enough to penetrate within the strands of the roving, and this will require control of the neat resin viscosity plus the filler particle size and filler level within the resin mix solution. The mat/roving composite will require a higher filler loading to prevent the resin mix from bleeding through the mat.

The resin cure properties are extremely important in developing a pultrusion system. The cure of the resin mix will be a function of both the cure properties of the neat resin plus the catalyst system employed. Many pultruders utilize the 180° SPI gel time as an indication of whether the resin would be suitable for pultrusion. More recently, some pultruders have been utilizing more sophisticated thermal analysis techniques such as the differential scanning calorimeter (DSC) to analyze the cure properties of pultrusion resin. The use of the DSC has been reported at an SPI conference and is also being studied by the ASTM subcommittee on pultrusion. For more detailed information on the possible use of DSC cure analysis techniques, consult the chapter references.

The choice of the appropriate viscosity and cure properties for the pultrusion resin is very much a function of the individual pultruder. Some pultruders favor lower viscosity resins and add more filler to obtain the proper resin mix viscosity, while others use a higher "neat" resin viscosity and less filler ("neat" is synonymous with resin only, no additives). Recently there has been some discussion concerning the maximum styrene levels permissible in production environments, and this may ultimately limit the amount of styrene that can be utilized with the resin. A shift toward higher resin mix viscosities probably can be anticipated. Similarly, the choice of cure properties is very much a function of the individual pultruder.

Pultruders utilizing minimal levels of lower temperature catalysts may require a shorter SPI gel time (or lower DSC onset temperature) from the neat thermoset polyester or vinyl ester resin. The resin reactivity required is a function of the thickness of a part being considered: thicker parts require a lower reactivity resin or a change in the catalyst system to prevent internal cracking. Resin matrices with higher percent tensile elongations tend to be more processible in thicker sections, although some high temperature property retention may be sacrificed.

6B.3.2 Reinforcements

Various grades of reinforcements are available for pultrusion processing. The most common reinforcement is glass roving, which is a glass bundle composed of 1000 or more individual filaments brought together in a single continuous strand and wound on a cylindrical package. This product is described in units of yield: "113 yield" means that there are 113 yards of the roving in 1 pound. This form of reinforcement is made from E-glass, and all roving suppliers use essentially the same chemistry for their glass.

The difference in rovings involves the binder, which is a substance attached to the roving to aid internal processing at the roving supplier's plant and to permit a better adhesion to the resin matrix. When binders are not compatible with the resin matrix being considered, a frequent result is poor mechanical properties, blisters, and/or generally poor wetout. Roving that processes well in one resin matrix may not process well in another, and all matrices to be used should be tested independently. Graphite is packaged as an "outside pull" product and is denoted by the number of filaments in the bundle; 12 k graphite implies 12,000 filaments.

Other roving issues include the formation of fuzz during processing. The fuzz represents real damage to the individual filaments of the roving package and may impart some additional weakness to the system. Binder spots (i.e., areas where the binder either has been heavily applied to the roving or has partially cured on the roving) constitute another potential problem associated with roving. Binder spots may represent areas of localized weakness, which may not affect static applications to the same degree as dynamic applications. Most pultruders today use what is known as single-end roving, where the filament bundle is held together. Early in the history of pultrusion, multiend roving was utilized, but because of processing problems this format is now in limited use.

Another form of reinforcement for pultrusion processing is continuous strand mat. Continuous strand mat appears as a sheet of glass in which the primary reinforcement is located in the transverse direction. This product is commonly supplied in units of ounces per square foot. The continuous strand mat must have enough longitudinal strength to be pulled through the resin bath and die. Continuous strand may be purchased either as A-glass or E-glass; only a few companies make any type of continuous strand mat.

The difference in chemistry between the A-glass and the E-glass is that the A-glass contains more sodium in the glass chemistry. The A-glass mat has finer filaments than the E-glass mat and produces a more attractive surface on the pultruded part. The E-glass mat currently produces better transverse (or crosswise) mechanical properties, but developments are being pursued among the A-glass mat suppliers to eliminate this discrepancy. There is some dispute on whether the A-glass and E-glass mats produce equivalent mechanical and electrical wet properties. This is a difficult topic to discuss in general because the wet properties are very dependent on the localized processing conditions of the individual pultruder. Among other parameters, wet properties are a function of reinforcement wetout, composite packing densities, shrink cracking, and part cracking due to abuse.

Another form of reinforcement consists of woven or stitched longitudinal reinforcement (both graphite and glass) in which the rovings are placed at specified orientations. Rovings that have been stitched parallel to the line of pull are called 0° rovings; 90° rovings refer to product that has been stitched perpendicular to the line of pull; and ± 45° rovings refer to product that makes 45° angles with the line of pull. Considerable variation in the stitching quality has been observed, with distortions from the desired angle and unwanted spacings between the reinforcement quite common. The pultrusion process itself may also impart some distortion; thus theoretical calculations based on the absolute orientations of these products must be made cautiously.

6B.3.3 Catalysts

The catalyst (more properly known as the initiator) is heat sensitive and initiates the cross-linking reaction in thermoset polyesters and vinyl esters. While one-component catalyst systems are often used in pultrusion, many pultruders have found it more efficient to utilize a cascading catalyst system consisting of a low temperature catalyst and a high temperature catalyst. In some instances a middle catalyst is also utilized. Peroxydicarbonate has gained a wide acceptance in pultrusion as a low temperature catalyst, while *t*-butyl perbenzoate is generally accepted as a high temperature catalyst. There are several different middle catalysts utilized in pultrusion, the only approximate rule of thumb is a 70–80 °C 10 hour half-life for the middle catalyst. As with other forms of the thermoset processing of fiberglass/filament reinforced plastics, different raw materials, such as pigments, may influence the choice of catalysts.

It should be noted that chemically the cure properties of the system will be a function of the catalyst, the inherent resin reactivity and the inhibitors present in the resin system. A secondary effect may be noticed with the use of high filler loadings or certain conducting reinforcements.

Epoxy resin matrices cure differently from thermoset polyesters and vinyl esters with amine or anhydride hardeners as the primary curing agents. The choice of hardener will dictate the high temperature properties, corrosion performance, and pot life of the resin mixture. Amine hardeners offer the superior higher temperature properties but are more difficult to process.

6B.3.4 Release Agents

As with other forms of plastic processing, the release agent used in pultrusion is a critical component of the resin mixture formulation. Stearates are one type of release agent. The generally accepted theory for the action of the internal release is that the release agent separates from the resin mix and coats the die surface, thereby preventing adhesion of the polymer to the surface. Chroming dies will permit the pultruder to reduce the amount of release agent required.

Release agents are not inert members of the resin mix formulation. In combination with the resin, filler, and catalyst system:

1. Some release agents may thicken the resin mix.
2. Some release agents will yield a higher gloss on the surface.
3. Some release agents may cause the part not to be paintable.
4. Some release agents may act as a corrosive agent onto the die surfce.
5. Some release agents may cause a discoloration in the part.

The pultruder must carefully research the release agent concept and determine which, if any, of these effects might be acceptable.

6B.3.5 Flame Retardants

Some resin systems contain a flame retardant inherently within the backbone of the resin. In another approach, the flame retardant is added to the resin mixture in a manner similar to the other ingredients (e.g., catalyst and filler). There are brominated and chlorinated additives; adjusting the resin mix will reduce the neat resin inventory.

Flame retardants should be understood as *retarding* but not completely preventing a fire.

6B.3.6 Fillers

Fillers are used to impart certain properties to the composite, one of which is reduced cost of the final profile. Aluminum trihydrate is traditionally used for flame retardants and arc tracking; clay often improves the surface finish; and, calcium carbonate is used because of its low cost. As with the release agents, different fillers may have certain negative side effects. For example, clay will produce an off-white part, and calcium carbonate is not recommended in corrosive applications.

6B.3.7 UV Inhibitors

UV inhibitors are substances that absorb ultraviolet radiation from the sun and reemit this radiation, often in the form of mild heat, preventing degradation. UV inhibitors are basically retardants rather than preventers of the UV radiation side effect.

6B.3.8 Surfacing Veils

Surfacing veils are synthetic fabrics, which are added to the exterior surfaces of the pultruded profile. The surfacing veils add corrosion protection, act as a slip plane protecting the die surface, prevent fiber blooming when the composite is exposed to UV radiation, and improve the surface quality. A well-made pultruded profile will always contain some form of surfacing veils. Thicker surfacing veils are often utilized in applications requiring excellent corrosion protection. It should be noted that different surfacing veils impart different surface appearances to pultruded composites.

6B.3.9 Pigments

As with release agents, pigments are not inert substances in the pultruded composites. Carbon black often affects the cure properties of the system; titanium dioxide may have some impact on the mechanical properties and the pigment often rapidly changes color when exposed to UV. Styrene, the monomer in most thermosets and vinyl esters, often discolors to a yellowish hue when exposed to UV radiation, and this can change the color of the pultruded composite. Carbon black and titanium dioxide white are often utilized when improved UV protection is required.

6B.3.10 Other Additives

Other additives are certainly possible in pultrusion. Wetting agents have been utilized for defoaming, foam cores have been added to the pultruded composite during the actual pultrusion process, and special additives are available to improve surface quality. The individual pultruder will undoubtedly catalogue a number of additives that appear to help a particular formulation and product.

6B.4 MECHANICAL PROPERTIES

Pultruded composites are often preferred over other forms of plastic processing because of the higher mechanical properties available from the pultrusion process. Reviewing individual suppliers' data sheets is often very critical in obtaining an estimate of the mechanical properties available. Although similar reinforcements are used, different pultruders utilize different thermoset resins and product forming techniques which may yield considerable variation in the performance of the pultruded part. The designer who works with the pultruded parts must be concerned with the minimum ultimate properties to be expected in the composite, not "typical" values. Typical values do not reveal the ultimate variation within the pultruder's system which may cause extremely low product performance. For example, pultruder A may advertise a typical value for tensile strength of 45,000 psi (310 MPa) while pultruder B advertises a minimum value of 30,000 psi (210 MPa). The use of typical values in pultruder A implies that values less than 30,000 psi are possible from pultruder A while pultruder B is stating that no values will be permitted less than 30,000 psi. The designer requires the more reliable presentation of the data.

Table 6B.1 compares aluminum, steel, a bulk molding compound, and one pultrusion company's advertised coupon values. These comparisons are useful proof

TABLE 6B.1   Properties of Metal Versus Filament/Fiberglass Reinforced Plastic

Property	Polyester BMC	Polyester pultrusion	Stainless steel	Wrought aluminum
Reinforcement, %	22	55	–	–
Specific gravity	1.82	1.69	8.03	2.74
Elongation, %	0.50	–	40.00	23.00
English Units				
Tensile strength, $\times 10^3$ psi	6.00	30.00	80.00	49.00
Tensile modulus, $\times 10^6$ psi	1.75	2.50	28.00	10.20
Compressive strength, $\times 10^3$ psi	20.00	30.00	80.00	49.00
Flexural strength, $\times 10^3$ psi	12.80	30.00	–	–
Flexural modulus, $\times 10^6$ psi	1.58	2.50	–	–
Metric Units				
Tensile strength, MPa	41.4	206.8	551.5	337.8
Tensile modulus, GPa	12.1	17.2	193.0	70.0
Compressive strength, MPa	137.9	206.8	551.5	337.8
Flexural strength, MPa	88.2	206.8	–	–
Flexural modulus, GPa	10.9	17.2	–	–

Source: EXTREN® Engineering Manual and *Guide to Engineering Materials,* 1988 (Advanced Materials and Processes).

223

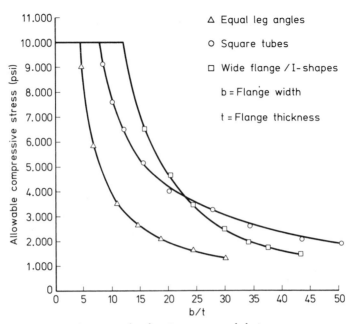

FIGURE 6B.4   Compressive strength of various structural designs.

tests for analyzing generalized material performances. They offer a minimum value to the designer who must estimate from these coupon tests what the possible mechanical properties of the completed structure might be. More valuable to the designer are graphic representations of full section properties under load. An example of this from the MMFG EXTREN® Design Manual (1) is the compressive strength of various structural designs as a function of geometry (Fig. 6B.4). From such data, with the application of an appropriate safety factor, the designer may successfully use the EXTREN® product in column (compression) loading. All pultruders do not have the same composite design for the same profile, and, coupled with resin/forming technique differences, such discrepancies mean that each pultruder must generate his own load characteristic curves.

Some pultrusion companies have studied mechanical property effects of arrangements from different reinforcement schemes. These studies are invaluable in aiding the pultruder to formulate composites satisfying a particular customer's specifications. These studies are often highly proprietary to the individual pultruder and are therefore not published; however, Reference 4 contains an example of different coupon properties obtainable by varying the ladder channel composite.

Other areas that should be well known to the competent pultruder involve (a) the effect of changing from polyester to vinyl ester resin, (b) the structural performance possible from the different graphites available, (c) a numerical estimate for the effect of mat-to-roving ratio, (d) a thorough knowledge of the differences in mechanical properties in utilizing A-glass versus E-glass mat, and (e) the effect of resin properties and cure on the ultimate high temperature properties of the composite. Competent pultruders should catalogue such information for use in applications.

Pultruders must also possess the capability of performing full section tests on the composite. Coupon tests constitute an excellent "proof test" but do not verify the full functionality of the composite. Full section bending tests to verify that the apparent modulus of elasticity has been maintained are excellent certification tests for the composite. Some sort of deflection machine is a must for all pultruders manufacturing structural grade profiles.

6B.5 ELECTRICAL PROPERTIES

The pultruded profile has found excellent acceptance in the market for filament/fiberglass reinforced plastic ladders because of its inherently good combination of electrical and mechanical properties. Most electrical property tests are described in ASTM D 149, ASTM D 495, and ANSI A 14,5. While many of the electrical test procedures require a dry specimen, some customers are beginning to require a water exposure. Some of these "wet" requirements request a saturation in room temperature water for periods of 24 or 48 hours, while other requirements request an immersion for 48 hours in 50 °C water. These immersion tests are often more difficult to pass because they highlight the potential for composite flaws to contribute to an electrical failure. Composite flaws that could cause an electrical failure include hairline cracks and inherent porosity.

The most critical attribute to a good electrical performance in pultruded shapes is the inherent wetout of the reinforcement by the polymer matrix and the prevention of cracking. The inherent reinforcement wetout will also strongly affect several of the reinforcement properties. The wetout can be stated functionally as follows:

$$\text{wetout} = f(\text{Br}, \text{V}, \text{Ti}, \text{S}, \text{T}, \text{Bi}, \text{Te}, \text{R})$$

where Br = the breaker bar system, which is dependent on the number of up/down cycles and the distance between the nodes of the up/down cycle

V = viscosity of the resin mix, which in turn is dependent on the viscosity of the neat resin and the filler type and particle size plus any release effect

Ti = the time in the resin bath

S = the squeeze-out system

T = the temperature of the resin bath; this is a very critical parameter because although higher temperatures will improve the wetout, higher temperatures will also decrease the pot life of the resin mix

Bi = the glass binder and its compatibility with the resin matrix chosen; the compatibility will be a function of the resin matrix chosen

Te = the tension in the reinforcement supplied by the guiding system; ultrasonic vibration may be utilized to spread the individual roving filaments that are subjected to tension

R = the resin matrix; some resin systems such as epoxies wet the reinforcements very well

225

It has been reported that the fiberglass continuous strand mat may have an impact on the electrical properties (6). Coarse mat may offer better electrical resistance than fine mat. This result must be verified for each application considered and for the different possible electrical tests. As with any of the factors potentially influencing electrical properties, their effect will be more pronounced in some electrical tests rather than others.

For example, aluminum trihydrate may have little impact on the dielectric strength but will have a substantial impact in arc tracking applications. The trapped water of hydration with the aluminum trihydrate will act as a heat sink, retarding the carbonization of the path in the arc tracking applications. The choice of composite will also affect the subsequent electrical performance. All longitudinal reinforced (nonstitched) composites tend to produce better electrical performance than when combined with continuous strand mat, regardless of the use of coarse or fine strands in the continuous strand mat.

Wicking is the placement of one end of a shape in a dye reservoir and determining the time required for the dye to penetrate a specified length of the composite; wicking has often been a specified electrical performance requirement. However, the data relating the wicking performance to subsequent electrical performance is currently being studied by an ASTM committee for a pultruded rod and has not been fully documented.

6B.6 CORROSION PROPERTIES

Pultrusion profiles are popular structural members in many chemical plants because of the presence of corrosive agents that would degrade metallic structures. For the purposes of this chapter, corrosion is subdivided into its UV (ultraviolet radiation) and chemical attack forms.

Ultraviolet radiation occurs when the structure is exposed directly to the sunlight. Despite the presence of UV inhibitors, all pultruded structures will show some fading of the pigmentation with time. Some colors such as blue and light gray will exhibit a more rapid change because of the instability of the pigment and/or the instability of the styrene monomer when exposed to UV radiation. The UV inhibitor only retards this reaction without fully terminating it. Some of the more stable pigments to UV radiation used in pultrusion are titanium dioxide, carbon black, and chrome yellow. Gel coats can be utilized to substantially improve the pigmentation life from the UV attack.

Another side effect from UV radiation is the appearance of glass fibrous materials on the surface of the part from the UV radiation attack. This phenomenon, known as "fiber blooming", is characterized by a prickly feeling when rubbing the hand across the surface of the part. This prickly feeling is due to actual exposed fiber filament, which can become embedded in the skin of the hand. The best way to inhibit fiber blooming is to cover the surface of the pultruded part with a surfacing veil material made from a synthetic fabric. This surfacing veil material may give a powdery appearance in time, but no fiber blooming will be observed nor any potential reduction in the physical properties as a result of the fiber blooming. Although the amount of physical property degradation experienced by the structure due to fiber blooming is not certain, there is no speculation on the potential damage from chemical attack with exposed fibers. The chemicals will be much more readily absorbed by the composite if it is allowed to wick down exposed reinforcement. The surfacing veil will not prevent color fading.

Chemical corrosion is not limited to attack from acids or bases; it can be caused by an attack from solvents, water, etc. As a highly generalized rule of thumb, polyesters will protect the system against most acids, while a vinyl ester resin matrix will protect the system against both acidic and basic (caustic, alkali) attack. The chemical corrosion protection from epoxy is dictated by the hardening agent employed. Surfacing veil often imparts additional corrosive protection to the system because it will force an additional resin layer 0.25 mm (0.10 in.) thick to protect the reinforcement. Gel coating the part with resins that normally are not pultrudable may be an additional way to impart some corrosive protection.

In reviewing a particular chemical for its corrosive attacking potential on the composite, three items are mandatory:

1. The exact nature of the corrosive chemical must be well known. Will this corrosive agent come in contact with the composite in a wet or dry environment?
2. The maximum temperature to be experienced by the composite must be well defined. Corrosive agents more aggressively attack the composite substrate at higher temperatures than at lower temperatures. In addition, different resin matrices will impart different maximum service temperatures to the composite. An application must be strongly considered if the maximum composite service temperature is being approached in the presence of a strong corrosive agent.
3. The percent concentration to which the composite will be exposed must be defined. Very low levels of a corrosive agent may be acceptable while higher levels may be unacceptable.

6B.7 QUALITY ASSURANCE

The continuous nature of the pultrusion process affords many opportunities for the application of quality assurance theory incuding statistical process control (SPC). The certification of a particular lot or run of parts based on specifications such as MIL-STD-105D is unacceptable in today's production environment, and SPC procedures are becoming more and more important to the pultruder. SPC is the acronym employed in this application as opposed to SQC (statistical quality control) because SPC implies a feedback loop in which information from an ongoing test is supplied to the pultrusion operator who, in turn, adjusts the process as required to conform to the specifications.

To properly operate a pultrusion quality assurance program, there must be a good understanding of the incoming raw materials. The primary raw materials for pultrusion are the resin and the reinforcement. The resin is important because it contributes to the wetout and the cure of the resultant pultrusion system while the reinforcement dictates the basic structural performance and is highly influential in other processing parameters. A successful SPC program with incoming raw materials must be based on a cooperative venture with the raw material suppliers. The pultruder must have the capability of performing all the incoming resin specifications as outlined by the SPI procedures in addition to whatever resin performance tests have been defined internally. Similarly, the reinforcement will require some means to analyze the incoming properties which will truly reflect upon the nature of the product being received. A sample inspection approach, such as found with MIL-STD-105D, may not be the best solution; and, each raw material customer must develop the approach best suited to his operation.

Dimensional inspections can be made on the pultrusion process at the initiation of the run. Many of the dimensions are "die struck" and remain relatively constant during the run. The individual pultruder should have the mathematical skill within his own organization to determine the process capability for each of the dimensions being considered. The statistical process capability can be formulated either from $\bar{x}/R$ (average/range) charts or from the more conventional $\bar{x}/s$ (average/standard deviation) process capability analysis.

Coupon mechanical property testing by ASTM procedures will be extremely difficult if it is to be performed in a timely fashion for feedback to the pultrusion operator. The testing of the full section modulus of elasticity by bending is a quick method to certify that the correct reinforcement is present and no structural defects exist within the composite. As required, this can be supported later by coupon testing verification for the mechanical properties. Many of the full section structural applications will be very dependent on the full section apparent modulus of elasticity.

Each company must decide on what kinds of quality control are to be used in its operations.

6B.8 ASTM PULTRUSION REFERENCES

Listed below are several references on pultrusion from ASTM. These are provided as an aid to understand the official terminology and test procedures associated with the pultrusion industry.

Specification Number	Description
D 2344	Test method for apparent inner laminator shear strength of parallel fiber composites by the short beam method.
D 3518	Procedure for in-plane shear strength–strain response of unidirectional reinforced plastics.
D 3647	Classifying reinforced plastic pultruded shapes according to composition.
D 3878	Standard definition of terms relating to high modulus reinforcing fibers and their composites.

Specification Number	Description
D 3914	In-plane shear strength of pultruded glass-reinforced plastic (GRP).
D 3917	Dimensional tolerance of thermosetting glass-reinforced plastic pultruded shape.
D 3919	Standard definition of terms relating to reinforced plastic pultruded products.
D 4385	Classifying visual defects in thermosetting reinforced plastic pultruded products.
F 1092	Standard specification for fiberglass (GRP) pultruded open weather, storm, and guard square handrails.

These specifications are in addition to the normal ASTM specifications for the determination of mechanical properties.

6B.9 APPLICATIONS

Four of the 10.6 m × 10.6 m × 7.6 m (35 ft. × 35 ft. × 25 ft.) high pyramid-shaped turrets shown in Figures 6B.5–6B.7 were fabricated using pultruded fiberglass shapes and then lifted to the top of the new Sun Bank Building in Orlando, Florida, USA, by helicopter. These turrets house communications equipment and were fabricated by Fibertron of Bessemer, Alabama, USA.

The tow target shown in Figure 6B.8 is fabricated using a thin walled 279 mm (11 in.) diameter pultruded fiberglass round tube shown in Figure 6B.9 for the casing. The target is used in military simulations for fighter pilots to practice firing on enemy aircraft. It is controlled by sending radio commands to a command receiver in the target; therefore, low radio interference is a requirement. Because the target is sometimes powered by a jetfueled combustion chamber located in the aft, the material must also be thermally non-conductive and fire-retardant. These targets are manufactured by Hayes International Corporation in Leeds, Alabama.

A 12.2 m (40 ft.) long skylight utilizing pultruded fiberglass shapes replaced a steel skylight in a Ciba-Geigy Chemical Plant in New York. The skylight, shown in Figure 6B.10, was fabricated of fiberglass louvers and structural shapes and is exposed to highly corrosive chemicals in the operating environment. It has lasted three times longer than steel in this application. The skylight was fabricated by Imco Reinforced Plastics in Moorestown, New Jersey.

A rectangular cover 11.68 m long × 8.43 m wide × 1.52 m high (38 ft. × 27.6 ft. × 5 ft.) for a sulfur pit was installed at a Shell Oil Refinery to contain fumes and meet the Environmental Protection Agency's regulations. The cover is shown in Figure 6B.11 and was fabricated using pultruded fiberglass reinforced shapes. The challenge was to design and fabricate a fiberglass cover to replace a steel cover that had failed. The cover had to enclose hydrocarbon emissions, meet Zone 4 UBZ (Earthquake Zone Building Codes) and meet air pollution control standards. The cover was designed and fabricated by Glass-Steel in Spring, Texas, USA.

FIGURE 6B.5–7   Pultruded fiberglass structural shapes.

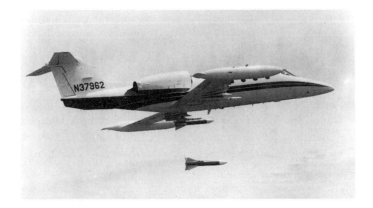

FIGURE 6B.8   Tow target used in practice firing.

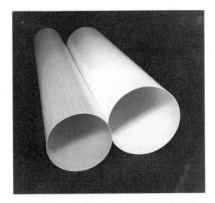

FIGURE 6B.9   Casing for the target made of pultruded fiberglass round tube.

FIGURE 6B.10   Skylight made of pultruded fiberglass sections.

FIGURE 6B.11   Rectangular cover for a sulfur pit.

231

Examples of the use of pultruded fiberglass shapes to replace metal ones are shown in Figures 6B.12 and 6B.13. Figure 6B.12 is a stile used in a pulp and paper plant which has a highly corrosive environment. It was fabricated by Fibertron in Bessemer, Alabama, USA. The handrails in Figure 6B.13 are in a wastewater treatment plant and were used because pultruded fiberglass shapes are corrosion-resistant and require minimal maintenance. This handrail system is manufactured by Glass-Steel in Spring, Texas, USA.

Pultruded fiberglass grating is used in a wide variety of corrosive environments which include wastewater treatment plants, pulp and paper manufacturing, chemical processing facilities, plating plants, and marine and oil production operations. The installation of fiberglass grating shown in Figure 6B.14 is taking place on a Shell offshore platform. The grating is manufactured by AFC in Chatfield, Minnesota.

A 10 m high × 3.2 m wide × 2.9 m deep (32.8ft. × 10.5 ft. × 9.5 ft.) all-fiberglass impulse generator was built by Maxwell Laboratories, Inc., in San Diego,

FIGURE 6B.12   All-fiberglass stile.

FIGURE 6B.13   Pultruded fiberglass handrail system.

FIGURE 6B.14   Pultruded fiberglass grating.

FIGURE 6B.15   Pultruded fiberglass structural members of this impulse generator dramatically illustrate the suitability of composite materials for high power electrical applications.

California, USA. Pultruded fiberglass shapes were used because of their superior dielectric strength, structural integrity and non-conductivity. Wood was previously used in the fabrication of these generators. The generator, shown in Figure 6B.15, is used by power companies to simulate lightning for testing switch gears and other heavy electrical equipment.

The Graph-Lite™ one piece composite driveshaft (Fig. 6B.16) pultruded by Morrison Molded Fiber Glass Company for Spicer Universal Joint Division, Dana Corporation, has significant advantages over two-piece driveshafts. These include a 60% reduction in weight, providing fuel economy; elimination of several parts, saving in assembly and inventory costs; reduced noise and vibration; and improved

233

FIGURE 6B.16   Graph-Lite™ one-piece composite driveshaft.

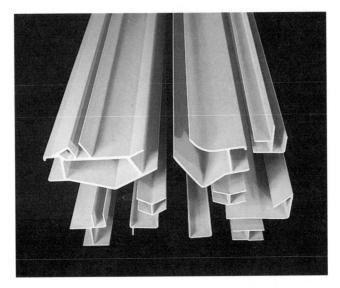

FIGURE 6B.17   Pultruded fiberglass shapes.

corrosion resistance. The Graph-Lite™ is manufactured by pultruding a fiberglass–graphite–proprietary resin composite over a seamless aluminum tube. An isolation barrier between the aluminum and graphite eliminates galvanic corrosion, and the composite reinforcement of the tube eliminates the need for a center bearing. By 1992 more than 500,000 Graph-Lite driveshafts for vans and light trucks, are expected to be sold annually.

Some of the eighteen different pultruded fiberglass shapes used as panel joiners and door framing for modular lavatories, unusually complex and angular, are shown in Figure 6B.17. These are among the first fiberglass pultrusions used in commercial aircraft replacing aluminum shapes. They are lightweight and resist the corrosive chemicals used in the lavatory operation and cleaning.

234

REFERENCES

1. *EXTREN*® *Design Manual* 1988 edition.
2. John D. Tickle, George A. Halloday, Joe Lazarou, and Brian Riseborough, "Designing Structures with Pultruded Fiberglass Reinforced Plastic Structural Profiles as Compared to Standard Steel Profiles," 33rd Annual Technical Conference, 1978 Reinforced Plastics/Composites Institute, The Society of Plastics, Inc., 1978.
3. Robert I. Werner and Zbigniew Kusibab, "Pultrusion Process Engineering Using Thermal Analysis," 38th Annual Conference, Reinforced Plastics/Composites Institute, The Society of Plastics Industry, Inc., 1983.
4. Robert I. Werner, "Properties of Pultruded Sections of Interest to Designers," 34th Annual Technical Conference, Reinforced Plastics/Composites Institute, The Society of Plastics Industry, Inc., 1979.
5. G. A. Hunter, "Pultruding Epoxy Resin," 43rd Annual Conference, Composites Institute, Society of the Plastics Industry, Inc., 1988.
6. L. P. Connors and Nelson H. Douglas, "The Effect of Material Composition on Electrical Properties of Pultrusions," 43rd Annual Conference, Composites Institute, Society of the Plastics Industry, Inc., 1988.

7 · Long Fiber Thermoplastic Composites

Narasimhan Raghupathi

Contents

N. Raghupathi, PPG Industries, Inc., Pittsburgh, Pennsylvania, USA

7.1 INTRODUCTION

Extensive efforts in the aircraft and aerospace industry over a number of years have led to the advancement of long and continuous fiber composites based on thermoset resin systems from state-of-the-art to science. This technology has proliferated into other areas such as the pipe industry (for high and low pressure pipes), the electrical industry (for printed circuit board substrates), sporting goods, and to a limited extent in the automative industry (for leaf spring and driveshaft applications). The resins used range from phenolics and epoxies in the aircraft industry to vinyl esters and unsaturated polyesters in the marine and automotive areas. Recently, a number of new thermoset resin systems have been developed for structual applications using the reaction injection molding/resin transfer molding process. In this process, the reinforcement is preshaped or preformed and placed in the mold before the resin is injected. Some of these resins include polyurethanes, polyurea, polyisocyanurate, and polydicyclopentadiene. The long fiber thermoset composites can be included in the category of materials used for structural and semistructural applications. Thermoplastic composites, on the other hand, were limited to a large degree to short fiber reinforcements. These composites were used mainly to fabricate nonstructural or, at the best, semistructural components via the injection molding process. The impetus for the long fiber thermoplastic composites mainly comes from the rapid advancements made in thermoplastic resin technology over the past decade, coupled with the recent innovations in fiber impregnation techniques using high viscosity thermoplastic resins. This has led to a family of long fiber reinforced thermoplastic composites capable of giving performances comparable to structural thermoset composites.

The key advantages of the advanced thermoplastic composites over the current generation of thermoset composites can be listed as follows: increased toughness, higher impact resistance, ease of repair, higher service temperatures with certain polymers, unlimited shelf life, no special storage conditions, applicability of metal stamping process for certain thermoplastic forming, material quality consistency, and anticipated higher economic return in recycling thermoplastic resins. The drawbacks, on the other hand, include the following aspects: higher processing temperature and pressure, difficulty in fiber impregnation (due to very high resin viscosity) leading to high voids in the composite, higher coefficient of thermal expansion for a given fiber volume, poor solvent resistance with certain thermoplastics, and susceptibility to environmental stress cracking. No one material is expected to dominate the structural composite arena. The material of choice is going to be determined by its ability to meet specific performance requirements for a given application coupled with the relative ease of processibility. The overall economics will also play a role, probably more significantly in applications outside the aircraft and aerospace areas.

Long fiber thermoplastic composites can be divided into three categories. The first is a family of long-and-short fiber materials suitable for injection molding. Typically, the reinforcement is longer than 6.3 mm (0.25 in.) but less than 25.4 mm (1 in.) long in the compounded pellets. This results in injection-molded parts with fiber length distributions in the range of 2–10 times greater than those obtained with the conventional short fiber process. The second category is a class of materials with discontinuous fibers in which the reinforcement length is at least 12.7 mm (0.5 in.) in the final molded part. Typically, such composites are based on chopped fibers, chopped strand mat, and nonwoven fibers. The last class of materials is based on continuous fiber reinforcements, including continuous fiber nonwoven as well as woven materials, preimpregnated tapes, and preimpregnated or in situ impregnated continuous fibers suitable for the filament winding or pultrusion processes.

This chapter addresses the current and emerging status in different segments of the long fiber thermoplastic composite technology, types of reinforcement and resin, various processes for resin impregnation, composite properties, molding and other fabrication methods, design guidelines to a limited extent, and finally, typical applications.

7.2 MATERIALS

Three types of fiber have been extensively used; namely, glass, graphite–carbon, and aramid. The fiber form, however, can vary widely depending upon process and performance requirements. The range of thermoplastic resins used extends from the low cost workhorse polypropylene at one end to high cost, high performance liquid crystal polymers at the other end.

7.2.1 Reinforcements

Typical properties of reinforcing fibers are given in Table 7.1. E-Glass fiber typically costs slightly under a dollar a pound; Kevlar 49, between $15 and $20 per pound; and graphite–carbon fibers, $30 or more per pound. As a rule, the price of the fiber increases exponentially as a function of specific tensile modulus and specific tensile strength. The focus in automotive and other high volume applications is toward glass fiber composites, whereas aramid and graphite fibers are considered in the aircraft/ aerospace, sporting goods, and other markets where costs are of secondary concern relative to performance.

The fiber forms vary widely but essentially start from single-filament building blocks. Single filaments, ranging anywhere from 400 to 12,000 in number, are combined into fiber bundles during the fiber forming process. These fiber bundles, known as strands, rovings (fiber glass industry), or tows (carbon fiber industry), are further processed in secondary operations into various fiber forms. One exception is the glass fiber used in a wet slurry process to make composite sheets. Typically, the fibers are chopped wet directly under the bushings (analogous to spinnerets) and shipped with a moisture content of 5–10% by weight. Various fiber forms used in the long fiber thermoplastic composite arena are given in Table 7.2.

TABLE 7.1 Properties of Reinforcing Fibers

Property	E-Glass	Kevlar 49	Carbon (Hercules)	Graphite (Hercules HMS)
Specific gravity	2.58	1.44	1.88	1.83
Tensile strength, MPa	3450	3600	3790	2200
, ksi	500	520	550	320
Tensile modulus, GPa	72.4	125	425	340
, Msi	10.5	18 (axial)	62	50
Tensile modulus, GPa		6.9		
, Msi		1.0 (transverse)		
Elongation, %	4.8	2.5	0.75	0.58
Coefficient of thermal expansion, $10^{-6}/K$	5.0	-5.2 (axial) 41.4 (transverse)		-0.99 (axial) 16.8 (transverse)
Thermal conductivity				
,W/m K	0.87		104	
,Btu/ft.h. °F	0.5		60	

TABLE 7.2 Fiber Forms

Filaments
↓
Fiber bundles
↓

↓ Discontinuous	↓ Continuous
Chopped fibers [> 12.7 mm (0.5 in.)long]	Roving
Chopped fiber mat[a]	Fiber tow
	Continuous fiber mat[a]
	Woven fabrics

[a] May be chemically or mechanically bound.

7.2.2 Thermoplastic Polymers

The choice of thermoplastic matrix resins is much broader than that for the fibers. The polymers, similar to the fibers, can be ranked on a performance/cost basis, the main performance aspect being the heat deflection temperature. At the low end of the spectrum are the polyolefins. Intermediate performance range resins include polyamides, thermoplastic polyesters, polycarbonates, and alloys of polyesters and polycarbonates. At the high performance end, typical resins include polyetherimide, polysulfones, polyphenylene sulfide (PPS), polyetherketones (PEK), polyetheretherketones (PEEK), and the like. Table 7.3 illustrates various resins, their transition temperatures,

TABLE 7.3 Thermoplastic Polymers

Polymer[a]	Type	T_g (°C)	T_m (°C)	Processing temperature (°C)[b]
Polypropylene	Semicrystalline	−20	168	200–250
Polyamides				
Nylon 6	Semicrystalline	70	220	230–275
Nylon 6,6	Semicrystalline	50	260	270–325
Polyesters				
PBT	Semicrystalline	20	240	240–270
PET	Semicrystalline	73	257	280–320
Polycarbonate	Amorphous	140–150		270–300
Polyphenylene oxide (GE's NORYL)	Amorphous	115–142		240–330
Polyetherimide (GE's ULTEM)	Amorphous	217		325–400
Polyphenylene sulfide (Phillip's RYTON)	Semicrystalline	88	290	315–330
Polyetherketone (BASF's PEK)	Semicrystalline	172	372	420–450
Polyetheretherketone (ICI's PEEK)	Semicrystalline	143	343	360–400

[a] PBT = polybutylene terephthalate; PET = polyethylene terephthalate.
[b] Temperature used in injection molding/extrusion.

and the temperature at which they need to be processed for obtaining effective impregnation of the fibers in the composite (1, 2).

Polypropylene resin based long fiber composites are extensively used in certain automotive applications where energy absorption between −40 °C and 80 °C is important. The typical processing temperature for polypropylene is around 200 °C, thus making it one of the easiest thermoplastics to process by conventional means. Fiberglass is the predominant reinforcement used with polypropylene. The heat deflection temperature of the composite is around 150 °C. Hence, polypropylene has temperature use limitations and is not recommended for continuous use above 80 °C. Intermediate performance range resins extend the application temperature range to 125 °C but require higher processing temperatures—in the neighborhood of 250–300 °C. The high performance thermoplastic resins usually require more severe processing conditions. For example, semicrystalline polymers are typically hot melt impregnated at one stage or another and require very high consolidation temperatures —in excess of 300 °C. Amorphous polymers may be impregnated either by softening them well above their glass transition temperatures or from solution, which will subsequently require a solvent removal process.

241

7.3 INJECTION-MOLDABLE LONG FIBER COMPOSITES[1]

Extrusion-compounded fiberglass-reinforced thermoplastic pellets have been traditionally used in part fabrication via the high speed injection molding process. Fiber lengths were typically limited to 6.3 mm (0.25 in.) or less in the compounding process. However, fibers degraded in length during the compounding and molding operations, resulting in inefficient use of the reinforcement in the final composite. Recent technological advances made in fiber impregnation techniques have led to a family of long fiber injection-moldable composites. The injection-moldable pellets in this case contain wetted fibers equal in length to the pellet, typically 12.7 mm (0.5 in.). Although some fiber degradation occurs during the molding operation, the fiber lengths after molding are typically 10 times larger than those obtained with short fiber composites. As a consequence, improved properties are obtained in the molded composite. The theory of short and long fiber reinforcements and the mechanisms accounting for the differences in their behavior are discussed later.

7.3.1 Impregnation Methods

A number of patented fiber impregnation processes have been developed commercially in the past decade. These processes do not follow the conventional extrusion compounding operation. In one variation, fibers are pultrusion compounded as illustrated in Figure 7.1. The continuous fibers in one form or another are pulled through a crosshead die by the puller. As the fibers go through the die, they are impregnated by the molten polymer from the extruder. It is essential that the fiber bundle open up in the operation so that the molten polymer can penetrate into the bundle to provide complete wetting of the filaments. In the fiberglass arena, the bundle opening can be manipulated to a degree by appropriate sizing of the fibers. The fiber opening can be further enhanced by subjecting the bundle to ultrasonic vibrations prior to entry to the die. The resin-impregnated fiber bundle is then pulled through a cooling bath into a pelletizer, which cuts the pellets to the desired length. The pultrusion-compounded pellets are typically much longer than the extrusion-

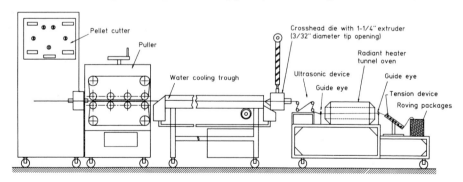

FIGURE 7.1   Pultrusion compounding process for long fiber reinforced pellets.

[1] The reader is also referred to Chapter 8 for additional information.

compounded pellets. Standard lengths are 12.7 mm(0.5 in.), although lengths up to 38.1 mm (1.5 in.) are not uncommon.

Another proprietary process uses a fluidized bed of resin powder through which the fiber bundle is pulled. The fiber bundle should, again, have the ability to open up, to permit the powdered resin to penetrate into the bundle. Electrostatic charging can help retain the resin powder in the interstices of the fiber bundle. The powder-impregnated bundle is then consolidated in a melting operation, solidified, and cut into pellets. This process has the disadvantage of requiring the polymer to be in a fine powder form. Polymers that are naturally produced in a powder form during the polymerization process are good candidates. Otherwise, some form of grinding (including cryogenic grinding) may be required, thus rendering the process expensive.

7.3.2 Molding

The long fiber compounds can be molded on conventional injection molding equipment without major modifications. Modifications are required only if maximum properties are required in the molded part. Otherwise, standard operation will still result in enhanced properties, although not maximum, by virtue of the fact that the thoroughly wetted fibers resist attrition during the molding operation. As a general guideline, the following conditions are necessary to obtain maximum performance: free-flowing valves, adequate runner sizes and shapes (to prevent clogging and log jam effects), nonrestrictive gates, and proper location of the gates (3). It is preferable to run the molding process, including the mold, at temperatures slightly higher than those used in conventional injection molding. This reduces the viscosity of the material entering the mold, thus reducing the shear, which in turn reduces the fiber attrition. A recent study (4) has indicated that the screw tip design influence the fiber length distribution of the long fiber composite. Conventional injection molding equipment with a general-purpose reciprocating screw was used in this study. A standard screw tip and a "free flow" (fluted channel) screw tip were evaluated. The major finding was that the free-flow screw tip resulted in higher overall mechanical properties (up to 20%) and fiber length (up to 30–50%) than with a standard screw tip. The short fiber composites were relatively unaffected by processing conditions or the kind of screw tip employed with the general-purpose reciprocating screw. Fiber length measurements for nylon 6, 6 molded composites showed typical lengths in the range of 0.64–2.9 mm (0.025–0.114 in.) and maximum fiber length up to 12 mm (0.472 in.) for the long glass fiber composite, versus average fiber lengths of 0.15–0.33 mm (0.006–0.013 in.) for the short fiber composite.

In an attempt to understand the differences in the fracture behavior of long and short fiber injection-molded nylon–glass fiber composites, the fiber length distributions were evaluated in another study (5). The fiber length distribution of the long fiber composite was influenced by the flow field. In the shear flow field close to the mold walls, more fiber attrition was observed relative to the core region (at the center away from the walls). But in both regions, the fiber length distributions were much higher than those obtained with short fiber composite (Fig. 7.2) (5).

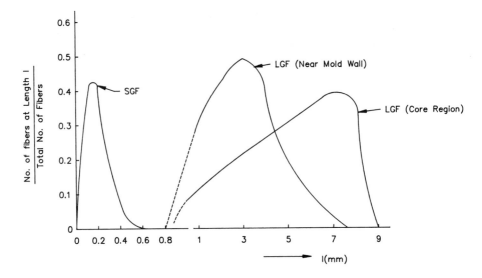

FIGURE 7.2   Fiber length distribution for short glass fiber (SGF) and long glass fiber (LGF) reinforced nylon 6,6 composites (fiber content, 30% by weight).

The explanation for the preservation of higher lengths in the molded composite using long fiber pellets is as follows. The long pellets produced by the pultrusion process contain fiber bundles that have been thoroughly wetted out (i.e., resin has penetrated into the bundle) and yet maintain their bundle integrity. During injection molding, these bundles undergo minimum length degradation, especially in flow regions where shear stresses are low. At the entrance region in the mold, a more extensional type of flow behaviour is observed with the long fiber composite. This enables the fiber bundles to align in the direction of flow and get through the constriction with minimum damage. The fiber morphology of long fiber composites shows more fiber bunching in the core region and fiber bending in the surface layer close to the mold wall. With the short fiber pellets, the fiber bundle integrity is lost during the extrusion compounding step. Thus, the filaments are thoroughly dispersed in the melt during injection molding. The resistance to attrition under shear flow field of these individual filaments is far less compared to the bundles which maintain their integrity. Typically, a five- to tenfold increase in mean fiber length is obtained when the long fiber composites are being compared with the short fiber counterpart.

7.3.3 Properties

The performance of the long fiber composites is significantly higher than their short fiber counterparts. Typical property enhancements are obtained in modulus, impact strengths, creep resistance, dimensional stability, fatigue resistance, and surface finish. Although most of the work has been with glass fibers, similar results have been obtained with graphite and aramid fibers.

The properties of short fiber composites can be predicted using the rule of mixtures theory. Tensile modulus and tensile strength of the composite based on this theory are represented by eqs 7.1 and 7.2 respectively (6):

$$E_c = E_f V_f C_0 \left(1 - \frac{l_c}{2l}\right) + E_m V_m \tag{7.1}$$

$$\sigma_c = \sigma_f V_f C_0 \left(1 - \frac{l_c}{2l}\right) + \sigma_m V_m \tag{7.2}$$

where E, σ and V denote modulus, strength, and volume fraction, respectively. The subscripts c, f, and m represent composite, fiber, and matrix, respectively. The critical fiber length, the mean fiber length, and the orientation factor are represented by l_c, l, and C_0, respectively. For randomly aligned planar fiber distribution, the value of C_0 is approximately 0.33. The critical length is defined as the length at which the tensile force on the fiber equals the interfacial shear force. At this length, the tensile stress at the midpoint of the fiber reaches its ultimate tensile strength (Eq. 7.3):

$$\left(\frac{l}{d}\right)_c = \left(\frac{\sigma_f}{2\tau}\right) \tag{7.3}$$

In this equation, l and d are fiber length and diameter, respectively; $(l/d)_c$ is the critical aspect ratio; σ_f is the fiber tensile strength; and τ is the interfacial strength between the fiber and the matrix. A schematic of the stress distribution is shown in Figure 7.3. The interfacial strength is strongly influenced by the effectiveness of bonding between the fiber and the matrix via coupling agents. As a general rule, for random short fiber composites, the maximum degree of reinforcement is achieved when the average fiber aspect ratio is approximately 10 times the critical aspect ratio. This implies that the fiber with 10 times the critical length is subjected to its ultimate load-bearing capability over 90% of its length. In other words, the fiber is used effectively at 90% efficiency. In a short fiber injection-molded composite, a portion of the fibers have aspect ratios near or below the critical value. These fibers do not effectively contribute to the overall composite strength. To begin with, the fibers have lengths of 3.2–6.3 mm (0.125–0.25 in.). During the compounding and molding operation, the fibers undergo attrition, resulting in a broad distribution in length, a portion of which is below the critical length. A typical fiber length distribution curve for injection-molded short glass fiber nylon 6, 6 composite is shown in Figure 7.4. The critical fiber length in this case was 0.6 mm as measured by the single fiber pullout test. The mean fiber length from this distribution curve was 0.76 mm.The curve clearly shows a significant fraction of fibers with lengths below the critical length (38% in this specific case). These fibers do not provide efficient reinforcement value. In the long fiber injection-molded composite, the mean fiber length is typically 10 times that of the short fiber composite for the reasons explained in connection with the compounding and molding operations. Thus, a higher portion of the fibers becomes effective in load bearing compared with the short fiber composite. Also for

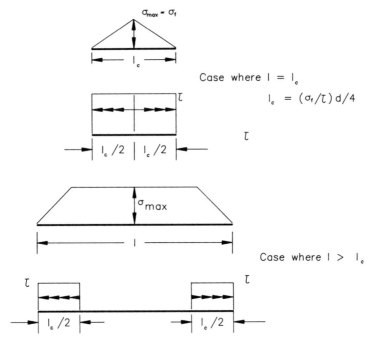

FIGURE 7.3 Stress distribution in a single filament: critical length concept. d = fiber diameter; l = fiber length; l_c = critical fiber length; σ_f = fiber tensile strength; σ_{max} = maximum tensile stress; τ = shear stress.

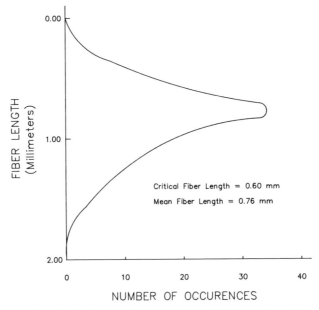

FIGURE 7.4 Fiber length distribution for injection-molded short glass fiber nylon 6,6 composite.

246

TABLE 7.4   Properties of Injection-Molded Composites at 40% Glass Loading[a]

Properties	Nylon 6,6 (3)		PBT[b] (10)		Polycarbonate (10)	
	Short	Long	Short	Long	Short	Long
Specific gravity	1.45	1.45	1.63	1.63	1.52	1.52
Tensile strength, MPa	223	223	129	134	160	160
, ksi	32	32	18.5	19.3	23	23
Flexural strength, MPa	281	343	204	225	190	260
, ksi	40	48.8	29	32	27	37
Flexural modulus, GPa	11.6	11.6	8.2	9.6	9.6	13
, Msi	1.7	1.7	1.2	1.4	1.4	1.9
Notched Izod strength, J/m	135	520	94	364	130	463
, ft–lbs/in.	2.6	10	1.8	7	2.5	8.9

[a] Short = input fiber glass length 4.76 mm (0.1973 in.) or less
 Long = input fiber glass length 12.7 mm (0.5 in.) or more
[b] PBT = polybutylene terephthalate

the same volume fraction of the reinforcement, the number of ineffective fiber ends in the long fiber composite is significantly reduced compared with the short fiber composite. Thus based on the theory, one would expect the long fiber injection-molded composite to have higher strength than its short fiber counterpart. Also, the composite tensile strength is more strongly dependent on the interfacial bond strength than the tensile modulus. This is evident from the fact that the particulate fillers (aspect ratio = 1 for spherical fillers) contribute to the increase in modulus but generally decrease the strength of the composite.

A comparison of properties of short fiber and long fiber injection-molded composites at 40% glass loading in three resin systems is given in Table 7.4. Improvements can be seen in notched Izod, flexural strength, and flexural modulus. Recent reports indicate typical property enhancements for nylon 6, 6 at 50% glass loading with long fibers (7). For example, flexural modulus and flexural strength show a 70% increase at 149 °C over their short fiber counterparts. Similarly, a 35% increase in flexural strength and a 56% increase in flexural modulus were obtained at 204 °C. At 60 °C, creep resistance of the long fiber composite was found to be equivalent to the short fiber composite at room temperature. Also, creep behavior of the long fiber composite at 100 °C was found to be equivalent to that of the short fiber composite at 60 °C. Similar results were also obtained in fatigue endurance tests. Finally, the enhancement in surface appearance of the long fiber composites is attributed to fewer fiber ends being exposed to the surface relative to the short fiber composites at equal fiber loading. Similar trends have been observed with carbon fiber reinforced PEEK composites. The use of a long carbon fiber provided additional strength and stiffness when compared with standard length carbon fiber (8).

7.3.4 Applications

Numerous applications are under development using the long fiber injection-moldable composites. The markets include automotive, electrical, sporting goods, aerospace, appliances, and other industrial components. The technology is especially attractive for applications that require the high performance of continuous fiber composites but cannot afford the high costs associated with it.

7.4 RANDOM FIBER STAMPABLE SHEET COMPOSITES

Another family of long fiber thermoplastic composites can be classified as random fiber stampable sheet composites. These materials have fibers oriented randomly with fiber lengths in excess of 12.7 mm (0.5 in.) in the molded composite. Generally, these materials are available in sheet form and can be converted to finished parts by a conventional stamping or high speed compression molding process. Typical advantages over the thermoset molding compounds include rapid molding cycle, indefinite material shelf life, improved impact strength, and recyclability.

A number of such materials are commercially available worldwide under various trade names. Companies that supply such materials include AZDEL, Inc.; Phillips; Exxon; BASF; Symalit (Switzerland); Arjomari (France); and Idemitzu and Ube-Nitobo (Japan). Products are based on glass and carbon fiber reinforcements and a variety of thermoplastic matrix resins.

7.4.1 Materials and Lamination Process

Various fiber forms are used in the lamination process. These include chopped fibers, random chopped fiber mat, and continuous fiber mat. Chopped fibers are typically between 12.7 and 38.1 mm (0.5–1.5 in.) long. Chopped fiber mats are comprised of fiber lengths in the same range as chopped fibers, and the fibers are held together by chemical binding agents. Continuous fiber mats have random swirl fibers either bound chemically or intertwined mechanically.

Two different lamination processes are used, depending on the fiber form. The chopped fibers are processed by a wet slurry process (similar to the paper process under high speeds). The polymer for this process should be in a powder form. Typically, the fibers, polymer powder, surfactants, and other processing aids are mixed together with water in a large mixing tank to form a slurry. The fiber bundles filamentize effectively in the mixing process. The slurry is then pumped onto a high speed conveyor where most of the water is removed by means of a vacuum. The result is a wet, nonwoven web of intimately dispersed fiber reinforcement and matrix resin. The web is further dried in a convective, continuous drying oven to remove the remainder of the moisture. The consolidation of the web is accomplished in a continuous double-belt laminator under heat and pressure. A schematic of the process is shown in Figure 7.5.

The advantage of the wet slurry process is that it enables the polymer powder to be in intimate contact with the fiber before melting, thus resulting in a well-impreg-

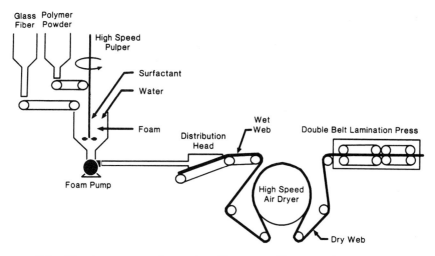

FIGURE 7.5   Wet slurry process for random fiber stampable composites.

nated composite after lamination. The process has the flexibility to handle very high viscosity polymers, limited only by lamination temperatures and the availability of the polymer in powder form. Higher fiber loadings (up to 60–70%) can be effectively achieved by this process. The process also has the capability to produce laminates in fully or partially consolidated forms. The partially consolidated laminate provides a lofted, low density material similar to structural foam and can be used in building and construction applications. Fully consolidated laminate up to a thickness of 5 mm can be made by this process.

The lamination of random fiber mats with the matrix is accomplished by a conventional melt extrusion process, as illustrated schematically in Figure 7.6. Continuous double-belt lamination equipment is used for this purpose. The polymer is delivered in molten form between two layers of mat from an extruder via a sheet die. The impregnation takes place in the heating zone of the laminator. Temperature, pressure, and residence time are the key factors that must be controlled to obtain good wetting. Laminating pressures are fairly low [in the neighborhood of $3.5–14\,\text{kg/cm}^2$ (50–200 psi)]. The material is cooled in the cooling zone and emerges as a continuous sheet at the exit end. The material must be cooled below the glass transition

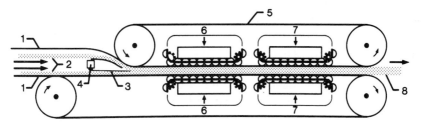

FIGURE 7.6   Continuous double-belt lamination process. (1) Thermoplastic resin sheets; (2) fiber glass mat; (3) thermoplastic resin extrudate; (4) extruder; (5) double belt laminator; (6) hot laminating zone; (7) cold zone; (8) finished sheet product.

temperature (in the case of amorphous polymers) and below the crystallization temperature (in the case of semicrystalline polymers) to develop enough rigidity in the material prior to exit. On-line slitting is used to cut the sheet into desired dimensions. In a secondary operation, the sheets can be further cut into desired blanks for molding purposes. The pretailored blanks eliminate the need for a slitting operation at the molder's end.

Variations of the same process have also been used for lamination. A series of sheet extruders can be used on-line, enabling the introduction of more than two layers of mat and different polymers between layers. In another variation, a calendering type of process has been used for lamination. A batch press lamination process is used for polymers that require excessively high temperatures for melting (i.e., temperatures that cannot normally be obtained in continuous belt laminators).

7.4.2 Molding/Stamping Process

The conversion of the long fiber thermoplastic composites to parts can be accomplished by stamping or high speed compression molding techniques depending on the nature of the polymer. Each process has its own advantages and disadvantages. Stamping, also referred to as solid state forming, is generally used with semicrystalline polymer matrices. Semicrystalline polymers can be permanently deformed at temperatures between glass transition and melting. The extent of stretching that can be obtained is limited by the narrow draw ratio of the polymer under these temperature conditions. The typical stretching is between 5 and 10%, depending on the type of polymer. Amorphous polymers, on the other hand, are generally too stiff to be rapidly formed into stable parts below their glass transition temperatures T_g. They have to be heated to temperatures substantially above their glass transition to be formed. They are no longer considered to be solids under these conditions. Exception to this are amorphous polymers that show yield behavior below T_g. However, such stamped parts will have poor thermal stability due to stress relaxation or strain recovery phenomenon. As a rule, solid state stamping is not recommended for parts with complex shapes, deep draws, deep ribs, and bosses. The major advantage of solid state stamping is that parts can be produced at rapid rates, typically in 10–20 seconds. Parts with a good surface finish can be obtained, since the fibers do not have the opportunity to come to the surface of the part during the stamping operation.

High speed compression molding, also known as flow forming, is not limited to semicrystalline polymers, since the matrix has to be melted before forming. The flow forming process has four primary steps: designing blanks, heating the material, compression molding, and solidifying the material. Four basic pieces of equipment are required: a heating oven which can be either an infrared or a hot air convection type; a matched metal compression molding die; a temperature control unit; and a high speed compression molding press. A typical schematic of the molding process is given in Figure 7.7.

Blank design is one of the important steps in producing quality parts. Proper blank sizes, temperature and placement in the mold need to be established during the prototype part development. The charge pattern can consist of single or multiple

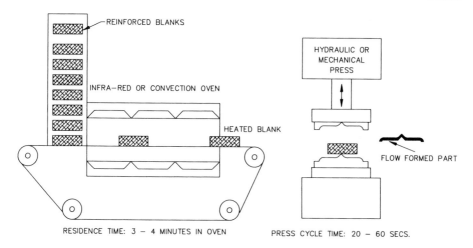

FIGURE 7.7   Molding process for sheet laminates.

blanks. It is dictated by the amount of material needed to mold a part at desired thickness, part complexity, and the extent of flow that can be obtained with a given polymer. Flow to the extent of 50–70% (i.e., charge coverage 30–50%) can be obtained in high speed compression molding, although this may induce preferred fiber orientation in the molded part. To minimize this effect, usually the extent of flow is limited to 30%. Blank heating needs to be optimized in terms of rate of heating, uniformity of temperature through the thickness, and residence time in the oven, coupled with the molding cycle. Infrared heating provides an effective, fast method of heating, but caution should be exercised not to overheat the surface in attempting to get the interior blank temperature hot. Excessive surface temperature can cause thermal degradation of the polymer. Circulating hot air ovens give more uniform temperatures through the blank thickness but are usually slow. Figure 7.8 shows typical heating cycles. It is recommended that the hot blanks be transferred to

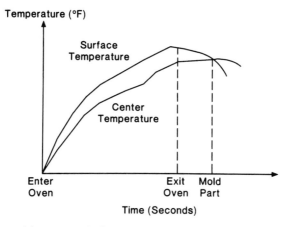

FIGURE 7.8   Typical heating cycle for molding sheet laminates.

251

the mold within 15 seconds or less of their emergence from the oven to reduce heat losses prior to molding.

The difference between stamping press and high speed compression molding press needs to be explained. Mechanical stamping presses operate at a single stroke speed rated in strokes per minute (SPM) and stroke length. The fastest mechanical presses used to date rate at 25 SPM with a 508 mm (20 in.) stroke. Mechanical presses are usually recommended for solid state forming only. High speed compression molding presses have the capability to vary the closing and pressing speeds. The closing or approach speed is typically in the range of 508–1270 cm/min (200–500 in./min). At the end of this stroke, the hot blanks come in contact with the top mold. Then the press switches over to the pressing speed, typically in the range of 76–203 cm/min (30–80 in./min). Pressing speed is a key aspect of the flow forming process. Pressing speeds should be fast enough to prevent the freezing of the hot blank surfaces when they come in contact with a relatively colder top mold surface. Suggested minimum pressing speed is 76 cm/min (30 in./min). Studies indicate that the mold filling takes place in less than 3 seconds, usually in the last inch of the pressing cycle. Mold cycle time for the flow forming process is influenced by a number of factors but is typically in the range of 20–60 seconds.

The mold temperature is another key factor in determining the cycle time and part quality. The mold temperature is largely dictated by the type of the polymer being processed. For amorphous materials, the temperature is typically maintained below the glass transition temperature of the polymer to facilitate removal of the molded part. For semicrystalline polymers, the mold temperature is maintained generally between the polymer crystallization temperature and the glass transition temperature. It is imperative that a rapid rate of crystallization and desired degree of crystallinity be obtained under these conditions. For example, polypropylene can be molded with ambient die temperatures, whereas a polyester (such as PET) will require a die temperature of 121–138 °C. The pressures required for flow forming are generally high because of the high viscosity of the thermoplastic resins. Typical pressures range from 100 to 200 kg/cm^2 (1500–3000 psi). Flows up to 50% are possible, although they are generally limited to 30% to obtain uniform fiber distribution throughout the composite.

Like the injection molding process, flow forming allows one to mold parts with varying cross-sectional thickness in the same part. Molded-in bosses, metal inserts, and molded-in threads are also possible. Since no fiber attrition occurs in this low shear molding process, excellent mechanical properties are obtained in the composite relative to injection molding. Also, cycle times are often competitive with injection molding. Flow forming also enables one to mold large parts not usually possible by injection molding. Injection molding of large parts requires excessive clamp pressures, hence expensive and large injection molding machines.

7.4.3 Properties

The properties of the random long fiber stampable composites can be predicted using the rule of mixtures equations (7.1 and 7.2). The fiber aspect ratio, however, is much greater than the critical aspect ratio; hence the factor $(1 - l_c/2l)$ essentially becomes 1. The modulus and strength equations are then modified as follows:

$$E_c = E_f V_f C_0 e + E_m V_m \tag{7.4}$$

$$\sigma_c = \sigma_f V_f C_0 e' + \sigma_m V_m \tag{7.5}$$

The symbols e and e' represent the modulus and strength reinforcement efficiency factors, respectively. The value of C_0 for in-plane random orientation of the fiber, as previously stated, is around 0.33. For three dimensional random orientation, the value is 0.18. The values fall in between for distributions that lie between planar and three-dimensional random orientation. The modulus and strength reinforcement efficiency factors for this type of composite are largely dictated by the effectiveness of the resin penetration into the fiber bundle. It is important that all the filaments in the fiber bundle be wetted out to achieve efficient reinforcement, assuming that the interfacial compatibility is excellent. This is largely dictated by the lamination or impregnation process used in the fabrication of the sheet composite. Table 7.5 gives values of strength reinforcement efficiency factors for composites with various fiber lengths (9).

TABLE 7.5   Strength Reinforcement Efficiency Factor (9)

Matrix	Reinforcement (glass fiber)	Fiber volume fraction	Tensile strength [MPa (ksi)]	Reinforcement efficiency factor
Polypropylene	Random short fiber	0.19	103.4 (15.0)	0.17
Polypropylene	Random long chop	0.19	120.0 (17.5)	0.20
Polypropylene	Swirl mat	0.19	120.0 (17.5)	0.20
Polypropylene	Twill weave	0.33	236.0 (34.3)	0.27
Polybutylene terephthalate	Random long chop	0.16	120.6 (17.6)	0.20
Polyetheretherketone	Random mat	0.31	227.5 (33.1)	0.21
Epoxy	Random mat	0.43	250.4 (36.4)	0.22
Epoxy	Twill weave	0.39	251.3 (36.6)	0.26
Sheet molding compound	Random long chop	0.35	158.6 (23.1)	0.13

Long fiber random mat thermoplastic composites offer a combination of properties unavailable with other thermoplastic materials. They possess not only high modulus but also excellent impact strengths. The impact resistance, in fact, is considerably higher than that obtained from long fiber injection-molded composites. The combination of stiffness and impact resistance, coupled with the high heat resistance of certain resins, provides a new class of materials for structural and semistructural applications.

The properties of random mat glass fiber reinforced AZDEL polypropylene composites are given in Table 7.6 (10). The impact strengths of polypropylene and

TABLE 7.6   Polypropylene Swirl Mat Composite Properties

Property	30% Glass	40% Glass
Specific gravity	1.10	1.19
Tensile strength, MPa	97	87
, ksi	14	12.6
Flexural strength, MPa	165	132
, ksi	24	19.2
Flexural modulus, GPa	5.52	4.62
, Msi	0.80	0.67
Izod impact strength, J/cm	7.48	6.51
, ft-lb/in.[a]	14	12.2
Dynatup impact strength [3.3 mm (0.130 in.) thick], J	22	20.5
, ft-lb[b]	16	14.9
Heat deflection temperature at 1.82 MPa (264 psi), °C	154	154
Coefficient of thermal expansion (-40 to 93°C), m/m/°C	2.7×10^{-5}	2.7×10^{-5}

[a] 1 ft-lb/in. = 0.534 J/cm
1 ft-lb = 1.356 J

polybutylene terephthalate (PBT) as a function of fiber forms are compared in Table 7.7 (9). It can be clearly seen that the impact strengths increase as one goes from short fiber to long fiber reinforcements. Properties of fiber glass mat composites based on various thermoplastic resins are illustrated in Table 7.8. Random fiber mat (including carbon fiber) composite properties based on Phillips's polyphenylene sulfide are illustrated in Table 7.9 (11). In addition to high strength, stiffness, and toughness, these composites provide excellent short- and long-term thermal properties, excellent chemical resistance, and inherent flame resistance.

7.4.4   Applications

A unique combination of strength, toughness, and easy processibility has made random fiber mat based thermoplastic composites strong competitors for applications in the automotive, electrical, appliance, furniture, power equipment, chemical process, aircraft, aerospace, military hardware, and sporting goods industries.

TABLE 7.7   Impact Strengths

Matrix	Reinforcement (glass fiber)	Volume fraction	Impact strength J/cm (ft-lb/in.)[a]
Polypropylene	Swirl mat	0.14	6.5 (12.1)
Polypropylene	Swirl mat	0.19	7.8 (14.5)
Polypropylene	Random long chop	0.19	3.0 (5.6)
Polypropylene	Twill weave	0.45	11.4 (20.7)
Polypropylene	Random short chop	0.19	0.74 (1.4)
Polybutylene terephthalate	Swirl mat	0.21	6.99 (13.0)
Polybutylene terephthalate	Random short chop	0.18	0.85 (1.6)

[a] 1 ft-lb/in. = 0.534 J/cm
1 ft-lb = 1.356 J

TABLE 7.8   Properties of Thermoplastic Swirl Glass Mat Composites[a]

Property	Polypropylene	PBT	PET	PBT/PC
Glass content, wt%	40	35	35	35
Specific gravity	1.19	1.59	1.67	1.52
Tensile strength, MPa	97	103	97	103
, ksi	14.0	15.0	14.0	15.0
Tensile modulus, GPa	5.52	8.27	6.9	6.9
, Msi	0.80	1.2	1.0	1.0
Flexural strength, MPa	165	193	196	186
, ksi	24.0	28	28.5	27
Flexural modulus, GPa	5.52	8.27	7.58	7.58
, Msi	0.80	1.2	1.1	1.1
Izod impact strength, J/cm,	7.52	6.99		2.69
, ft-lb/in.	14.0	13.0		5.0
Dynatup impact, J	22	29		
, ft-lb.	16	22		
Heat deflection temperature at 264 psi, °C	154	218	248	166

[a] PBT = polybutylene terephthalate; PET = polyethylene terephthalate; PBT/PC = alloy of PBT and polycarbonate.

TABLE 7.9   Polyphenylene Sulfide Based Composite Properties; Random Mat Reinforcement

Property	30% Glass	40% Glass	20% Carbon	30% Carbon
Tensile strength, MPa	124	158	138	200
, ksi	18.0	23.0	20.0	29.0
Tensile modulus, GPa	11.0	12.4	11.7	17.2
, Msi	1.6	1.8	1.7	2.5
Flexural strength, MPa	220	234	214	317
, ksi	32	34	31	46
Flexural modulus, GPa	11.0	12.4	11.7	17.2
, Msi	1.6	1.8	1.7	2.5
Izod impact strength				
notched, J/cm	6.44	7.52	0.38	1.34
, ft-lb/in.	12	14	1.3	2.5
unnotched, f/cm	10.2	13.4	2.68	3.78
, ft-lb/in.	19	25	5	7
Heat deflection temperature at 264 psi, °C	273	274	270	

255

FIGURE 7.9   Passenger car bucket seat (polypropylene Azdel composite).

FIGURE 7.10   Bumper backup beam (polypropylene Azdel composite).

Typical automotive applications include battery trays, seats (Fig. 7.9), bumper beams (Fig. 7.10), station wagon load floors (Fig. 7.11), engine oil pan and valve covers, and sound shields. Among the nonautomotive uses are office furniture (Fig. 7.12), modular flooring, microwave food trays, corrosion liners, and impact-resistant liners. Research and development efforts are also in progress to produce composites capable of giving a Class A surface finish for the exterior automotive horizontal panel application. These composites have to meet the typical paint bake temperatures in

FIGURE 7.11   Ford Taurus station wagon load floor (polypropylene Azdel composite).

FIGURE 7.12   Typical office furniture (polypropylene Azdel composite).

excess of 149 °C. This means the polymer should have essentially no shrinkage coming out of the paint bake cycle. Even a small amount of shrinkage can cause unacceptable fiber prominence on the surface of the molded part. It is a challenge to tailor such a polymer system and yet maintain adequate processing characteristics in terms of viscosity. Crystalline polymers with a crystallization temperature range near or below 149 °C will not be suitable due to resin shrinkage which occurs as a result of crystallization. Ideally, an amorphous polymer with glass transition temperature above 149°C will be required to overcome the shrinkage issue. However, such a polymer usually will have a high molecular weight, resulting in high melt viscosities. The temperature during lamination has be to sufficiently high—315 °C for a polymer with a T_g of 149 °C—to have a workable polymer viscosity. The fiber form, including the strand geometry (filament diameter and number of filaments in a bundle), also has a dominant effect in obtaining a Class A surface. Small bundles and fine filaments are ideal from the surface viewpoint. However, they may provide resistance to resin penetration, especially if the reinforcement is in the form of a mat. Optimization of fiber geometry and form with respect to surface and wettability is essential. In addition to materials, the molding process needs to be optimized. It is preferable to keep the mold temperature below T_g for amorphous polymers and near crystallization temperature for semicrystalline polymers to minimize postmolding shrinkage. However, this may cause the surface of the molten charge to freeze as soon as it comes in contact with the mold surface during the molding cycle. This, in turn, may cause surface defects. Thus, numerous technical challenges must be met to make Class A stampable thermoplastic material a commercial reality. A systems approach, namely, combination of materials and molding process coupled with integration or consolidation of part design, will be required to meet the performance and economic needs of the automotive industry in this class of application.

7.5 CONTINUOUS FIBER THERMOPLASTIC COMPOSITES

The product forms used in thermoplastic composites are similar to those available in thermoset systems. Unidirectional tow and tape, woven forms, and three-dimensional knitted preforms that use carbon, glass, and organic fibers are available to meet various application requirements. Product forms can further be classified as preimpregnated or postimpregnated (12). In the former case, the fibers are completely wetted and fully impregnated by the resin in one step before part fabrication. In the latter case, the fibers and resin are brought in close physical proximity to each other without any adhesion or bonding. Impregnation takes place during the part fabrication step. Unidirectional tape and tow, woven fabric, and woven prepreg tow represent preimpregnated forms, while cowoven carbon (or glass)/thermoplastic fibers, commingled carbon (or glass)/thermoplastic fibers, woven fabric with stacked films, and powder impregnated strand/tow or woven fabric represent examples of postimpregnated forms.

The choice of resins used in each of the processes is dictated by the physical properties of the polymers. Resins that can be converted to film, powder, or fibers can be used in both processes. However, resins such as pseudothermoplastics (12) are limited to preimpregnation. Pseudothermoplastics are distinguished from conventional thermoplastics as follows. In conventional thermoplastics, the chemistry is complete and processing involves reversible physical change, such as heating, melting, and cooling to solidify. In pseudothermoplastics, the chemistry continues during processing, increasing molecular weight and/or expelling volatiles to achieve target physical properties. Polymers that are categorized as pseudothermoplastics include polyamide-imide (Torlon) and polyimides (Aramid KIII, LARC-TPI). The intrinsic properties of the polymer, such as compatibility or solubility, and melt viscosity, are major factors in dictating the type of prepreg process. Resins with low melt viscosities can be used in the melt impregnation process, whereas resins that are soluble in selective solvents are suitable for solution impregnation. Other resins may require monomeric or prepolymer impregnation.

7.5.1 Preimpregnation: Process and Forms

As stated earlier, preimpregnation is done using melt or solution impregnation. Melt impregnation uses an extruder and a crosshead die as shown in Figure 7.1. The fiber bundle has to be opened up by suitable means prior to impregnation so that the molten polymer can be made to penetrate into the bundle. The die can be configured with a round opening for fiber tows or bundles or a slit opening for tapes. To obtain uniform fiber distribution in the prepreg, it is important to maintain appropriate tension, alignment, and spacing of the fibers in the tape process. In the solution impregnation process, the polymer is dissolved in an appropriate solvent. The viscosity of the solution can be manipulated by modifying the polymer concentration and the solution temperature. The fibers pass through the solution and then through suitable squeezing devices to remove excess polymer. The squeezing devices can be in roll or orifice form. The impregnated fibers then pass through a solvent recovery system where sufficient heat and time are provided to remove all of the solvent. Of course, the solvent is recovered and reused.

Carbon fiber single tows (3 k, 6 k, and 12 k) and unidirectional tapes up to 25 mm (12 in.) wide are available with almost all thermoplastic resin systems (12). The typical weights range from 80 to 190 g/m². The typical industry standard is 145 g/m² and 0.125 mm (0.005 in.) thick. Single tows are available in lengths up to 914 m (3000 ft), whereas tapes are typically 46 m (150 ft) long. The conventional thermoplastics produce tack-free prepregs that are like thin cardboard, whereas pseudothermoplastics produce slightly tacky prepregs because of trace solvent still present at this stage. Fiberglass forms are available, but to a limited extent due to shortcomings in certain interfacial performances, such as thermal and hydrolytic stability. Aramid fiber has been successfully preimpregnated via the solution impregnation technique, but the high temperature required in melt impregnation with polymers such as PEK and PEEK may cause some thermal degradation of the aramid fiber.

Preimpregnated fabrics in broad widths are available with carbon, glass, and aramid fibers and resins that are amenable to the solvent impregnation process. As a result of the high temperature, high viscosity requirements for semicrystalline and liquid crystal polymers, melt impregnation of fabrics has been limited to narrow widths.

7.5.2 Postimpregnation: Process and Forms

As stated previously, postimpregnation can be done via film, commingled fibers, or powder technology. The processing ease and the ultimate properties are dictated by

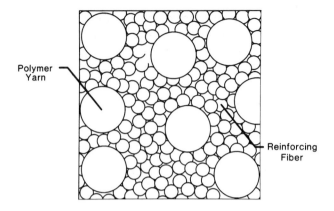

FIGURE 7.13   Commingled polymer yarn and reinforcing fiber.

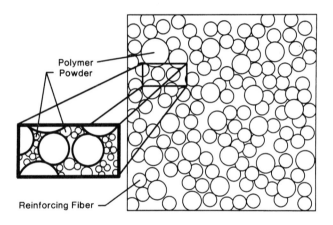

FIGURE 7.14   Commingled polymer powder and reinforcing fiber.

how intimately the resin and the fibers are in contact during the consolidation step. In the film stacking, alternate layers of polymeric films and fiber (usually fabric) are subjected to high temperatures and pressures to achieve consolidation. This requires excessively long times (sometimes on the order of hours). The resulting wetting and properties are less than optimum.

The commingling fiber process enables one to bring the monofilament or multifilament forms of the reinforcement and the polymer fibers in improved contact with one another. Commingling fine multifilaments of the polymer (20 μm) with the reinforcing fiber can result in excellent wetting and properties close to those obtained from preimpregnated forms. Figure 7.13 shows schematically a commingled yarn with high reinforcement fiber volume.

The powder impregnation technology uses a fluidized-bed technique in which the polymer powder is trapped between the filaments of a bundle by electrostatic charge. Some partial fusing of the powder may be required to provide integrity in the

260

handling of the fiber bundle after impregnation. An alternate technique includes covering the fiber-impregnated bundle by a thin sheath of the same polymer (preferably lower molecular weight) to prevent any powder loss during handling. The properties obtained by this technique improve as the powder diameter approaches micrometer range. Figure 7.14 schematically illustrates a powder commingled composite with high fiber loading.

The postimpregnated forms from commingled fibers and powder technology lend themselves to be processed through braiding, knitting, and weaving processes. This allows fabrication of a wide range of styles, widths, thicknesses, and shapes of fabrics for the manufacture of parts. These fabrics provide excellent drapability for complex part fabrication, especially when deep draws are encountered.

7.5.3 Properties

Properties of the unidirectional composites can be predicted using the rule of mixtures. Properties of the quasi-isotropic and other forms can be calculated using the laminate theory. The actual values obtained relative to theoretical prediciTons determine the efficiency of the impregnation process and the effectiveness of the interfacial bonding between the polymer and the reinforcement.

Properties of continuous glass fiber thermoplastic composites fabricated via the pultrusion process are given in Table 7.10 (13). Properties of carbon fiber based composites obtained via the commingling process are given in Table 7.11. The elevated temperature property retentions are normally dictated by the type of resin in the composite.

TABLE 7.10   Unidirectional Glass Fiber Composite Properties[a]

Property	Nylon 6	Polypropylene	PETG polyester
Fiber, wt %	50	65	55
Specific gravity	1.58	1.40	1.79
Tensile strength, MPa	637	624	762
, ksi	92	90	110
Tensile modulus, GPa	28.9	26.2	37.0
, Msi	4.2	3.8	5.1
Flexural strength, MPa	624	637	693
, ksi	90	92	100
Flexural modulus, GPa	23.9	17.9	36.3
, Msi	3.3	2.6	5.0
Interlaminar shear, MPa	41.6	25.6	37.4
, ksi	6.0	3.7	5.4
Notched Izod strength, J/cm	48.3	24.2	40.3
, ft-lb/in.	90	45	75
Unnotched Izod strength, J/cm	56.4	32.2	48.3
, ft-lb/in.	105	60	90

[a] Composites fabricated via pultrusion process; PETG = Copolyester of terephthalic acid, ethylene glycol and cyclohexane dimethanol

TABLE 7.11   Unidirectional Carbon Fiber Composite Properties[a]

Property	PPS	PEEK
Tensile strength, MPa	1829	1850
, ksi	264	267
Tensile modulus, GPa	117	174
, Msi	17	21
Elongation, %	1.29	1.27
Flexural strength, MPa	1781	2286
, ksi	257	330
Flexural modulus, GPa	131	131
, Msi	19	19
Interlaminar shear, MPa	62.4	90.1
, ksi	9	13

[a] Composites fabricated via commingling process: PPS = polyphenylene sulfide; PEEK = polyetheretherketone

7.5.4 Fabrication

The fabrication process for advanced thermoplastic composites is dictated by part shape, the type of resin (true thermoplastic or pseudothermoplastic), and the product form (preimpregnation or postimpregnated). The true thermoplastics, in which chemistry is complete, are amenable to automated processing at rapid rates because they require only melting, shaping, and consolidating during the fabrication step. The pseudothermoplastics, on the other hand, require precise control of heating rates, temperature, pressure, and time to reach desired molecular weights or to remove trace solvents. The fabrication processes are similar to those used for the advanced thermosets (such as bagging and autoclaving).

Preimpregnated forms lend themselves to high speed automated processing. However, the product forms are stiff, lacking the drapability of the thermoset or postimpregnated forms. This limits their application to the fabrication of flat or simple contoured parts. Postimpregnated forms, although flexible and drapable, enabling complex part fabrication, require higher temperature, pressure and time and are generally processed by matched metal die or pultrusion techniques.

Preconsolidated sheets can be heated rapidly using infrared quartz heaters. Typical heating times are 1–2 minutes when the prepreg is heated from both sides. The cooling rates for semicrystalline polymers must be controlled appropriately to obtain suitable morphology in the matrix, which, in turn, controls properties such as toughness and solvent resistance. Hydro-forming and rubber block forming can be used for a sheet stock. The process requires a single-sided mold. The preheated sheet is shaped by applying pressure from a hydraulically pressurized rubber diaphragm or by mechanically deforming the rubber block.

Roll forming can be used to form long structural shapes with hat or Z sections. Pultrusion through a hot die can be used to reshape thermoplastic prepreg tape to desired forms. Die design is critical to the part quality. The die design must permit heating while gradually changing the cross section as the tape progresses through the die. Some compliance may have to be built into the die to provide constant pressure for consolidation to compensate for lack of compliance of the composite (with high fiber loading) during processing. Postimpregnated forms are not generally suitable for pultrusion because they require excessively long times for melting, shaping, and consolidating.

The automated tape laying process can be used for larger planar parts. Heating, followed by rolling on the mold surface, is done for each layer. Consolidation pressure is maintained for a few seconds after rolling down each layer to get adhesion to the previous layer. Filament winding, similar to the tape laying process, can use unidirectional material in the tape or tow form. The tape or the tow is heated before being wound on the mandrel. The mandrel can be heated to aid consolidation. Pressure required for consolidation can be applied by maintaining adequate winding tension or through the use of pressure rollers. In both processes, the adjacent layers of the material must be in the molten form when they come in contact with each other to obtain good consolidation.

7.5.5 Economics

Continuous fiber thermoplastic composites are in their infancy. Commercial success will depend on economic routes in making structural parts and assemblies. A systems approach integrating material, process, fabrication, and part design will be necessary to realize the full benefits of using these composites. The current raw material costs reflect the developmental stage of the technology. The costs of the products forms are dictated, in addition, by the number and complexity of the process steps required to produce specific forms. The processing steps can add between 1.5 and 2.5 times the cost of the raw materials to the overall cost of the product form. In addition, design, fabrication, and assembly techniques significantly influence the total cost of the finished part. The costs are bound to go down as technological advances are made, and these composites can then be expected to become competitive with other structural materials.

7.6 CONCLUSION

Long fiber thermoplastic composites provide alternate material forms for semi structural and structural applications and mostly lend themselves to high speed fabrication processes used in the metal and thermoset industries. These advantages, combined with indefinite material shelf life, ease of part repair, increased durability and toughness, corrosion resistance, and recyclability provide tremendous opportunity for the growth and advancement of this segment of fiber reinforced composite technology.

Different segments of the long fiber thermoplastic composites technology are in different stages of advancement. The segment that has probably made excellent commercial progress is stampable sheet composites. Substantial research and development efforts are taking place in short/long fiber injection-moldable composites.

Extensive commercial use of these materials can be expected in the next 5 years. Continuous fiber thermoplastic composites will require significant development and optimization of the materials and process technologies before they can make substantial inroads into the reinforced composite marketplace. Even then, they will be used in select applications and niche markets. Initial penetration will be in the aircraft industry and, as the technology matures, proliferation into other industries, such as automotive, can be anticipated.

ACKNOWLEDGMENTS

The author thanks the following companies for providing technical information: AZDEL, Inc.; BASF Industries; LNP Corporation; PPG Industries, Inc.; Phillips Petroleum Company; and Polymer Composites, Inc. Special thanks to PPG Industries, Inc., and AZDEL, Inc., for providing photographs for illustration.

REFERENCES

1. R. B. Gosnell, in *Engineered Materials Handbook*, Vol. 1, *Composites*, ASM International, Metals Park, OH, 1987 pp. 97–104.
2. *Materials & Process Report*, Massachusetts Institute of Technology, May 1988.
3. C. R. Gore, G. Cuff, and D. A. Cianelli, *Mater Eng.*, March 1986.
4. D. A. Cinanelli, J. E. Travis, and R. S. Bailey, *Plast. Technol,* pp. 83–89, April 1988.
5. J. Karger-Kocsis and K. Friedrich, *Compos. Sci. Technol.*, 32, (4), 293–325 (1988).
6. R. L. McCullough, *Composite Design Guide*, Vol 2, Center for Composite Materials, University of Delaware, 1981.
7. J. E. Travis, D. A. Cianelli, and C. R. Gore, *Machine Des.*, p. 3, Feb. 12, 1987.
8. K. R. Quinn and G. S. O'Brien, *Plast. Eng.*, 44(4), (1988).
9. D. M. Biggs, D. F. Hiscock, D. C. Schiltz, J. R. Preston, and E. J. Bradbury, *Plast. Eng.*, 44(3), (1988).
10. R. L. Stadterman, 43rd SPI Annual Technical Conference, Session 3-F, Composite Institute, February 1988.
11. D. G. Brady and T. P. Murtha, Society of Plastic Engineers, ANTEC 1985, pp. 1178–1180.
12. M. T. Harvey, in *Engineered Materials Handbook*, Vol. 1, *Composites*, ASM International, Metals Park, OH, 1987, pp. 544–553.
13. A. Youngs, Technical Conference on Composites in Manufacturing, EM 85-102, Society of Manufacturing Engineers, January 1985.

8 · Injection Molding of Thermoplastic Composites

Bohuslav Fisa

Contents

●

B. Fisa, Department of Mechanical Engineering, École Polytechnique de Montréal, Montreal, Quebec, Canada

8.1 INTRODUCTION

Injection molding of thermoplastic materials is a cyclic process in which a molten polymer is injected into a closed, cold mold where it solidifies, taking the shape of the mold cavity. The mold is then opened, the molding is removed, and the cycle is repeated. Since the first hand-operated injection machines were introduced more than 60 years ago, injection molding has evolved into a complex, sophisticated process particularly suitable for large production runs—many thousands or millions of nominally identical parts. Of all plastics processing methods, injection molding is second only to extrusion: in 1987, in the United States, more than 4 million tons of thermoplastics (about one-fifth of the total plastics production) was injection molded (1). Pails, milk crates, pipe fittings, TV and radio housings, transparent drinking cups, telephone housings, and fans are some of the most common examples of injection-molded products.

Virtually every thermoplastic resin is injection molded, and most resins are also available in several filled and/or reinforced forms. The terms "filled" and "reinforced" indicate that a second, discontinuous, usually more rigid, phase has been blended into the polymer. When the aspect ratio (i.e., the ratio of the largest to the smallest dimension) is close to 1, the second phase is referred to as "filler." If the aspect ratio is much larger than 1, as is the case with fibers, the term "reinforcement" is used to describe the second phase. Filled and reinforced thermoplastics are used because they allow a significant and easy modification of the base resin properties. This can be seen in Table 8.1. Introduction of calcium carbonate into polypropylene improves room temperature stiffness significantly, whereas high temperature stiffness, as measured by deflection temperature under load, is only slightly improved. Glass fibers provide a much higher room and high temperature rigidity than the unfilled polypropylene (2). Talc, a plateletlike material, provides a reinforcement intermediate between those of calcium carbonate and glass fibers. Similar changes in properties are observed with other commodity and engineering resins.

The process of incorporation of the second phase into the polymer matrix (compounding) is expensive and can be justified only if better properties are provided

TABLE 8.1 Effect of Calcium Carbonate, Talc and Glass Fibers on Selected Properties of Polypropylene (2)

Material	Flexural modulus (GPa)	Notched Izod impact strength (J/m)	Deflection temperature under load (°C)
Polypropylene/ homopolymer	1.7	45	65
Polypropylene, 40% (wt),/ calcium carbonate	2.8	53	75
Polypropylene, 40% (wt),/ talc	4.2	35	97
Polypropylene, 30% (wt),/ glass fiber	6.5	100	148

267

TABLE 8.2   Cost of Selected Injection Molding Resins (3)

Material	Resin cost	
	$/kg	$/1000 cm^3
Polypropylene		
Unfilled	1.09	1.00
40% mineral filled	1.84	2.28
40% glass fiber	1.85	2.28
Polyamide 66		
Unfilled	4.11	5.03
Mineral filled	2.77	4.03
30% glass fiber	3.81	5.43
Polycarbonate		
Unfilled	3.55	4.36
30% glass fibers	4.59	6.92
Polyetheretherketone		
Unfilled	50.60	66.70
30% glass fibers	45.10	67.12

by the filler or by the fibrous reinforcement. This is illustrated by Table 8.2, which lists costs of four resins in unfilled, filled, and reinforced varieties. For example, a typical filler costs less (per unit weight) than polypropylene resin, yet the cost of filled polypropylene is almost double that of unfilled polymer. In fact, fillers and reinforcements extend the range of properties available from a given base polymer toward values that would otherwise require the use of a different, more expensive, resin rather than reduce its cost. Thus glass fiber reinforced polypropylene can in certain applications replace unfilled polyamide or polycarbonate. While it is possible to reinforce efficiently any thermoplastic resin, only certain fiber–resin combinations have found widespread use. For example, glass fiber reinforced high density polyethylene is little used because the properties sought (mainly high temperature stiffness) can be much more easily obtained with a similarly priced but higher melting polypropylene. Table 8.3 lists the production of reinforced thermoplastics in the United States in 1987. Polypropylene, thermoplastic polyesters, and polyamide each have about 25 % of the

TABLE 8.3   Use of Reinforced Thermoplastics in the United States in 1987 (1)

Material	Amount used (tons/year)
Polyamides	70,000
Polypropylene	65,000
Polycarbonate	21,000
Thermoplastic polyesters (PET, PBT)	56,000
Styrenics (SAN, ABS, SMA, etc.)	22,000
Other (MPPE, sulfones, PVC, etc.)	16,000
Total	250,000

market. In Europe, the situation is different: polyamides occupy more than half the reinforced thermoplastics market (4).

Reinforcing material can be either of fibrous or planar shape. In practice fibrous reinforcements are almost exclusively used, with the glass fibers dominating this market segment. Carbon or aramid fibers provide higher stiffness at lower weight, but their use in large volume applications has been limited by their high cost. Planar reinforcements such as talc, mica, or glass flake are also available and can be used where stiffness and isotropy are required. Mica has found applications as an efficient low cost stiffener of certain commodity plastics where lower impact strength can be tolerated.

Strengths and limitations of reinforced thermoplastics have the same origin: they can be processed on the same equipment as unreinforced resins. Fibers or fillers can be easily blended into the molten resin using a single-screw or twin-screw extruder; the resulting compound is processed on an ordinary injection molding machine at high production rate. Unfortunately, the necessity to make the matrix–fiber mix flow during processing imposes serious limitations on the product. In particular, fiber length and maximum fiber loading are far less than optimal. Moreover, fiber orientation in the molding is determined by flow. Both these phenomena affect and control the ultimate properties available from this class of reinforced plastics.

This chapter deals with injection-molded thermoplastic composites. According to a relatively broad (but not the broadest) definition, any two-phase material can be considered to be a composite. In practice, however, there are difficulties in establishing a clear boundary that distinguishes between filled and unfilled polymers, since the latter very often contain a certain amount of a second phase (pigments, antioxidants, etc.). In addition, filled thermoplastic composites (i.e., those in which the aspect ratio of the rigid, dispersed phase is close to 1) process and behave more like unfilled polymers than those containing higher aspect ratio reinforcing particles. For this reason, this chapter covers mainly fiber reinforced thermoplastics.

We will explore relationships indicated in Figure 8.1. As with other composite materials, properties of a short fiber reinforced thermoplastic composite depend not only on the properties of its constituents but also on the way they are put together. There has to be a certain compatibility between the materials and the process. High molecular weight of the polymer and high fiber length and/or concentration will hinder flow and limit the geometry of molded parts to simple shapes. Both process and material characteristics influence composite microstructure, which in turn determines, together with the properties of its constituents, the ultimate properties.

We begin with a brief description of the injection molding process. The section that follows describes how a molded part is formed in the mold, in particular how the combination of flow and rapid cooling affects the microstructure of the finished part. We then proceed to a discussion of the properties available from these materials. The chapter concludes with a short description of some recently developed related materials (long fiber injection-moldable composites, liquid crystal polymers, reinforced structural foams) and processes (coinjection, injection–compression, and lost metal core injection molding).

269

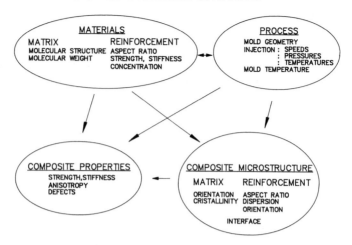

FIGURE 8.1   Structure–processing–properties relationships in injection molded thermoplastic composites.

8.2 INJECTION MOLDING PROCESS

An injection molding machine (Fig. 8.2) consists of two principal parts: an injection unit and a clamping unit. The injection unit performs two functions. It converts the thermoplastic material into a uniform, continuous melt and injects it into the mold. During the initial stages of cooling, it holds the molten polymer in the mold under pressure to compensate for the thermal contraction the material undergoes on cooling. The clamping unit holds the mold and keeps it closed while it is being filled and the material solidifies. It also operates the opening and closing mechanism and ejects the molded part from the mold.

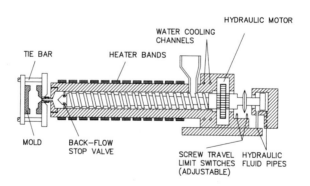

FIGURE 8.2   A reciprocating-screw injection molding machine.

8.2.1 Injection Unit

There are several configurations of injection units in use today. The simplest, first-generation plunger and torpedo machines are still being made, but this chapter describes only the most widely used type, the reciprocating single-screw injection unit. Some of the more complex machines, which may combine a screw with a plunger or two independently operated screws, are discussed in several recently published books on injection molding (5–8). The screw in reciprocating-screw injection unit acts in two stages. First, it acts as an extruder; that is, it plasticates a sufficient amount of the thermoplastic resin required for one injection cycle. The raw material, typically supplied in the form of pellets from the hopper, is conveyed forward by the rotating action of the screw. In the first zone of the screw the solid material is compacted, and the air from spaces between pellets is expelled and escapes back to the atmosphere through the hopper. Compacted polymer gradually melts as it advances along the screw. The energy necessary to melt the polymer is provided by a combination of heat from the heating bands and mechanical heat generated by the rotating screw. While the screw rotates, the front end of the barrel is closed, the pressure generated by the melt pushes the screw rearward, and the molten polymer accumulates in the space ahead of the screw. The pressure the screw must overcome before it starts moving back is called *back pressure*: it is a measure of the mechanical work done on the polymer. Optimum back pressure produces well mixed, thermally homogeneous melt. Excessive back pressure may lead to polymer and fiber length degradation. When a sufficient amount of polymer melt has accumulated in front of the screw, molten plastic is injected into the mold by pushing the screw forward. Here the screw acts as a plunger. To prevent the melt from flowing back into the screw, the screw tip is equipped with a nonreturn valve. The screw rotation may be stopped before the screw starts moving forward or only when the mold is full.

The most important process parameters controlled by the injection unit are:

- *Melt temperature*: The temperature of the melt when it penetrates into the mold is controlled by the temperature control system of the injection unit but may also be affected by the injection speed (see Section 8.3.1.4) and by the level of back pressure (5).
- *Injection speed*: This is the speed at which the screw advances during the mold filling step. Modern machines are equipped with variable injection speed control—a profile of speeds rather than a single constant value is used to fill the mold. Typical mold filling starts at a slow speed to prevent jetting (see Section 8.4.2); speed is increased during the middle part of filling and reduced again toward the end to prevent flashing or air trapping and to allow smooth and accurate transition to pressure control (5,9), which takes over when the mold is full.
- *Injection pressure*: The pressure exerted by the screw on the melt is not constant during the mold filling stage. Injection pressure builds up as the mold is filled and as the resistance to flow increases. It is only when the mold is full that a transfer from speed control to pressure control takes place. Injection pressure is the principal variable during the holding stage (i.e., when the melt in the mold is being

271

compressed to control shrinkage) (5). The transfer from speed to pressure control is initiated by one of the four modes: time, screw position, mold cavity pressure, or hydraulic pressure (9).

The injection units are rated by their *injection capacity*, the maximum volume of melt that can be injected into the mold during one molding cycle. Injection capacity, which depends on the screw diameter and on the maximum screw stroke, traditionally has been expressed in terms of weight of polystyrene. Thus a 500 g machine will be able to inject 500 g of a polymer having the same melt density as polystyrene. The smallest injection units have an injection capacity of a few tens of grams; the largest one ever built has a capacity of 350 kg (5). Other important characteristics of an injection unit are:

- *Maximum injection pressure* the screw can exert on the melt: in a typical injection unit it may be between 100 and 200 MPa.
- *Maximum injection rate* at which the melt can be injected into the mold (expressed in cubic centimeters per second).
- *Recovery rate* is the measure of the plasticating capacity of the injection unit (expressed in cubic centimeters per second).

8.2.2 Clamping Unit and Mold

The mold is mounted and operated by the clamping unit. The principal function of the clamping unit is to keep the mold closed during an injection cycle. The clamping units are rated in terms of *clamping force*, which is the maximum force available to resist opening of the mold by the melt. Small injection machines have clamping unit capacities of less than 100 tons. The above-mentioned largest machine ever built has a clamping unit rated at 10,000 tons. Compared with other plastics processes the clamping forces required are very high. In a study reported by Morton-Jones (10), a 5 kg molding of large flat shape requires 50 tons clamp force when molded by reinforced reaction injection molding (RRIM), 800 tons by sheet molding compound (SMC), and between 2500 and 3500 tons in thermoplastics injection molding. For this reason the injection mold construction must be substantially more sturdy and molds are therefore more expensive than for other processes.

The mold installed in the clamping unit contains one or more cavities, which hold the injected plastic until it has solidified sufficiently to be ejected from the mold. The simplest type of injection mold is illustrated in Figure 8.3. It consists of two halves, one of which is mounted on the stationary part of the clamp unit and contains the sprue bush (the channel connecting the injection unit to the mold and the runner system). Runners are channels located between two mold halves which connect the sprue bush to the mold cavities. In the *runner* there is usually a restriction immediately before the cavity. This restriction, which may assume several shapes, is called *gate*. The other half is mounted on the mobile part of the clamp unit. It houses the ejector system. The cavity surface temperature is maintained constant by means of a water- or oil-filled cooling system. *Mold temperature* is one of the most important molding

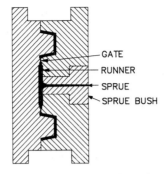

FIGURE 8.3   A simple two-plate mold. Cooling channels and ejectors are not shown.

parameters. Today complex and intricately shaped parts are made by injection molding requiring accordingly complicated molds. Several monographs deal exclusively with the art and science of injection mold design (11–13).

8.2.3 Molding Cycle

The operating sequence of an injection molding machine is illustrated in Figure 8.4. We start (stage 1) with the open mold and the injection unit located some distance from the mold to prevent heat transfer between the sprue bush and the hot injection nozzle. The injection unit contains enough melt for one cycle. The mold is closed and the injection unit is thrust against the sprue bush (stage 2). The screw acting as a plunger moves ahead (stage 3) and injects the melt into the mold. When the mold is almost full, the control is transferred from injection speed to injection pressure. With the mold completely filled, the injection unit maintains pressure on the polymer melt. This pressure is called *holding pressure*. Stage 4 begins when the material in the gate has solidified. At this moment it is no longer possible to compress the melt by

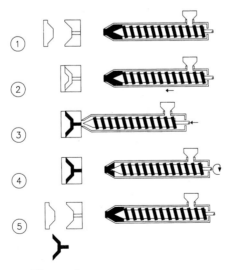

FIGURE 8.4   Injection molding cycle.

the action of the screw; the nozzle is then closed, and the entire injection unit backs off and starts plasticating the polymer required for the next cycle. Finally, the mold is opened, the part is ejected (stage 5), and another cycle can be initiated. Depending on the part size and thickness, an injection molding cycle can last from a few seconds to several minutes.

8.2.4 Molding Reinforced Thermoplastics

There is no one set of rules universally applicable to reinforced thermoplastics. The principal areas of concern when working with these materials are machine wear and fiber length degradation. Machine wear can be reduced by manipulating the temperature profile along the screw. The idea is to melt the polymer early, thus reducing the mechanical work done on the reinforced melt as it progresses along the screw. In the mold the runners and gates are larger than those used with unfilled polymers. Both these factors also reduce the fiber length degradation during molding. Molding conditions recommended for a great variety of reinforced plastics are given in Reference 14.

8.3 PART FORMATION IN THE MOLD

The principle of injection moldings is simple: fill the mold cavity with molten polymer and wait until it solidifies. However, the events taking place in the mold during filling and cooling are very complex and not all of them are fully understood at this time. These phenomena may have a determining influence on the properties of the molded product.

Considering only the mold, an injection molding cycle (Fig. 8.5) can be divided into three stages: filling, packing, and cooling (6). As the polymer is injected into the mold, a pressure sensor located in the mold close to the gate registers a surge in pressure when the melt front reaches it ($t = 0$) but, since the mold is almost empty

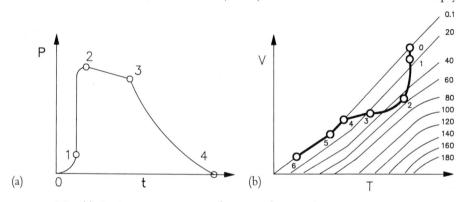

FIGURE 8.5   (a) Cavity pressure, P, as a function of time t during one injection cycle. (b) An injection molding cycle on a pressure–volume–temperature (PVT) diagram of an amorphous polymer. Numbers on the right indicate the pressure in mega pascals (see text).

274

at this stage, the pressure is low. As the melt spreads in the mold, the pressure gradually grows until the mold has been cmpletely filled ($t = 1$). At this moment, the mold packing stage begins and (between 1 and 2), the melt is compressed to compensate for the volume shrinkage the material will undergo during cooling (Fig. 8.5B). At this time ($t = 2$), the bulk of the material in the mold did not have time to cool substantially and the average melt temperature in the area near the gate is close to the initial one ($t = 0$). During the second part of the packing stage (between 2 and 3), when the gate and runners have not yet completely frozen (i.e., when there is still a pathway of the molten polymer to the nozzle), additional melt is injected to compensate for shrinkage caused by cooling. At $t = 3$, the material in the gate has frozen and the thermal shrinkage can no longer be compensated by bringing in more material from the injection unit. The pressure now decreases rapidly to reach atmospheric pressure at $t = 4$. At this time, the partially solidified object may separate from the mold wall. The molding is ejected from the mold at $t = 5$, and additional shrinkage occurs before reaching ambient temperature at $t = 6$. It should be understood that while the simplified picture of the molding cycle as presented in Figure 8.5 is qualitatively correct, it is by no means complete. In fact, each point of the molding has a different flow–temperature–pressure history, and consequently the microstructure and properties will not be uniform throughout the part.

In this section, we are concerned with the generation of the molded part in the mold. We start with a brief discussion of flow behavior of reinforced melts.

8.3.1 Flow of Reinforced Melts

8.3.1.1 Viscosity

High molecular weight polymers in molten state have high viscosity, particularly at slow speeds of deformation. As the rate of deformation is increased, the viscosity is reduced. This behavior is represented by curve A in Figure 8.6. The viscosity η is shown as a function of shear rate $\dot{\gamma}$. At low shear rates the viscosity is nearly constant and independent of the shear rate. At high shear rates the viscosity decreases rapidly with the increasing shear rate. This behavior is called shear thinning or pseudoplastic. Injection molding, being a high speed process, takes place at high shear rates: $10^3 < \dot{\gamma} < 10^7 \mathrm{s}^{-1}$ (15), where the melt is considered to obey the power law (16):

$$\eta = A\,\dot{\gamma}^{n-1} \tag{8.1}$$

where A and n are constants dependent on material and experimental conditions. For shear-thinning fluids, the power law index n is smaller than 1. One important aspect of the flow in the mold with both neat and filled or reinforced melts is the shape of the velocity profile (Fig. 8.7). For example, for a Newtonian fluid ($n = 1$) flowing through a tube, the velocity profile is parabolic. For power law fluids ($n < 1$), velocity profiles become flatter, and when $n = 0$, the material would flow as a plug through the tube, with all the shearing taking place at the wall (16).

Introduction of solid particles into the polymer matrix generally leads to an increase in viscosity of the resulting suspension. The magnitude of the increase will

275

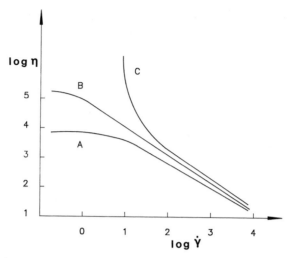

FIGURE 8.6   Shear viscosity η, as a function of shear rate $\dot{\gamma}$ of an unfilled polymer (curve A); a polymer filled with short glass fibers (curve B), and a polymer filled with a finely divided solid such as carbon black (curve C).

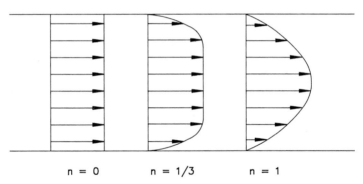

FIGURE 8.7   Velocity profiles of power law fluids flowing in a circular channel for three values of power law index n.

depend on several factors: particle concentration and aspect ratio, as well as the state of dispersion, are among the most important ones. It is convenient to consider the viscosity of suspensions using the Mooney equation (17). This equation is one of the many proposed to relate the viscosity of concentrated suspensions to the particle concentration and shape:

$$\ln \eta_r = \frac{k_E \Phi}{1 - \Phi/\Phi_m} \tag{8.2}$$

where η_r represents the relative viscosity of the suspension obtained by dividing its absolute viscosity by the viscosity of the matrix, Φ is the volume fraction of the filler, and k_E is the Einstein coefficient, which depends on particle shape. For spheres, k_E is 2.5. For particles other than spheres, k_E is larger and increases with the aspect ratio.

276

The remaining parameter, Φ_m, represents the maximum packing fraction. It can be viewed as the maximum concentration the suspension will be able to accommodate; above Φ_m there would not be sufficient quantity of the matrix to wet and embody all the particles. According to Nielsen (18), Φ_m can be defined as follows:

$$\Phi_m = \frac{\text{true volume of the filler}}{\text{apparent volume occupied by the filler}} \qquad (8.3)$$

This definition shows the importance of the spatial arrangement when the "filler" is nonspherical. Consider the case of very long fibers. At random, their apparent volume is very high and Φ_m approaches 0. The same fibers in close hexagonal packing would have a Φ_m of about 0.9.

Packing of particles of various shapes and combinations of shapes was studied by Milewski (19–21). Recently an empirical equation relating Φ_m to the aspect ratio p for randomly oriented glass fibers was proposed (22):

$$\Phi_m = \frac{1}{1.38 + 0.0376p^{1.4}} \qquad (8.4)$$

For example, when p is 100, Φ_m is about 0.04. For a polymer having a melt density of $1\,g/cm^3$ this would be equivalent to a weight concentration of 10%. This suspension would have a nearly infinite viscosity; to make it flow, sufficient stress would have to be applied to break some of the fibers or to orient them. In practice, these phenomena occur simultaneously. Fragile fibers (glass and carbon) or flakes (glass and mica) break (23–25), whereas tough, flexible fibers such as aramid or cellulose orient and bend but do not break (26). Orientation and reduction of aspect ratio both lead to an increase in Φ_m and to a reduction of k_E. As a result, the relative viscosity is reduced.

Rheology of fiber filled melts over a wide range of shear rates was studied by Crowson et al. (27–29). The effect of glass fibers on viscosity was found to be large at low shear rates, decreasing rapidly with the increasing shear rates (curve B, Fig. 8.6). The slope of the curve for $\log \eta$ versus $\log \dot{\gamma}$ was found to be close to -1, indicating that the value of the power law index n is almost zero and that the velocity profile in the channel is very flat and consequently the material flows almost as a plug. The effect of fiber length was also investigated (20): at low shear rates, longer fibers give higher viscosities; at high shear rates, the differences become insignificant.

A somewhat analogous situation occurs with suspensions of particles that have a tendency to aggregate, as do many finely divided fillers and pigments, especially those without surface treatments by processing aids and/or coupling agents. In this case the aggregates themselves can be considered to be large particles with their interstitial spaces filled with immobilized polymer; that is, the system behaves as if the filler concentration were higher than what was actually added. This type of behavior is represented by curve C in Figure 8.6. At low shear stresses the material does not flow and its viscosity is infinite (it exhibits yield stress). As sufficient stress is applied, aggregates break, leading to an even more rapid drop in viscosity with the increasing shear rate than is the case with the pure polymer matrix.

8.3.1.2 Melt Elasticity

Chains of a molten thermoplastic polymer form an entangled network that when deformed in one direction, as is the case in flow, generates internal tensions normal to the flow direction. The existence of normal stresses is at the origin of melt elasticity. In injection molding the melt elasticity causes the melt to expand in the mold as it emerges from the gate and facilitates mold filling. An incorrect combination of injection speed with the melt expansion is a potential source of defects in injection-molded parts (see also Section 8.4.2).

One of the easiest methods for characterizing the melt elasticity is to measure its die swell ratio. This is done by extruding the polymer through a circular die and measuring the extrudate diameter. The die swell ratio B is then defined as follows:

$$B = \frac{d_c}{d_d} \tag{8.5}$$

where d_c and d_d are the diameters of the extrudate and of the die, respectively. With unfilled polymers, die swell increases strongly with shear rate as illustrated in Figure 8.8, which was obtained with polycarbonate (30). The presence of even a small amount of glass fibers reduces the die swell substantially. A similar phenomenon (viz., substantial reduction or supression of the die swell) was observed in many other polymer systems with particles of all shapes (30–32).

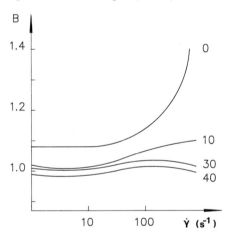

FIGURE 8.8   Die swell ratio B as a function of shear rate $\dot{\gamma}$ of polycarbonate filled with different amounts of short glass fibers. Numbers on curves indicate the glass fiber concentration in weight percent (30).

8.3.1.3 Particle Orientation in Flow

Both shear and elongational flows play an important role in mold filling. Shear flows involve velocity fields in which velocity is constant in the direction of flow but varies in the perpendicular direction. In elongational flows velocity gradients are in the direction of flow rather than orthogonal to it. Understanding and control of flow-generated orientation of anisometric particles in these flows is clearly of great importance for the performance of composites formed in processes involving these flows. This subject, as it applies to polymer processing, has been recently reviewed by Utracki (33,34) and by Goettler (35). The theoretical treatments developed for infinitely diluted suspensions of rods or platelets (36) in Newtonian liquids are not directly applicable to reinforced polymer melts for at least two reasons. First, the suspending medium is not Newtonian but viscoelastic. The second reason has to do with the particle concentration, which in composites under consideration is usually sufficiently high for the particle–particle interactions to become significant.

An excellent example of the state of the art is provided in the work of Vincent and Agassant (37,38). These authors have studied the orientation of glass fiber reinforced polyamide melts in steady flows. To quantify the orientation, they use the fiber orientation function f for fibers belonging to a plane and defined as follows:

$$f = 2 \langle \cos^2\theta \rangle - 1 \tag{8.6}$$

where

$$\langle \cos^2\theta \rangle = \frac{\sum\limits_{\theta=0}^{\theta=\pi/2} n_\theta \cos^2\theta}{\sum\limits_{\theta=0}^{\theta=\pi/2} n_\theta} \tag{8.7}$$

and n_θ is the number of fibers inclined at an angle θ to the flow direction. The orientation function can assume values between $+1$ (perfect orientation parallel to flow) and -1 (perfect orientation orthogonal to flow). For random-in-plane orientation, $f = 0$. In Couette flow (simple shear with the shear rate constant across the gap), starting with an initially random orientation (at $t = 0, f \simeq 0$), the authors varied the shear rate and time of rotation. The results are shown in Table 8.4. The results show that the fibers become strongly oriented in the direction of flow, but the transition from the initial to the final state requires a significant amount of time. For Poiseuille flow in a circular channel (shear rate decreases from a maximum at the wall to zero in the center) with the same material, the results shown in Figure 8.9 were obtained. In these experiments the capillary ($L = 50$ mm, $D = 6$ mm) was fed from the larger diameter cylinder, giving a strongly converging flow at the capillary entrance. As discussed in the next paragraph, the converging flow orients the fibers in the direction

TABLE 8.4 Values of Fiber Orientation Function f with Respect to the Flow Direction, in Couette Flow as a Function of Shear Rate $\dot{\gamma}$ and Time of Rotation t[a]

$\dot{\gamma}$ (s^{-1})	t (s)	f
40	10	0.76
40	90	0.92
180	10	0.96

[a] Glass fiber reinforced polyamide (30 wt%), at $t = 0$ and $f \simeq 0$ (38).

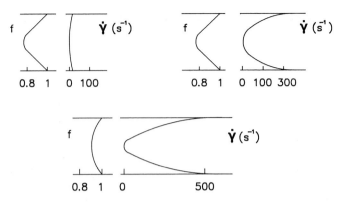

FIGURE 8.9   Fiber orientation function f across the diameter of a circular tube for three values of shear rate $\dot{\gamma}$. Tube length and diameter are 50 and 6 mm, respectively. Polyamide 66 matrix containing 30% wt of short glass fibers (38).

of flow and, consequently, the initial fiber orientation can be considered to be parallel to flow. The results show that the value of f increases with the shear rate; but even at low shear rates ($\dot{\gamma} < 15\,\mathrm{s}^{-1}$) there is a high degree of orientation in the direction of flow. The high degree of orientation in the center of the capillary where shear rate $\dot{\gamma}$ is zero should be considered in view of the initial state of orientation in the capillary entrance; flow along the capillary did not change it precisely because of the low shear rates in the central region.

In elongational flows, orientation behavior of fibers appears to be quite well understood and can be summarized as follows: converging flow results in high degree of fiber alignment in the flow direction, whereas diverging flow causes the fibers to align perpendicular to the major flow direction. The orientation process in elongational flow can be described by Eq. (8.8) (33,35):

$$\frac{\tan \psi}{\tan \psi_0} = \left(\frac{A}{A_0}\right)^{3\chi/2} \tag{8.8}$$

Thus, a particle having an aspect ratio p, oriented at an angle ψ_0 to the local streamline, will reorient to an angle ψ in flow characterized by the convergence ratio A/A_0, $\chi = (p^2 - 1)/(p^2 + 1)$.

8.3.1.4 Temperature and Pressure Effects

The dependence of viscosity on temperature is usually expressed by an Arrhenius-type relation:

$$\eta_T = K e^{\Delta H/RT} \tag{8.9}$$

In this equation K is a constant, characteristic of the polymer and its molecular weight, R is the universal gas constant, and ΔH is the activation energy of the flow process. Two possible values of ΔH may be obtained: $\Delta H_{\dot{\gamma}}$ by comparing viscosities at constant shear rate or ΔH_τ when viscosities at constant shear stress, τ ($\tau = \dot{\gamma}\,\eta$) are considered. For power law fluids with power law index n, the two values are related by:

$$\frac{\Delta H_{\dot{\gamma}}}{\Delta H_{\tau}} = n \qquad (8.10)$$

The effect of fibers or other fillers on the flow activation energy is still somewhat unclear. It was recently studied by Akay (39) at shear rates of up to $10^6\,\mathrm{s}^{-1}$ with glass fiber reinforced polypropylene and polyamide and with calcium carbonate filled polypropylene. In this work the activation energies $\Delta H_{\dot{\gamma}}$ of base (unfilled) resins were found to be independent of shear rate over the entire range of shear rates. With glass fiber reinforced polypropylene, $\Delta H_{\dot{\gamma}}$ remained constant at about $3\,\mathrm{kcal/mol}$ up to $\dot{\gamma} = 2 \times 10^5\,\mathrm{s}^{-1}$; between $\dot{\gamma} = 2 \times 10^5\,\mathrm{s}^{-1}$ and $\dot{\gamma} = 10^6\,\mathrm{s}^{-1}$ it decreased by more than half. For glass-reinforced polyamide, a very strong decrease of activation energy is observed. These results suggest that the shear rate dependence of activation energy is stronger than that of unfilled polymers.

Pressure dependence of viscosity can be also treated in terms of free volume theory of polymers (40), which leads to the relation:

$$\eta_p = \eta_0\, e^{\beta p} \qquad (8.11)$$

where η_p and η_0 are viscosities at pressure p and zero, respectively, and β is the pressure coefficient of viscosity. For polypropylene in the range of shear rates used in injection molding, β was estimated to be only $5.8 \times 10^{-9}\,\mathrm{Pa}^{-1}$ (29). This means that the viscosity triples as the pressure is increased from 0 to 190 MPa, the latter value being at the high end of pressures encountered in injection molding.

In injection molding the flow does not take place at constant pressure and temperature. The pressure in the polymer in the process of being injected varies from the injection pressure in the nozzle entrance to zero at the melt front. Moreover, the mechanical work expended on the melt to make it flow into the mold is converted into heat, causing a significant temperature rise (15,29). The problem was analyzed by Crowson et al. (29) who, assuming adiabatic flow, propose the following relation for the temperature rise ΔT resulting from flow:

$$\Delta T = \frac{\Delta P}{\rho c_p}\,(1 - \alpha T) \qquad (8.12)$$

where ΔP represents the pressure drop, ρ is the polymer density at temperature T, c_p its specific heat at constant pressure, and α is the volume coefficient of thermal expansion. The first term of this equation $(\Delta P/\rho c_p)$ corresponds to the dissipative heating during flow, whereas $\alpha T\,(\Delta P/\rho c_p)$ is the temperature decrease due to the decrease of pressure along the flow path. In an injection machine, during one molding cycle, the temperature regime will be as follows (29). The melt reservoir in the front of the screw is filled with melt at temperature T at a low pressure ($P \simeq 0$). As the screw moves forward, the melt is pressurized to the pressure ΔP and its temperature will rise by an amount given by $\alpha T\,(\Delta P/\rho c_p)$. For polypropylene this compressional heating produces a temperature rise of 16 °C for ΔP of 140 MPa. The melt will flow

281

and the pressure along the flow path will gradually fall to reach zero at the melt front. The decompression will be accompanied by a decrease of temperature. However, during the flow, the temperature will rise by an average amount throughout the melt given by the first term $\Delta P / \rho c_p$. For polypropylene with ΔP at 140 MPa, the dissipative heating was measured to be 73 °C (29).

The pressure and temperature effects on flow of filled and reinforced melts are more significant when compared with their unfilled counterparts principally because of the higher viscosity caused by the fillers and reinforcements. In particular, the temperature rise due to viscous dissipation may be high enough to cause significant polymer degradation.

8.3.1.5 Fiber Length Degradation

The underlying reasons for reduction of fiber length in flow were explained in Section 8.3.1.1. Starting with undispersed fibers, a very significant length degradation occurs when the fibers are being blended into the melt and continue to break at a slower rate during subsequent flow (24,41,42). Recently developed compounding processes circumvent the breakage associated with the dispersion of fibers into the melt (see Section 8.5.1). Fiber fracture in a flowing, well-dispersed melt is usually interpreted in terms of two types of interaction:

- Stresses induced on the fiber by the flowing melt without any concurrent interaction from other fibers;
- Stresses resulting from fiber–fiber interactions (collisions, spatial hindrance, friction, etc.).

According to Czarnecki and White (26), the most frequent fiber fracture caused by fiber–melt interactions occurs by bending. Bending stress on a fiber in shear flow is approximately proportional to $\dot{\gamma} \eta L^2$, where L is fiber length. It suggests that a reduction of shear rate (large flow channels) or of viscosity minimizes the length reduction. The probability of fiber–fiber collision increases with the fiber concentration and fiber length but must also be influenced by fiber reorientation in flow (convergence, divergence, and sudden change of flow direction). Unfortunately, flow-induced fiber length degradation is part and parcel of any processing method used with these materials; it can be reduced but not avoided.

8.3.1.6 Particle Segregation in Flow

In laminar shear flow, suspended particles tend to move away from the wall toward the center (i.e., to the region of highest fluid velocity). In a process like injection molding, this is transformed into a concentration gradient along the flow direction, the concentration increasing with the distance from the gate. It appears that significant gradients are observed only with large isometric particles such as glass beads larger than roughly 100 μm in diameter, whereas with fibers or fine fillers the phenomenon is not important (43–45). A set of rules governing the polymer–filler redistribution in injection molding applicable to spherical particles is given in Reference 46.

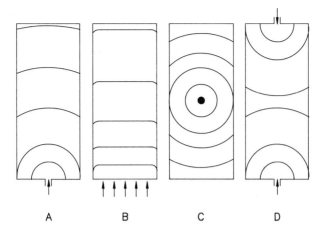

FIGURE 8.10   Mold filling patterns in a rectangular cavity of uniform thickness: (A) edge gate, (B) film gate, (C) centrally located gate, and (D) double-edged gate.

8.3.2 Flow in the Mold

8.3.2.1 Mold Filling Patterns

When the polymer melt leaves the injection nozzle and flows into the mold, the first polymer to contact the cold mold wall solidifies instantaneously, forming an immobile "frozen layer." Additional flow then takes place within the envelope created by the frozen layer. To illustrate the phenomena taking place during mold filling, let us first consider flow in a simple rectangular cavity of constant thickness. An edge[1] gate is located along the shorter side (Fig. 8.10A). As the melt penetrates into the mold, it starts spreading around in a semicircular fashion until the longer side is reached. The principal flow direction in the melt front then changes to follow the longer axis until the mold is almost completely filled. For a given mold shape, the filling pattern can be affected by the location and/or the type of the gate. For example, with a film gate (Fig. 8.10B) the initial shape of the melt front would no longer be semicircular but would be almost straight throughout the whole mold filling stage. Use of a centrally located gate (Fig. 8.10C) would reduce the flow path length approximately by half; pressure required to fill the mold would be smaller (assuming equal gate size). A similar effect could be obtained by using two gates (Fig. 8.10D). In this case, the zone where the two melt fronts meet (weldline) might have different, usually lower, properties. In practice, the cavity geometries are of intricate shapes and melt front advancement during mold filling stage is very complex. The knowledge of mold filling patterns is important for several reasons. Bangert (47) identified some of the more important questions that can be answered if the positions of melt front during mold filling are known. They are:

[1]There does not appear to be a generally accepted, unified terminology applicable to gate types. As a result, there are discrepancies in gate designation among books dealing with mold design. In this chapter the terminology used is that of Reference 12.

283

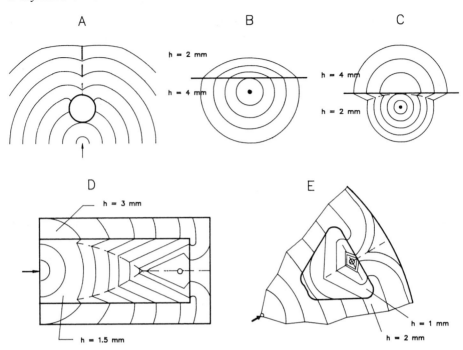

FIGURE 8.11   Mold filling patterns in (A) mold with a circular insert and (B–E) molds with sections of different thickness. Dashed lines in A, C, D, and E indicate weldline locations. Small circles in D and E show the location of entrapped air pockets (47).

- Does the injection unit have sufficient capacity to fill the mold?
- What is the location of weldlines (and where are gates placed to prevent the weldlines from being in a critical area of the molding)?
- Are air entrapments to be expected?
- What is the minimum number of gates to ensure adequate filling of a given molding, and where are these gates to be located?

Bangert then presents a simple method, derived from the theory of wave propagation, to predict the melt flow progression as the mold is being filled. Figure 8.11 shows a number of his results, some of which were validated by experiment. In Figure 8.11A (flow around a circular core), the melt front is divided in two by the core and, when it recombines behind the core, a weldline is created. Presumably, after a certain additional flow, the polymer will "forget" the core and the weldline will disappear. However, for a certain distance behind the core, differences in microstructure and properties may be expected in the weldline area compared to a molding without the core. Figures 8.11B and 8.11C illustrate the flow in two molds, each having two sections of different thickness h_1 and h_2. Flow patterns were constructed assuming that flow lengths l_1 and l_2 are related to the cavity thickness by:

$$\frac{l_1}{l_2} = \frac{h_1}{h_2} \tag{8.13}$$

This relation, based on the assumption of isothermal Newtonian flow, means that the melt front velocity is proportional to the cavity thickness in these molds. Thus, the thicker section will fill faster. While the thicker section is being filled, the melt front in the thinner part advances at a correspondingly slower rate. Only when the thicker section is full can the melt front accelerate in the thinner section. The discrepancies in melt front velocities can lead to pockets of entrapped air and to weldlines. In figure 8.11C, where the gate is located in the thinner section, the melt flows into the thicker section and then partly back again.

Examples of applications to slightly more complex molds are also shown in Figure 8.11. For the molding of Figure 8.11D, a pocket of entrapped air will form if the gate is located in the thinner area as indicated and if the difference in thickness is too great. The example in the Figure 8.11E represents a section of a cable reel with heart-shaped depression. Again mold filling patterns indicate the position of weldlines and of a trapped air pocket.

The mold filling patterns of Figure 8.11 illustrate the general nature of the process. For flow in complex molds when the nonisothermal, non-Newtonian nature of the flow cannot be neglected, one of the several mold filling programs may be used (48–50). Some of these programs are commercially available and can be very useful, provided the rheological behavior of the polymer is sufficiently well known.

8.3.2.2 Fountain Flow

Another very important aspect of the mold filling process that has a major effect on the microstructure and properties of all injection-molded objects is the "fountain flow" in the cold mold cavity (51), as illustrated in Figure 8.12. The melt flows from left to right between two immobile frozen layers. Because of the shape of the velocity

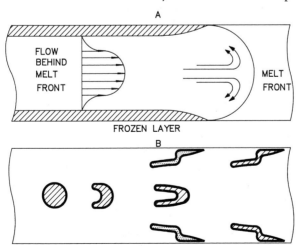

FIGURE 8.12   Fountain flow. (A) Formation of the frozen layer from the expanding melt front showing. Idealized shape of the velocity profile in the core behind the melt front (51). (B) Deformation of an initially spherical fluid element at successive positions of the advancing melt front (see text) (52, 53).

profile (Fig. 8.12A), the fluid elements in the core will move at a faster than average velocity until they reach the melt front zone. There, assuming a constant shape of the melt front profile, they will expand and flow to the wall, where solidification takes place. As the streamlines indicate, a considerable stretching takes place in the melt front area and the fluid that has stretched most is the one to contact the cold mold wall. Because of the viscoelastic character of the fluid, the orientation generated by stretching is frozen, forming an oriented skin that is part and parcel of all injection-molded articles. Schmidt (52,53) has used a color tracer technique to demonstrate the fountain flow. His results are summarized in Figure 8.12B. The tracer particle, which enters the mold as a sphere, gradually deforms into a circular arc with a U-shaped cross section. The flow and deformation continue until the tracer reaches the melt front, where the material flowing on the midplane at the highest velocity splits the tracer and forces the leading edges to the mold wall. The first tracer particles to enter the mold will solidify closest to the gate. Similar techniques were also applied to more complex molds (54).

An injection-molded article can then be said to consist of a skin, which originates from the continuously regenerated melt front surfaces, and a core, which is a result of a more or less fully developed flow behind the melt front. Two questions must be asked: how does the flow in the mold affect the orientation of fibers in the finished molding and, more importantly, to what extent can it be influenced by the material and process parameters available? Clear answers are not yet known; only a general qualitative understanding has been reached. The principal difficulties lie in the insufficient understanding of the rheology of concentrated fiber suspensions in visco-elastic melts and also in the approach used to study the flow-generated fiber orientations. This method consists of studying finished moldings and assuming that observed fiber orientation patterns had been generated during mold filling. This, as some recently published evidence suggests (55,56), may not be justified, and certain features of fiber orientation in finished molding may be attributed to flow during packing and holding. However, since it is not possible to separate *a posteriori* the contributions of flow during filling and packing, let us review the situation and assume, as most authors do, that the fiber orientation has been principally determined by flow during mold filling.

8.3.2.3 Fiber Orientation in Molded Objects

Fiber orientation has been mostly studied in relatively thin, flat moldings where fibers are usually found to be oriented parallel to the plane of the molding. In general, fiber orientation in the skin is very different from that in the molding core and depends on the mold and gate geometries, the molding process parameters, and the resin used. Most of the work published dealt with glass fiber reinforced polymers; carbon fibers, aramid fibers and platelet reinforcements have been studied much less.

Three elementary cavity–gate configurations have been studied by several authors: centrally gated discs and rectangles injected either through a film gate or through an edge gate located in the center of the shorter side.

For the disc, working with polypropylene containing 20% glass fibers, Woebcken (57) identified a four-layered structure:

- Skin, with random-in-plane fiber orientation.
- Under the skin, where the fibers are oriented in the radial direction, parallel to flow.
- Intermediate layer, with random orientation.
- Core layer, with fibers oriented perpendicular to flow.

Akay (58) reports similar observations. While the division into layers may be convenient for visual interpretation of micrographs, it is probably more correct to view the structure as continuously changing throughout the thickness. This can be done by using some kind of fiber orientation function. For example, Vincent (37), working with polyamide containing 30 wt% glass fibers, determined the fiber orientation function defined in Eq. (8.6). His results are shown in Figure 8.13 for two mold temperatures. Delpy and Fisher (59) report that in glass fiber reinforced polyamide, in the core the value of the orientation function decreases; that is, fiber orientation becomes more perpendicular to flow direction with the distance from the gate. This effect is more pronounced for thick discs. Similar results were obtained by other authors (60–63).

In the centrally gated disc, the perpendicular to flow fiber orientation in the core can be easily interpreted in terms of the tangential stretching of the melt as it flows outward. It can be shown (64) that the tangential component of the velocity at the melt front is 2π times larger than the radial component. Surprisingly, very similar fiber orientation is found in rectangular cavities with film gates (Fig. 8.10B), including the case of a gate cross section that is optimized to provide uniform speed of melt front during filling. Hegler (65) found exactly the same layers in this type of mold as those described above for centrally gated discs. Several other reports consistent with Hegler's results have been published. Wright and Whelan (66) worked with needle-shaped wollastonite reinforcement having much smaller particle size and reported similar orientations.

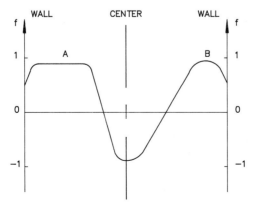

FIGURE 8.13 Fiber orientation function f across the thickness of a centrally gated, circular molding made of wt% glass fiber–polyamide 66 at two mold temperatures: 150 °C (A) and 45 °C (B) (38).

287

Menges and Geibüsh (67) propose a mechanism whereby the fibers adapt this orientation as a result of fountain flow upon the entrance into the mold and, because the fiber-filled melt is highly pseudoplastic (flat velocity profile), fiber orientation acquired in the gate region persists as the melt flows away from the gate. Malzahn and Schultz (55) disagree with this mechanism. Comparing the fiber orientation in short shots (moldings from partially filled mold) to filled and packed moldings, they attribute the core orientation to the flow during mold packing. However, results of similar work with short shots (68) are at variance with those of Malzahn and Schultz.

Flow in rectangular cavities into which the melt enters via a small edge gate located at the shorter side (Fig. 8.10A) shows features of the two geometries just described. First, the polymer spreads radially; then, having reached the sides, it flows along the longer axis of the mold. Fiber orientation reported for this geometry is similar to those described above: more or less random in plane in the skin, gradually changing to perpendicular to flow in the core (60,69–73).

Fiber orientation in molded parts depends on a number of parameters in addition to the mold and gate geometries; these include rheology of the matrix material and injection speed. Bright and Darlington (60) and Darlington and Smith (63) compared polyamide to polypropylene, the latter being more pseudoplastic than the former. As a result of the flatter velocity profile and low shear rates in the center, glass fiber reinforced polypropylene will tend to conserve more of the perpendicular-to-flow orientation acquired in the gate area. The core layer will be thicker than with polyamide, which exhibits a more developed velocity profile. The intermediate layer, where the shear is large enough to reorient the fibers in the direction of flow, will be less pronounced in polypropylene than in polyamide. Injection speed has an effect on the thickness of the frozen layer: slow injection will favor it; fast injection with the higher amount of flow generated heat and shorter duration will play against it (74).

To summarize, the fiber orientation in injection-molded parts depends principally on the mold and gate geometry and on the matrix flow behavior. Three distinct regions with different fiber orientations can be identified:

- A skin originating from the expanding melt front.
- An intermediate layer where the shear flow oriented the fibers in the direction of flow.
- A core that keeps the fiber orientation it acquired in the gate area.

It is clear that the fiber orientation will vary throughout the molding, but tools that would allow us to fully characterize and predict the fiber orientation in the molding have yet to be developed. The limitations of currently available fiber orientation prediction software are discussed in Reference 75. At this time only a qualitative understanding of the mold filling process with fiber filled polymer melts has been attained.

8.3.2.4 Mold Packing

The packing stage of an injection molding cycle starts when the mold has been filled with the melt and ends when, somewhere in the gate or runner system, solidified polymer completely obstructs the path between the mold and the injection unit. In the beginning of the packing there are pressure gradients: the last space to fill is close to the atmospheric pressure. Additional material is injected into the mold to compress the melt to the desired level of holding pressure. This is relatively fast (76,77). The initial pressurization is followed by a phase in which more material is added as the polymer shrinks on cooling. Very little is known about packing of unfilled polymers and virtually nothing about filled and reinforced polymers. It is likely that the flow during packing affects fiber orientation to some degree but the extent is not known except for one special situation dealing with weldline strength (54) (see Section 8.4.2).

8.3.3 Solidification and Microstructure

Properties of any thermoplastic material can be varied, over a great range of values, by controlling the orientation of polymer chains and, in crystallizable polymers, the concentration and structure of the crystalline phase. For example, strength and rigidity similar to those of carbon or aramid fiber can be achieved by orienting such an obsequious polymer as polyethylene. In finished injection-molded parts, orientation is caused by freezing the polymer before polymer chains, deformed from their equilibrium shape by flow, can relax. Again the local differences in thermomechanical history of the melt lead to nonuniform distribution of many properties. Moreoever, in crystallizable polymers the temperature and the pressure, as well as the chain orientation in the melt when it crystallizes, will affect the polymer morphology, hence many of its technological properties. When discussing the matrix properties in injection-molded thermoplastic composites, it is convenient to consider separately amorphous polymers, semicrystalline polymers, and the effect of the presence of fibers on the matrix microstructure.

8.3.3.1 Amorphous Polymers

Several methods of polymer physics are available to characterize orientation in amorphous polymers. The simplest and most widely used one consists of relating the optical anisotropy as measured by differences in refractive indices (birefringence) to the orientation of macromolecular chains in the polymer. The results of the published work on the subject can be summarized as follows.

1. If the flow in the mold is unidirectional (e.g., Fig. 8.10B), orientation, averaged over the sample thickness, decreases with the distance from the gate (78). In a more typical molding with a small edge gate (e.g., Fig. 8.10A), the orientation pattern shown in Figure 8.14 is obtained. Orientation is highest in the gate area. This is probably due to the packing flow. It decreases very rapidly and at a short distance from the gate starts growing again. A plateau occurs approximately when the melt

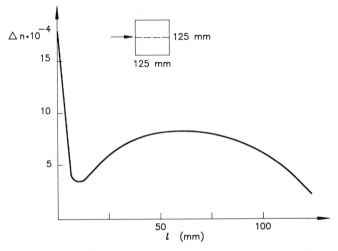

FIGURE 8.14   Birefringence Δn as a function of distance from the gate, l along the line shown in the inset. Polystyrene molded in a square mold with an edge gate (78).

front has reached mold sides. Several process parameters have a marked influence on orientation. Increasing melt and mold temperature (79–82) and injection pressure (78) can significantly reduce the overall level of orientation in the molded article. In general the thicker parts will be less oriented than thinner ones.

2. Orientation is always much more important in layers close to the surface than in the core. This is shown in Figure 8.15, redrawn from the work of Kamal and Tan (83). The first layer (from the surface to point B) is the "frozen layer," which solidified instantaneously when the expanding melt front came in contact with the wall. This layer is relatively thin (40–120 μm), and this may explain why other

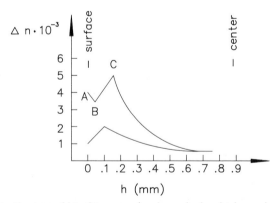

FIGURE 8.15   Distribution of birefringence Δn through the thickness h of a rectangular polystyrene molding. The mold dimensions were 127 × 63 × 1.8 mm. An edge gate was located in the center of the shorter side. Upper and lower curves were obtained by measuring the birefringence in the directions parallel and perpendicular to the longer side at a distance of 11 mm from the gate (83).

290

authors have not detected its presence. The second layer (B–C) is believed to be the result of solidification during mold filling. The core solidified during quiescent cooling, with the orientation decreasing rapidly toward the center of the molding. In accord with this interpretation, increasing the mold temperature should lead to a reduction of both highly oriented layers (A–B and B–C). This was in fact observed by Kamal and Tan (83).

3. Großkurth (84) studied the morphology of injection-molded polystyrene by ion etching and related the morphological features observed to those found in hot stretched polymer. Transition from highly oriented "line" to unoriented globular structure is observed as one moves from the surface to the core. The influence of injection molding parameters on orientation and on mechanical properties of polystyrene was investigated by Hoare and Hull (85). The fracture was found to initiate in the core when the material was stressed in the injection direction and in the skin when the stress was applied perpendicular to injection direction. This is of particular importance for objects loaded in flexure where the stress is highest at the external fiber.

For polycarbonate, molded in a centrally gated disc mold, an orientation pattern consistent with the picture presented above was reported (64).

8.3.3.2 Semicrystalline Polymers

Formation of a crystalline phase involves two distinct steps: nucleation [i.e., formation of an initial primary crystalline embryo (nucleus)] and subsequent crystal growth from the nucleus. It is generally accepted that in most systems nuclei consist of foreign bodies such as catalyst residues, submicrometer dust particles, and antioxidants. In this case the nucleation is said to be heterogeneous, as opposed to homogeneous nucleation, which occurs as a result of local density and chain conformation fluctuations in the melt (86). Homogeneous nucleation can be observed only when special care is taken to eliminate the heterogeneous nuclei from the system (87). A polymer's ability to crystallize depends principally on its molecular structure: flexible, regular chains such as those of linear polyethylene and isotactic polypropylene will crystallize easily, whereas irregular polymers such as atactic polystyrene or polymethlymethacrylate will not crystallize at all. Keeping in mind the injection molding process, the following comments about the crystallization behavior of thermoplastics can be made:

1. For each crystallizable polymer, there is a range of temperatures in which the rate of crystallization will be significant (Fig. 8.16). At temperatures below but close to the crystalline melting point, the rate of crystallization is low because of the slow rate of nucleation. At low temperatures, the rate of crystallization is low because of the insufficient mobility of the polymer chains. It follows that in the system that undergoes rapid cooling, the ultimate degree of crystallization (crystallinity) will be related to the time the polymer spent in the range of temperatures where the rate of crystallization is significant. This will be shortest in the skin and longest in the core.

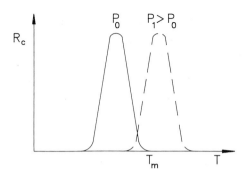

FIGURE 8.16   Rate of crystallization R_c as a function of temperature T, T_m represents the melting point at pressure P_0 (88).

2. The temperature at which a molten polymer will start to crystallize increases with pressure; for example, highly crystalline polyethylene is produced at 230 °C and 500 MPa (88). Many other polymers exhibit the same behavior. During the packing stage of the injection molding cycle, pressures of 100 MPa are common, and significant upward shift in the temperature occurs. For isotactic polypropylene molded under typical conditions, onset of crystallization is predicted at a temperature higher by 20–35 °C than what is observed at atmospheric pressure (89).

3. Deformation of the polymer melt in shear or elongational fields leads to a decrease of conformational entropy of macromolecules and to an enhanced rate of nucleation.

The combined effect of temperature, pressure, and melt orientation was considered by Trotignon and Verdu (89), who divided the solidification process of a crystallizable polymer (polypropylene) melt into four elementary steps.

- *Mold filling*: During this step the melt temperature is above the crystallization range except in the immediate vicinity of the mold wall, where a highly oriented layer of low crystallinity has formed. The pressure is low.
- *Filling–packing transition*: When the mold is filled, the melt velocity drops but the pressure remains low. The flow-induced orientation begins to relax in the subskin layer until the temperature at which the polymer starts to crystallize is reached. The rate of crystallization is high because it is favored by the chain orientation. A crystalline (α) form of polypropylene known to crystallize at low pressures predominates.
- *Packing stage*: A sudden increase in pressure during the packing stage produces an effect similar to quenching (fast cooling). Thus the front of crystallization (position of significant rate of crystallization) jumps toward the core, leaving behind a layer of low degree of crystallinity (see Fig. 8.16). A second consequence of the packing pressure concerns the orientation distribution across the thickness. The melt compressed by the packing pressure has lower free volume and, as a result, the rate of orientation relaxation is considerably reduced. This contributes to higher levels of orientation close to the gate where the pressure is highest.

• *End of molding cycle*: Once the front of crystallization has reached the unoriented core ($\simeq 500\,\mu$m from the surface for polypropylene), morphology associated with quiescent crystallization is observed.

The concentration dependence of the β form of crystalline polypropylene, which crystallizes only at high pressures from oriented melt, quickly cooled to 120–125 °C (i.e., below the range of maximum rate of crystallization of the α form), gives credence to this mechanism (89).

Microscopic examination of injection-molded semicrystalline polymers confirms the layered structure. In the skin, which must have resulted from the instantaneously cooled melt front, the chain orientation is high and parallel to the skin. The crystallinity will depend on the crystallization behavior of the polymer. This layer is relatively thin ($\simeq 50\,\mu$m). In the intermediate layer, the crystallization is dominated by matrix orientation and high temperature gradients. Nuclei are formed in the zone of low temperature near the skin and crystallization proceeds toward the core. The term "transcrystallization" is used to describe the crystal growth in one direction. Finally, the core contains randomly nucleated structures, which grow in all directions.

The layered structure described above has been reported for a wide variety of semicrystalline polymers (90–102), although sometimes additional layers are claimed to be present at the boundaries between the three principal zones (skin, intermediate-sheared, core). Obviously the classification of observed morphological features into layers is a useful mental tool, but it probably would be more accurate to view the structure as continuously changing from the surface toward the core. Relative thickness of layers can be affected by the process and material parameters. In general, slowing the crystallization process by using higher melt or mold temperatures will favor the central, core layer. Overall degree of crystallinity will also be increased. Using a polymer of lower molecular weight (shorter chains, fast relaxation) will reduce the thickness of the intermediate layer. Moreover the thickness of the intermediate layer typically decreases with the increasing distance from the gate. An example of layer thicknesses observed in a simple polypropylene molding is shown in Figure 8.17 (103).

Mechanical properties of injection-molded semicrystalline thermoplastics are affected by the variations in crystallinity and orientation. Typically, the oriented but

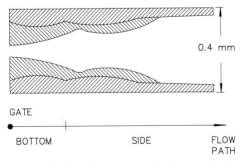

FIGURE 8.17 Thickness of the three layers in a injection-molded polypropylene drinking cup as a function of the distance from the gate (103).

293

less crystalline skin and the unoriented, albeit more crystalline, core will be more ductile than the intermediate layer (99,100,104). As a result, fracture will often initiate in this intermediate layer in objects loaded parallel to the injection direction (105).

8.3.3.3 Fiber–Matrix Interactions

The presence of solid particles in the melt may have an important influence on the matrix morphology. The subject has been recently reviewed by Burton and Folkes (106) for fiber-reinforced thermoplastics. The presence of the second dispersed phase may affect both matrix orientation and crystallization behavior.

The fiber surface may provide a large number of nucleation sites, and the growth of the crystalline phase will start at the surface and proceed into the melt. Such effects were observed in glass–polypropylene (107), aramid–polyamide 66, carbon—polyamide 66 (108), and carbon–polypropylene (109) but not in glass–polyethylene terephthalate (110) and glass–polyamide 66 (108) systems. The local flow field around a fiber immersed in the melt is different from the macroscopic field. Chain alignment created by the shearing and stretching of the melt may in itself be the source of crystal nucleation (111,112).

The effect of glass fibers on the matrix orientation for a glass fiber–high density polyethylene system is shown in the Figure 8.18 (70). The peak associated with the highly oriented sheared layer observed in unfilled polyethylene becomes less pronounced as the fiber concentration increases. In fact the matrix orientation becomes dominated by the fiber orientation. The effect of modified crystallization behavior of the matrix in the presence of fibers is obviously of great interest. Not only may the matrix have properties different from those in the unreinforced state, but the adhesion at the fiber–matrix interface must also be related to the matrix structure near to it. Unfortunately, little is known about this subject at this time.

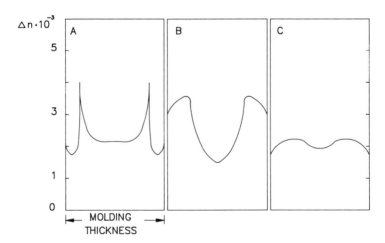

FIGURE 8.18 Birefringence Δn through the thickness of a rectangular plaque of high density polyethylene containing (A) 0, (B) 20 wt%, and (C) 50 wt% of glass fibers (70).

8.3.3.4 Surface Texture

For a number of applications, smooth surface is required. With unfilled thermoplastics, the surface texture is rarely a problem and is essentially determined by the mold surface. Introduction of glass fibers into the matrix, however, has a deleterious effect on the surface smoothness. Depending on the processing conditions, surface varying from dull and porous to relatively smooth can be obtained. Microscopic examination of the dull surface reveals the presence of ridges, which appear to be the fibers embedded with the polymer. It is likely that the expanding melt front contains fibers protruding from the surface, as is the case in extrusion of fiber-reinforced melts (30,113) and, when deposited on the cold mold wall, solidification occurs before a continuous film had time to form. Increasing the injection speed (73,114) or the mold temperature (115) improves the surface finish significantly; but as demonstrated by Cox and Mentzer (114), even when the surface appears "smooth," microscopic depressions are present in fiber-reinforced, injection-molded thermoplastics.

8.4 PROPERTIES AND APPLICATIONS

8.4.1 Mechanical Properties

Among the many factors influencing the mechanical behavior of short fiber reinforced composite materials, fiber aspect ratio, concentration, orientation, and adhesion at the fiber–matrix interface are of great importance. In Section 8.3, we saw that the nature of the injection molding process imposes limitations on three of these factors: maximum fiber length and concentration are limited by the necessity to make the material flow in the mold using reasonable pressures; and orientation is to a large extent determined by the flow itself. Parts made of reinforced thermoplastics will exhibit many of the general characteristics of their unreinforced injection-molded counterparts (postmolding shrinkage, thermal residual stresses, etc.). However, the flow-generated fiber orientation leads to a degree of anisotropy greater than that of unfilled or filled (when the filler aspect ratio is close to 1) plastics. Anisotropy of some materials with respect to linear coefficient of thermal expansion leads to warpage, and local fluctuations in fiber orientation are at the root of weak weldlines.

Mechanical properties are generally quite inferior to what could be expected from the rule of mixtures. This is true even when the reinforcement is perfectly oriented. Blumentritt et al. (116,117) studied the strength and stiffness of a series of thermoplastic resins filled with a variety of short fibers of a series of thermoplastic resins filled with a variety of short fibers used in volume fractions of 0.1–0.5. Two different orientations were studied: unidirectional (116) and random in plane (117). These authors calculated the "fiber efficiency factors" for modulus K_E and strength K_σ using the rule of mixture expressions:

$$\sigma_{uc} = K_\sigma \, \sigma_{uf} \, \Phi_f + \sigma_m' \, (1 - \Phi_f) \qquad (8.14)$$

$$E_c = K_E \, E_f \, \Phi_f + E_m \, (1 - \Phi_f) \qquad (8.15)$$

295

TABLE 8.5 Fiber Efficiency Factors for Glass Fiber Reinforced Polyethylene and Polycarbonate: Unidirectional vs. Random-in-Plane Orientation (116, 177)

Φ_f	Polyethylene				Polycarbonate			
	Unidirectional		Random-in-plane		Unidirectional		Random-in-plane	
	K_E	K_σ	K_E	K_σ	K_E	K_σ	K_E	K_σ
0.1	0.49	0.44	0.18	0.23	0.51	0.43	0.21	0.14
0.2	0.59	0.48	0.18	0.12	0.41	0.33	0.17	0.13
0.3	0.46	0.36	0.20	0.17	0.48	0.34	0.25	0.18
0.4	0.47	0.29	0.21	0.15	0.46	0.30	0.26	0.13

where σ_{uc} is the ultimate strength of the composite, σ_m' is the matrix stress at the fracture strain of the composite, σ_{uf} is the ultimate strength of the reinforcing fiber, E_f and E_m are Young's moduli of the fiber and the matrix, respectively, and Φ_f is the fiber volume fraction; E_c represents the Young's modulus of the composite in the direction parallel to the fiber axis for the unidirectionally oriented composites and in the plane of the fibers for the random-in-plane composites.

Some selected results for polyethylene and polycarbonate reinforced with 6 mm long glass fibers are shown in Table 8.5. For unidirectionally oriented composites, values of K_E of about 0.5 were found; K_σ values are lower, particularly at higher loading. The deviation from the rule of mixtures ($K_E < 1$ and $K_\sigma < 1$) may be due to a variety of factors including nonuniform spacing and imperfect alignment; but the main reason is the finite fiber length. In a continuous fiber unidirectional composite, the average strain in the fiber is equal to that of the matrix. In a short fiber composite, the stress is transferred from the matrix to the fiber by shear at the interface and the strain will vary along the fiber. The efficiency of a given fiber–matrix system may be estimated by comparing the observed fiber length to the critical fiber length L_c defined as follows:

$$L_c = \frac{\sigma_{uf}\, d}{2\tau_y} \tag{8.16}$$

where σ_{uf} and d are the tensile strength and the diameter of the fiber, respectively, and τ_y is the shear strength of the matrix or of the interface (whichever is weaker). Ideal stress distribution at the fiber matrix interface in a ductile matrix–brittle fiber system is shown in Figure 8.19 for a fiber longer than L_c (118). Assuming good adhesion, the shear stress transferred through the interface will reach a limiting value represented by the yield shear strength of the matrix τ_y. Thus there will be a region toward the fiber ends where shear stress is constant and equal to τ_y. Tensile stress in the fiber increases from zero at the fiber end to reach a maximum value in the central region, where the stress borne by the fiber equals that which would be observed in a continuous fiber composite. When the stress in the fiber reaches the ultimate strength, the fiber will break. In fibers shorter than L_c the ultimate fiber strength is never reached and the fibers slip out in the event of fracture. Even when the fiber length

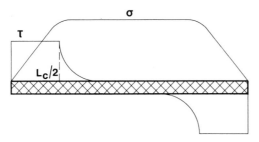

FIGURE 8.19 Idealized stress distribution at a ductile matrix–brittle fiber interface $L > L_c$, σ is tensile stress borne by the fiber and τ is shear stress at interface (18).

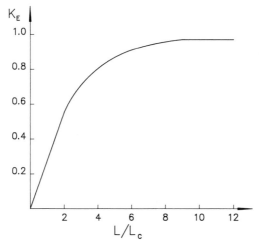

FIGURE 8.20 Reinforcement efficiency factor K_E as a function of relative fiber length L/L_c (119).

exceeds L_c, the reinforcing potential of short fibers is not fully utilized because of end regions. The question of how long is long enough was addressed by Bader and Bowyer (119).

Figure 8.20, which shows the fiber efficiency factor for modulus K_E as a function of fiber length, suggests that a fiber length of at least $5\,L_c$ is desirable. In short fiber reinforced thermoplastics, the majority of fibers are of subcritical length. For a commercial glass fiber reinforced polyamide 66, where L_c was estimated to be $270\,\mu m$, only about 20 % of fibers were longer than L_c and none achieved the desired $5\,L_c$ required for efficient reinforcement (119). This work and many others published since confirm that fiber length in short fiber reinforced thermoplastics is far from optimum and that even a modest improvement in fiber length could lead to significantly better mechanical properties. Darlington et al. (120) showed that attempts to preserve the fiber length by careful control over processing conditions results in improved strength and stiffness. Back pressure (i.e., intensity of shearing in the injection unit) was found to have the most influence on fiber length. Results obtained on glass fiber reinforced polyethylene terephthalate are shown in the Figure

297

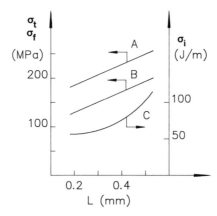

FIGURE 8.21   Flexural σ_f, tensile σ_t, and impact σ_i strength of injection-molded glass fiber reinforced (30 wt%) polyethylene terephthalate as a function of fiber length L (121).

8.21 (121). The effect of fiber length on impact strength is particularly significant. According to an analysis reported by Lunt and Shortall (42), observed variations in fiber length are not sufficient to explain the differences in mechanical properties. One possible explanation is that a change in fiber length produces a change in fiber orientation: the longer the fibers, the better they are aligned in the plane of the molding, giving a stronger and stiffer product.

The effect of fiber orientation on mechanical properties is equally important. In the preceding section we saw how the combined effects of flow and cooling affect the fiber orientation in finished moldings. Bright and Darlington (60) studied the relations between the mold geometry and mechanical properties of glass fiber reinforced polypropylene and polyamide 66 and concluded that mechanical properties are controlled by the relative thicknesses of layers of different fiber orientation. Some of their results are shown in Table 8.6. In the case of polypropylene, where the core layer is relatively thick moldings are stiffer and have higher ultimate strength in the direction perpendicular to flow. These results show that the usual assumption of higher stiffness and strength in the flow direction is not always correct. This latter assumption was found to be accurate for polyamide 66 molded in the same mold. In polyamide 66 moldings, the core layer is thin or absent.

It is relatively straightforward, albeit very tedious, to measure the fiber length and orientation distribution in a given molding and, using one of the available multidirectional short fiber composite models (122–124), to calculate the mechanical properties and compare them with those measured on the same moldings. However, the desired situation, whereby a mold designer could calculate the fiber orientation and mechanical properties of the part before building the mold, is yet to be achieved.

8.4.2 Weldlines

In polymer processing, weldlines occur whenever two molten polymer interfaces are brought into contact (125). The terms "knitline" (126) or "meldline" (127) are also used in the literature to designate essentially the same thing. In injection-molded objects, weldlines are found when two separate melt streams join, either in multigated molds or as a consequence of flow around obstacles in the mold (pins, inserts, zones of different thickness, as in Fig. 8.11). In unfilled amorphous polymers the properties of a weldline will depend on whether the polymer chains had enough time to diffuse across the interface to form a strong bond (128). In unfilled semicrystalline polymers, the morphology of the crystalline phase may be different in the weldline (129).

The subject of weldlines in injection molded parts was reviewed by Malguarnera (130) and by Moslé et al. (131). In brittle thermoplastics such as polystyrene or styrene–acrylonitrile copolymer, the weldline produces a rather severe reduction in tensile strength (50 % for a double-gated standard tensile bar mold). In semicrystalline (polyacetal, nylon) or ductile, amorphous (polycarbonate) polymers, the effect of weldline on mechanical properties is minor or negligible, principally because of the material's ability to yield. In addition, in unfilled polymers, the effect can be reduced by modifying processing conditions (e.g., increasing melt temperature, injection speed, or pressure) (132).

In general, weldlines are much weaker in multiphase systems, including incompatible polymer blends (133). While generally there are fewer data on the weldline behavior of filled and reinforced polymers, the consensus is that weldlines represent a source of weakness and that they are weakest (relative to the same composite tested away from the weldline) in composites containing highest aspect ratio reinforcements. Savadori et al. (134) studied the weldline strength of polypropylene filled with calcium carbonate, talc, galss fibers, and glass spheres. Results obtained with a double-gated mold are shown in the Figure 8.22. More moderate drop in strength is

TABLE 8.6 Tensile Modulus and Ultimate Tensile Strength of Glass Fiber Reinforced Polyamide 66 (PA) and Polypropylene (PP) for Different Mold Configuations.[a]

Mold shape and dimension (mm)	Resin	Tensile modulus, E (GPa)			Ultimate tensile strength, σ (MPa)		
		E_0	E_{90}	E_0/E_{90}	σ_0	σ_{90}	σ_0/σ_{90}
End-gated bar, 3 × 40 × 250	PP	3.51					
Thin corner-gated plaque, 2 × 120 × 120	PP	2.80	3.20	0.87	48	55	0.87
Thick corner-gated plaque, 6 × 120 × 120	PP	2.65	3.75	0.71	45	60	0.75
Thin center-gated disc, 2 × 170 diameter	PP	2.85	3.48	0.82	42·	53	0.79
Thick center-gated disc, 6 × 170 diameter	PP	2.68	3.49	0.77	40	50	0.80
Thin corner-gated plaque, 2 × 120 × 120	PA	5.55	3.65	1.52	94	69	1.36
Thick corner-gated plaque, 6 × 120 × 120	PA	5.3	6.25	0.85	95	109	0.87

[a] Subscripts indicate direction parallel (E_0 and σ_0) or perpendicular (E_{90} and σ_{90}) to the flow (60).

299

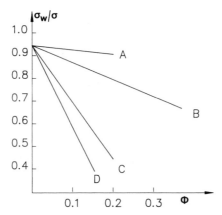

FIGURE 8.22   Relative weldline strength σ_w/σ, where σ_w and σ represent strength of double-gated and single-gated tensile specimens, as a function of filler volume fraction Φ. Curves: A, glass spheres; B, calcium carbonate; C, glass fibers; D, talc (134).

observed with low aspect ratio fillers (calcium carbonate and glass spheres) than with glass fibers and talc (plateletlike shape). Similar results were also reported by Cloud et al. (135). For example, these authors reported a 50 % loss of strength in 40 % glass fiber reinforced nylon weldlines.

Most of the results published on weldlines in reinforced plastics were obtained using standard double-gated tensile bar molds (e.g., ASTM D 647). With this mold, fibers are oriented almost perfectly in the direction perpendicular to flow. Assume a structure where the weldline zone has a lower modulus and lower strength than the rest of the sample. When the sample is loaded, the local strain in the weldline will be much higher than elsewhere leading to rapid fracture. These weldlines (double-gated bars) represent the worst situation that could be found in a molded reinforced part and as such are usually avoided by the mold designer.

The other type of weldline—formed, for example, where the melt stream is divided as it flows around an obstacle—is more common by far, although very little is known about fiber orientation in weldlines of this type. An example of an injection-molded dashboard made from fiber-reinforced acrylonitrile–butadiene–styrene (ABS) polymer is shown in Figure 8.23 (136). In this part seven weldlines of varying lengths were identified. We must ask what is the difference in properties

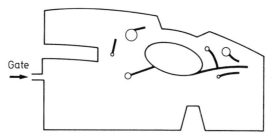

FIGURE 8.23   Weldlines (heavy lines) observed in an instrument panel molded of glass fiber reinforced ABS (136).

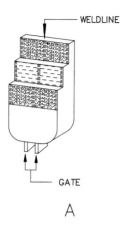
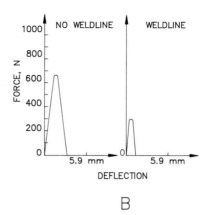

FIGURE 8.24 (*A*) Fiber orientation observed in a rectangular polypropylene molding where a weldline was produced using two edge gates as illustrated. (*B*) Impact load–deformation curves for samples without and with the weldline (69).

between the weldline zone and away from the weldline and how it changes with the distance. Akay and Barkley (69) studied weldlines produced in a rectangular mold with two edge gates placed side by side. Their results obtained with glass fiber reinforced polypropylene can be summarized as follows:

- The fibers in the weldline are oriented parallel to flow, whereas in the area without the weldline the fibers have random-in-plane orientation in the skin and perpendicular-to-flow orientation in the core (see Fig. 8.24A). Moreover, dynamic mechanical analysis suggests an essentially fiber-free zone in the weldline.
- Loss of impact strength in the weldline is greater than loss of tensile strength (Fig. 8.24B) and increases with the fiber length.

In another work (137), a rather slow change in properties along this type of the weldline was reported. However, in molds where the weldline formation is followed by laterally expanding flow, the change in particle orientation is more rapid, resulting in stronger weldlines. The effect of packing on weldline strengths was investigated by Hamada et al. (56). They used two experimental molds shown schematically in Figure 8.25 and worked with 20 wt % glass fiber reinforced polycarbonate. In the mold A, all flow is symmetrical; variations of the holding pressure do not affect the fiber orientation in the weldline zone and weldlines having only about half of the tensile strength of weldline free samples are produced. In mold B, pressure unbalances during the holding stage lead to unsymmetrical packing flow, which will distort the weldline (Fig. 8.25B). Linear increase in weldline strength as a function of holding pressure was recorded. At 120 MPa the weldline had 76 % of the strength measured in samples without weldline, versus 55 % at a holding pressure of 0 MPa. Application of oscillating holding pressure (138,139) can lead to analogous results.

A particular case of weldlines is produced by jetting. As mentioned in Section 8.3.1.2, filled and reinforced melts are less "elastic" and do not swell as much when injected in the mold. An incorrect combination of gate size and injection speed may

301

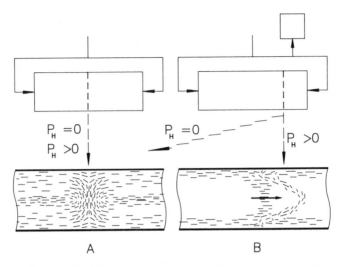

FIGURE 8.25 Influence of holding pressure P_H on the fiber orientation in the weldline zone for (A) symmetrical mold and (B) asymmetrical mold (56).

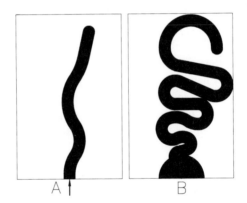

FIGURE 8.26 Jetting in the early (A) and later (B) stages of mold filling.

lead to the phenomenon illustrated in the Figure 8.26. The melt penetrates into the mold as a straight jet, which then deforms into a pattern similar to that in Figure 8.26B. It is only when the incoming melt is slowed by the material already in the mold that the mold can fill in the conventional fashion. However, the jetted melt remains essentially immobile; and when the melt front reaches it, its parts in contact with the wall have already frozen and remain visible in the finished molding. Jetting can be avoided by designing molds in which the melt stream immediately strikes the mold wall (13) or by using a gate diameter that obeys the relationship:

$$\frac{BD}{h} \geqslant 1 \tag{8.17}$$

where B is the die swell, D is the gate diameter, and h is the cavity thickness (140).

8.4.3 Residual Stresses and Warpage

Residual stresses in an injection-molded article originate from two sources. First, there are thermal stresses resulting from nonisothermal solidification: during the cooling step, the skin solidifies first around a still molten core. Subsequent cooling and shrinkage of inner layers exert compressive stresses on the skin while the core is maintained under tension. Then there are orientation stresses due to the flow-induced orientation of macromolecules. We have seen that depending on material and process parameters, the polymer is oriented to a different degree throughout the molding. The subject of residual stresses was reviewed by Isayev and Crouthamel (141), who concluded that thermal stresses are generally much more important than those caused by frozen macromolecular orientation. In simple injection-molded articles (e.g., flat plate) made of unreinforced polymers, the residual stress–depth curve assumes a symmetrical, nearly parabolic shape with the maximum (tensile) stress located in the center when the rate of cooling is identical on both sides. Nonuniform cooling (faster on one side) will produce nonsymmetrical residual stress distribution, which will lead to dimensional distortions called warpage (142,143). In more complex objects such as those with ribs, corners, and sections of different thickness, uniform cooling is impossible: inevitably some sections will solidify sooner, restraining the shrinkage of other sections. As a result, complex residual stress distribution patterns will be observed in such moldings (144).

The introduction of fillers into the matrix produces two types of effect:

1. Fillers and reinforcements have a higher thermal conductivity and lower coefficient of thermal expansion than polymers. Cooling will proceed at a faster rate and volume shrinkage will be less. Spherical and other fillers with aspect ratio close to 1 will reduce internal stresses (145) and warpage (146) as well as the cycle time (147).

2. Fibrous fillers will inevitably cause differential shrinkage as a result of directional dependence on linear coefficient of thermal expansion —(148). This in turn will lead to an increased tendency to warp (146). The higher warpage in fiber-reinforced materials is as unavoidable as is the mechanical anisotropy. It can to a certain degree be dealt with during mold design (149). However, most manufacturers of reinforced thermoplastics also sell low warp grades where, by using combinations of glass fibers with other mineral fillers such as mica or talc, the warpage is reduced but with some loss of strength (150). The most obvious solution—use of high aspect ratio, platelet-like fillers, which give isotropic moldings—failed to materialize: glass flakes break too much in the injection molding machine (151), and mica flakes do not have sufficient internal cohesion; they cleave when stressed (152).

Very little is known about residual stresses in fiber reinforced thermoplastics. According to Hindle et al. (153), who worked with glass fiber reinforced polypropylene, near the gate and parallel to flow the stresses are compressive at the surface, with a maximum value some distance below the surface while the core is in tension. Far from the gate and perpendicular to flow, residual stresses were compressive at the surface, tensile below the surface, and compressive again in the center of the sheet.

303

Annealing and aging have a significant effect on residual stress distribution in both reinforced (154) and unreinforced polymers, with potentially far-reaching consequences for in-use performance over the long term (155–160).

8.4.4 Design Considerations

The need to have simple yet relatively accurate design methods that would take into account the viscoelastic behavior of thermoplastic materials has led to the development of "pseudoelastic design method" (161–163). In this method values of time- and temperature-dependent properties such as creep (or relaxation) modulus are determined and substituted into equations normally used with elastic materials. If the value chosen for the modulus takes into account the required service life of the part as well as the maximum allowable strain, this approach can give satisfactory results. Suppliers of plastics intended for engineering applications provide creep curves from which appropriate values of creep modulus can be extracted.

Design methods used with thermoplastics in general can be applied to reinforced thermoplastics as well, provided the flow-induced fiber orientation is taken into account. The simplicity of the preceding statement is unfortunately misleading. Consider the set of requirements put forward recently (164), which must be met before an engineering design using these materials can be carried out with confidence.

- A quantitative method to convert mold filling and rheological data into an accurate description of fiber orientation.
- A quantitative method to convert the fiber orientation to an expected distribution of properties throughout the part.
- Nondestructive test methods to follow the quality of moldings (in terms of fiber orientation, location and nature of defects, etc.).

As discussed in the preceding section, we cannot expect either of the first two requirements to be satisfied in the forseeable future. However, probably the largest potential source of error is to consider the material as being stable in time. Plastics are generally used at temperatures where their microstructure undergoes significant changes with time, even in the absence of any aggressive chemical agent (165,166). For example, consider compression-molded, annealed (i.e., relatively stress free) polypropylene aged for 6000 hours at room temperature in the dark under inert atmosphere; the room temperature impact strength of this specimen decreases by more than 50 %, while the creep modulus increases by 75 % as a result of changes in its amorphous phase (165). In composite materials, the state of the interface may be affected by stresses generated during temperature and humidity cycling. In injection-molded parts, complex relaxation of residual stresses coupled with physical and chemical aging usually take place. The effects of these phenomena on in-use performance of the molded parts are largely unknown and therefore difficult to allow for in design. Uncertainties about material characteristics and their evolution upon aging make necessary a significant program for in-use testing of the product molded in the commercial environment before a design can be considered to be final.

Darlington and Upperton have investigated the applicability of design procedures for unreinforced thermoplastics to their fiber-reinforced counterparts (167–169). Given the uncertain state of fiber orientation in molded parts, they present a "bounds approach" to mechanical properties. It is based on the assumption that mechanical properties in a molded part may vary between two limiting values: upper and lower bounds. Upper bound values are those found in injection-molded standard test bar, where the fibers are aligned along the tensile axis. Data from the tests on these bars represent the best strength and modulus that can be found in an injection-molded part. Lower bound values are taken from a film-gated square mold in which a reasonable degree of fiber alignment is observed at least with some polymers. Tensile modulus and strength data found in samples cut in the direction perpendicular to flow are considered to represent the lower bound values. A third set of properties corresponding to random-in-plane orientation can be calculated from the upper and lower bound values. Data from these tests are compared with the mechanical behavior of several molded parts subjected to static loading. The authors conclude (167) that when designing for stiffness, the random-in-plane approach offers the designer a reasonable compromise between accuracy and ease of use. The reliability of this approach decreases where the component is of simple form and the stress field is also of simple form (e.g., uniaxial tension). Correction based on knowledge of fiber orientation may then be indicated. With respect to strength, they recommend the use of lower bound values unless the fiber orientation distribution is known with certainty to be more favorable.

8.4.5 Examples of Applications

Structural analysis is only one of the several important aspects of product design of injection-molded products. Processing considerations usually related to mold design and dimensional stability of molded parts (shrinkage and warpage) may impose solutions other than those that would be based on the results of structural analysis alone. Aesthetic factors or need for additional processing (painting, welding) may tip the scale toward one or another otherwise equivalent choices. The examples discussed below were taken from the technical literature. Most information originates from but is not published by the material or equipment suppliers.

8.4.5.1 Housings for Electrical Tools

One of the early applications in which fiber-reinforced plastics have replaced die-cast metals consists of housings for drills, jig and circular saws, and similar electrical tools. Smooth and aesthetically appealing on the outside, the internal faces are quite complex, with structural ribs, areas provided for tight insertion of motors and switches, metal inserts for screws, etc. (Fig. 8.27). The complex shape usually necessitates considerable mold prototyping. Compared with die-cast metals, reinforced plastics provide the advantage of safety (double-insulation) and a surface finish that does not require painting. Glass-reinforced polyamides (66 or 6) offer best cost/performance combination. The dimensional performance depends on mold

305

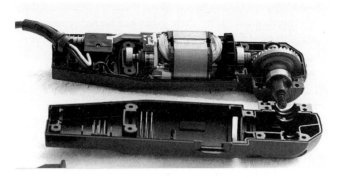

FIGURE 8.27   An orbital grinder. Housing is made of glass fiber reinforced polyamide 66.

design and molding conditions. Acceptable surface finish is achieved by formulation of the resin (170,171).

8.4.5.2 Automotive, Under-the-Hood Applications

Functional requirements include a combination of high temperature, chemical, and mechanical resistance. An excellent example is provided by the radiator tanks (parts attached to the radiator top and bottom: Fig. 8.28). Formerly these parts were made from copper alloys. Coolant pipe fittings had to be brazed or soldered to the radiator tank body. The plastic has to withstand the temperature of 130 °C and the effect of many chemicals (hydrocarbons, antifreeze, cleaning products, calcium chloride, etc.). Today glass-reinforced polyamide 66 is almost universally accepted. Some manufacturers also used glass–reinforced polyphenylene oxide (172), but gave it up because of the better stress-cracking resistance provided by the polyamide. Plastic radiator tanks can be used with aluminum radiators (no electrolytic corrosion). Large differences in coefficients of thermal expansion between the tank and the metal radiator required a redesign of the gasket (170), however. To date tens of millions of polyamide tanks have been successfully molded.

FIGURE 8.28   Radiator tank made of glass fiber reinforced polyamide 66.

Other under-the-hood parts made from glass-reinforced polyamide include fans, fan brackets, and gearbox covers. Some recently developed passenger car models contain as much as 6 kg of glass-reinforced polyamides and polybutylene terephthalate, the latter being mainly used in electrical applications (173). For noncritical applications (fan shrouds, horn shells, heater housing), glass-reinforced polypropylene is a material of choice because of its lower cost (172).

8.4.5.3 Plastic Drawers (174)

Plastic drawers for kitchen furniture (largest size $550 \times 435 \times 75$ mm) used to be made of high impact polystyrene, which is susceptible to stress cracking in contact with oils, fats, and household cleaners. Thus it was decided to replace this material with another low cost plastic—polypropylene. Because polypropylene is semicrystalline, the volume shrinkage during cooling from the melt is much more important than for the amorphous polystyrene. Despite incorporation of 20 wt % of talc, the linear shrinkage remains at 1.0–1.1 % (almost twice as high as that of polystyrene). As a result of the greater shrinkage and possible shrinkage differences caused by the material anisotropy, polypropylene drawers tended to warp much more than those made of polystyrene. The problem was solved by providing additional separate cooling circuits in the corner zones, which receive more cooling than other areas of the mold. Higher flexural modulus of talc-filled polypropylene over the specified temperature range (20–60 °C) led to a reduction of thickness from 3 mm to 2 mm, which more than outweighed the slightly higher cost of talc-filled polypropylene. This in turn enabled manufacturers to maintain a cycle time identical to that experienced with polystyrene.

The talc-filled polypropylene (40 wt %) has also found a major market in injection-molded lawn furniture, where its higher flexural stiffness ($E = 4$ GPa) makes it preferable to unfilled polypropylene.

8.5 RELATED PROCESSES AND MATERIALS

8.5.1 Long Fiber Reinforced Injectable Thermoplastics[1]

The need for longer fibers in injectable compounds has been recognized for a long time. Initially it led to the development of a compounding process derived from extrusion wire coating. Multifilament rovings are pulled through an extrusion die, coated with polymer, cooled, and cut into pellets. Fibers in the pellet are not dispersed in the matrix—they stay bundled together. In injection molding, good dispersion can be achieved by sufficiently shearing the material (e.g., by using high back pressure). This leads to a severe fiber length degradation and machine wear. Although average fiber length in the molding may be slightly longer than with conventional short fiber

---

[1] The reader is also referred to Chapter 7 for additional information.

compounds, they are of the same order of magnitude. Fiber length in these compounds can be preserved at the expense of good dispersion (175).

Recently developed compounding processes (176,177) circumvented these drawbacks and led to the introduction of well-dispersed long fiber pellets. In these processes, called pultrusion compounding, the roving consisting of several thousand continuous monofilaments is opened (i.e., filaments are separated one from another), and the molten thermoplastic is forced into the roving. Thus individual filaments are coated with the polymer. Cooled impregnated roving is then cut into pellets. With this process, fiber loadings higher than those obtainable with extrusion compounding are possible (60 vs. 40 wt %). Literature published on these materials usually deals with one or both of the following questions:

• How much of the initial very high fiber length can survive the injection molding process, and how can it be preserved?
• What advantages in finished moldings can be derived from long fibers?

A recent study by Cianelli et al. (178) in which a 50 wt % long glass fiber/polyamide 66 compound was compared with an equivalent short fiber compound, provides many answers. In these experiments the two compounds were molded side by side on the same equipment. Screw type (standard and low shear), screw tip (standard and free flow), and processing conditions (back pressure, speed of screw rotation, injection, or speed) were all varied. Different temperature profiles along the barrel were used for each compound. The principal conclusion is that long glass fiber compounds give significantly longer fiber length than their short fiber counterpart regardless of the processing conditions even though practically every fiber broke several times during processing. Under the most severe conditions used (high shear screw, standard screw tip) 55 % of all fibers (by volume) still were longer than 1 mm (initial fiber length $\simeq$ 12.7 mm). This is much longer than in conventional short fiber compounds, where most fibers are in the 100–400 μm range.

In another study (179) it was shown that the fiber length in the molded object is not uniform. Material in the skin was found to contain shorter fibers than the core region. This effect was attributed to shear during the mold filling process. These materials are more pseudoplastic than their unfilled or short fiber reinforced counterparts (180). Although at low rates of deformation their viscosity is much higher, differences become less significant in the range of shear of elongation rates characteristic of injection molding (see Table 8.7). Their moldability as determined by the spiral mold flow test is quite similar to that of identically loaded short fiber compounds (181). Considering the combination of high fiber concentration and the fiber length of these materials and comparing it with the maximum (random) packing fraction [Eq. (8.4)], it is obvious that the fibers must keep much of their initial alignment throughout the molding process. They may slide, but they do not rotate relative to each other.

Unfortunately, very little else is known about the rheology of long glass fiber compounds at this time (182,183). The range of mechanical properties offered by these materials exceeds that of short fiber compounds not only because of the fiber length

TABLE 8.7   Effect of Short and Long Glass Fibers on Viscosities of Polyamide 66[a]

Polyamide 66	Viscosities (Pa)		A	n	B	m
	Shear, η, at $\dot{\gamma} = 1000\,s^{-1}$	Elongational, λ, at $\dot{\varepsilon} = 1000\,s^{-1}$				
Unfilled	17	180	62	0.81	356	0.90
Short fibers	42	760	376	0.68	3820	0.77
Long fibers	74	10200	1420	0.57	221000	0.55

[a] Fiber loading is 50 wt% in both cases; $T = 285\,°C$. Parameters of power law equations for both shear $\eta = A\dot{\gamma}^{n-1}$ and extensional $\lambda = B\dot{\varepsilon}^{m-1}$ are shown (180).

but also as a result of higher fiber loadings possible with the long fiber compounds (see Table 8.8). At equal fiber concentrations the materials have approximately the same stiffness and tensile and flexural strength. Long fibers give much higher impact strength.

Part and mold design with these materials is more difficult because all the undesirable effects associated with flow-induced fiber orientation (warpage, weldlines) are dependent on fiber length.

Reported applications of long fiber reinforced thermoplastics include:

- *Chain saw gear wheel (184):* Offers better impact strength and fatigue resistance than short fiber compounds.
- *Aircraft fuel tank antisurge valve flap (181,184):* Used as a replacement of an asbestos-filled phenolic compound.
- *Golf putter head (178,185):* This product includes cavities into which molten lead is poured to achieve precise weight distribution; the material is not affected by the molten metal; impact strength is excellent.

TABLE 8.8   Typical Properties of Long and Short Fiber Reinforced Polyamide 66 Containing 50 wt% of Glass Fibers[a]

Property	Fibers	
	Short	Long
Tensile strength, MPa	155	165
Elongation at break, %	6	6
Flexural strength, MPa	240	260
Flexural modulus, MPa	9.6	11.2
Notched impact strength, J/m	13	37
Deflection temperature under load, °C	250	261
Linear coefficient of thermal expansion, $10^{-5}/K$ (dry)	5	5

[a] Results at room temperature and 50% relative humidity (177).

8.5.2 Liquid Crystal Polymers

Unlike ordinary semicrystalline polymers, which lose all their crystallinity when heated above their crystalline melting temperature, liquid crystal polymers keep a high degree of structural order in the liquid state. There are two classes of liquid crystals: lyotropic and thermotropic. The former exhibit a liquid-crystalline phase in solutions, the thermotropics can be molten and processed by conventional plastics processing methods (extrusion, injection molding, etc.). The second type is briefly discussed below.

Liquid crystal polymers have many outstanding properties, high temperature performance being the most important. For example, one commercial type based on p,p'-bisphenol, p-hydroxybenzoic acid, and terephthalic acid has a heat deflection temperature of 355 °C and a continuous use temperature of 240 °C (186,187). There is considerable analogy between injection molding of liquid crystal polymers (LCP) and that of fiber-reinforced thermoplastics. Molten LCPs contain elongated domains that orient in flow in a manner very similar to that of short fibers. In fact, structural studies on injection moldings reveal a multilayer skin–shear layer–core structure (188–192). Because of different orientation of these domains from layer to layer, the properties vary with part thickness as the skin/core ratio increases with the decreasing thickness: Anisotropy and weak weldlines represent a potential problem, and mold and part design are again critical (193).

LCPs are available unfilled (transparent) or in mineral, glass fiber, and carbon fiber filled grades. Applications include electric and electronic devices (connectors, chip carriers, bobbins), where the ability to withstand high temperatures and to exhibit flame resistance outweighs the high cost. They are also suitable for dual oven use (i.e., conventional and microwave) transparent cookware (193,194).

8.5.3 Structural Foams

The term "structural foam" refers to materials that consist of a foamed core surrounded by a solid skin. The core and the skin are from the same material and are molded simultaneously on a modified injection molding machine. There are two approaches to structural foam molding. The first one uses a chemical blowing agent that decomposes when the polymer is heated in the machine, giving gas. The second method uses an external supply of gas, usually nitrogen, which is injected into the molten polymer in the injection unit. The high pressure keeps the nitrogen in solution until the melt is injected into the mold, where the gas separates and foams the polymer. In the most common variant of this process, the volume of the polymer injected is less than the volume of the mold, the expanding polymer–gas mixture ensuring mold filling. Typical overall void content in structural foam moldings varies between 5 and 30 %.

Structural foam molding is used with both reinforced and unreinforced polymers. Relative advantages and disadvantages of the process with respect to conventional injection molding were recently reviewed by Morton-Jones and Ellis (195).

- Structural foams provide higher flexural stiffness to weight ratio. The modulus of the foam is related to density by (196):

$$E_f = E \left(\frac{\rho_f}{\rho} \right)^2 \tag{8.18}$$

where E_f and E represent the moduli and ρ_f and ρ the densities of the foam and the solid polymer, respectively. Since the stiffness S is proportional to the cube of thickness h_f, it follows that for a uniformly foamed panel of a given weight, the stiffness (load divided by deflection) will be inversely proportional to the panel density:

$$S \propto E_f h_f^3 \propto \frac{1}{\rho_f} \tag{8.19}$$

The actual structural foam moldings exhibit a density gradient across the thickness: a dense, hence stiffer skin, and a lower modulus core. This further improves the flexural stiffness of the system: models based on analogy with I-section beams have been used to calculate the stiffness of structural foams (161,162,195).
- Dimensional stability is good. The presence of expanding gas forces the skin against the mold wall, thus compensating for shrinkage. The structural foam products are less oriented.
- Cheaper molds can be used (a process advantage), since the mold filling is completed by expanding gas, which means that clamping force can be lower.
- Process disadvantages include very visible flow patterns and surface roughness due to the collapse of the foam cells against the mold wall and longer cycle time; poor conductivity of the foam requires longer cooling time, thus limiting the maximum thickness to about 15 mm. Minimum thickness is about 5 mm.

Properties of polypropylene glass reinforced structural foam were studied by Hornsby et al. (197,198). Some of the advantages of the foam process appear to be lost, however, because the fiber reinforcement causes considerable warpage in flat moldings. The fiber orientation in moldings is similar to that observed in conventional reinforced polypropylene (i.e., parallel to flow in the skin and perpendicular to flow in the core). During expansion in the mold, fibers remain within the polymer matrix, surrounding the cells rather than bridging them. Weldline zones consist of higher density or solid material and may have higher stiffness and impact strength than the weldline-free zone (198). Typical applications of reinforced structural foam are those in which product appearance is not critical. A detailed description of the design of a 6.3 kg washing machine tank molded of 30 % glass fiber–polypropylene structural foam is given in Reference 195. This part replaced 16 welded steel parts. Required rigidity was achieved by using thick sections and ribs. Steel bearing, motor mounting flange, and spring legs were all molded in. Other applications include automobile seat shells, material handling crates, and furniture (199).

8.5.4 Coinjection Molding

Coinjection molding is a bicomponent process: two injection units fill one mold to form a skin–core structure, and the skin performs functions not available from the core material. One version of the process (Fig. 8.29) usually involves a combination of sequential and simultaneous mold filling from both injection units. The first use of the coinjection process involved bicomponent structural foam molding where the skin is solid, thus eliminating the poor surface quality of conventional structural foam. Recently solid, relatively thin-walled products were introduced (200). Eckhard and Munschek (201) consider both possible combinations (reinforced/unreinforced) and present the following advantages.

- *Core reinforced/skin nonreinforced:* This combination provides products with the external appearance of unreinforced materials with most of the structural integrity of reinforced ones.
- *Skin reinforced/core nonreinforced:* By using this combination, products exhibiting higher flexural stiffness per unit of material cost can be obtained.

The flow behavior of both compounds needs to be carefully balanced to avoid flow instabilities, which may lead to the core striking through the skin. Akay (202) studied the coinjection process in a centrally gated disc mold with filled, glass-fiber reinforced, and unfilled polypropylene. Fiber orientation was found to be similar to that in conventional (one-component) injection molding. Important variables for obtaining uniform skin/core structure were injection speed and melt and mold temperatures. Interface instabilities were found to be more pronounced when the higher viscosity component occupied the skin and when the injection speed was high.

Typical applications of reinforced core/unreinforced skin are those characterized by a need to hide the reinforced material. One such example is that of a headlight reflector made of polybutylene terephthalate (PBT), the skin material provides the necessary smoothness to accept vacuum metallization, whereas the glass fiber reinforced PBT core provides the mechanical and thermal properties required. Stainless steel fibers for shielding from electromagnetic interference can be also incorporated into the core (200).

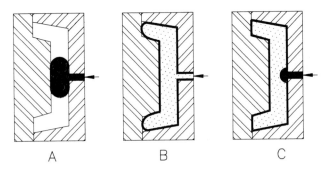

A   B   C

FIGURE 8.29   Coinjection molding: (*A* and *C*) injection of the skin material; (*B*) simultaneous injection of the core and the skin (201).

8.5.5 Injection–Compression Molding

Injection–compression molding is a variety of injection molding designed for production of flat thermoplastic parts with dimensionally stable surfaces (e.g., lenses, compact disc records). In injection–compression molding the thermal contraction of the cooling melt is compensated by a reduction of the mold cavity. The process is illustrated in Figure 8.30. The distance between the mold halves is significantly larger than in finished molding, and the cavity is only partially filled during the mold filling stage. The pressure required to fill the mold is low (point 1,2 in the PVT diagram, Fig. 8.30B). The gate is sealed by freezing or mechanically at point 3. With the beginning of the compression operation, the melt is first made to fill the cavity. This results in a slight increase of the pressure. The mold is full at point 4, and pressure increases steeply up to the maximum set value. From there (point 5) the cooling occurs at constant volume until atmospheric pressure has been reached (point 6). The advantages of this process over injection molding stem from the absence of packing, and the macromolecular orientation due to the holding pressure will be largely avoided. Strategies to minimize orientations and residual stresses are reviewed in Reference 203. With fiber-reinforced thermoplastics, injection–compression molding gives lower warpage and as such may be considered for large flat parts such as external body panels for passenger cars (204).

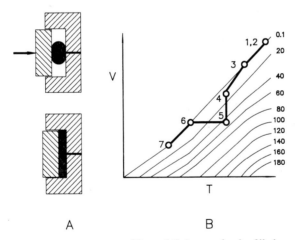

A   B

FIGURE 8.30   Injection–compression molding. (A) Incompletely filled mold (top) and compressed melt (bottom). (B) Pressure–volume–temperature diagram of an amorphous polymer during an injection–compression cycle (see text). (203).

8.5.6 Lost Metal Core Molding

One of the characteristics of injection molding is its unsuitability for the production of precision hollow parts. This difficulty can be circumvented by using a low melting metal core, which is used as an insert and is melted out upon completion of the injection molding process. This process has been successfully used for manufacturing

313

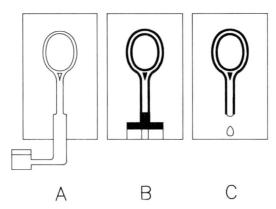

FIGURE 8.31   Lost metal core molding. (*A*) Die casting of the core. (*B*) Injection molding with the core as insert. (*C*) Melting out of the core (206).

of high performance tennis rackets from carbon fiber reinforced polyamide 66 (205,206). The first stage of the process illustrated in Figure 8.31 consists of die casting the metal core using a bismuth–tin alloy melting at 138 °C. The core is inserted into the injection mold and the reinforced polymer, which melts at around 270 °C, is molded around the core. Despite the high melting point of the plastic, the core, acting as a heat sink, does not melt. After the injection molding, the core is removed by heating to 160 °C for 15 minutes. The alloy is recycled. The body of the racket is then filled with polyurethane foam.

Clearly this process can be justified for relatively expensive products. Another reported application—namely, water pumps for passenger car engine cooling systems (206)—illustrates, together with other variations of thermoplastic injection molding, the almost endless possibilities offered by this process.

REFERENCES

1. Anon., *Mod. Plast*, 65(1), 95 (1988).
2. T. J. Brown and C. J. Hooley, *Int. J. Veh. Design. Technol. Adva. Veh. Design Ser*, SP6, 23 (1986).
3. Anon., *Plast. Technol.*, 34(5), 171 (1988).
4. L. Longoni, *Short Fiber Reinforced Thermoplastics*, International Conference Uxbridge, June 17–18, preprints published by Plastics and Rubber Institute, London, 1985.
5. A. Whelan, *Injection Moulding Machines*, Elsevier Applied Science Publishers, London, 1984.
6. F. Johannaber, *Injection Molding Machines, A User's Guide*, Hanser Publishers, Munich, 1984.
7. D. V. Rosato and D. V. Rosato, Eds., *Injection Molding Handbook*, Van Nostrand Reinhold, New York, 1986.
8. A. Whelan, *Injection Moulding Materials*, Elsevier Applied Science Publishers, London, 1983.
9. R. A. Hetzer, in E.C. Bernhardt, Ed. *CAE. Computed Aided Engineering for Injection Molding*, Hanser Publishers, Munich, 1983.

10. D. H. Morton-Jones, in A. Whelan and J. P. Goff, Eds., *Developments in Injection Moulding*, Elsevier Applied Science Publishers, London, 1985.

11. K. Stoeckhert, Eds, *Mold Making Handbook for the Plastics Engineer*, Hanser Publishers, Munich, 1983.

12. R. G. W. Pye, *Injection Mould Design*, George Godwin, London, 1983.

13. G. Menges and P. Mohren, *How to Make Injection Molds*, Hanser Publishers, Munich, 1986.

14. S. D. Gerbig in A. Whelan and J. P. Goff, Eds., *Developments in Injection Moulding*, Elsevier Applied Science Publishers, London, 1986.

15. F. Kurihara and S. Kimura, *Polym. J.*, 17, 863 (1985).

16. R. B. Bird, R. C. Armstrong, and O. Hassager, *Dynamics of Polymeric Liquids*, Vol. 1, *Fluid Mechanics*, Wiley, New York, 1977.

17. M. Mooney, *J. Colloid Sci.*, 6, 162 (1951).

18. L. E. Nielsen, *Polymer Rheology*, Dekker, New York, 1974.

19. J. V. Milewski, *Polym. Plast. Technol. Eng.*, 3, 101 (1974).

20. J. V. Milewski, *Composites*, 4, 258 (1973).

21. J. V. Milewski, *Ind. Eng. Prod. Res. Dev.*, 17, 363 (1978).

22. M. M. Cross, A. Kaye, J. L. Standord, and R. F. T. Stepto, *ACS Polym. Mater. Sci. Eng. Div., Prepr.*, 49, 531 (1983).

23. B. Fisa, B. Sanschagrin, and B. D. Favis, *Polym. Compos.*, 5, 264 (1984).

24. B. Fisa, *Polym. Compos.*, 6, 232 (1985).

25. D. Bigg, *Polym. Compos.*, 6, 20 (1985).

26. L. Czarnecki and J. L. White, *J. Appl. Polym. Sci.*, 25, 1217 (1980).

27. R. J. Crowson, M. J. Folkes, and P. F. Bright, *Polym. Eng. Sci.*, 20, 925 (1980).

28. R. J. Crowson and M. J. Folkes, *Polym. Eng. Sci.*, 20, 934 (1980)

29. R. J. Crowson, A. J. Scott, and D. W. Saunders, *Polym. Eng. Sci.*, 21, 748 (1981).

30. B. A. Knutsson, J. L. White, and K. B. Abbas, *J. Appl. Polym. Sci.*, 26, 2347 (1981).

31. V. M. Lobe and J. L. White, *Polym. Eng. Sci.*, 19, 617 (1979).

32. L. A. Utracki, B. D. Favis, and B. Fisa, *Polym. Compos.*, 5, 277 (1984).

33. L. A. Utracki, *Polym. Compos.*, 7, 284 (1986).

34. L. A. Utracki, *Rubber Chem. Technol.*, 57, 507 (1984).

35. L. A. Goettler, in D. W. Clegg and A. A. Collyer, Eds., *Mechanical Properties of Reinforced Thermoplastics*, Elsevier Applied Science Publishers, London, 1986.

36. H. L. Goldsmith and S. G. Mason, in F. R. Eirich, Ed., *Rheology*, Vol. 4, Academic Press, New York, 1967.

37. M. Vincent, *Etude de l'orientation des fibres de verre courtes lors de la mise en oeuvre de thermoplastiques chargés*, Doctoral thesis, Ecole Nationale Supérieure des Mines de Paris, 1984.

38. M. Vincent and J. F. Agassant, in J. C. Seferis and P. S. Theocanis, Eds., *Interrelations Between Processing Structure and Properties of Polymeric Materials*, Elsevier, Amsterdam, 1984.

39. G. Akay, in A. Whelan and J. P. Goff, Eds., *Developments in Injection Molding*, Vol. 3, Elsevier Applied Science Publishers, London, 1985.

40. J. E. Mark, A. Eisenberg, W. W. Graessley, L. Mandelkern, and J. L. Koenig, *Physical Properties of Polymers*, American Chemical Society, Washington, DC, 1984.

41. J. M. Lunt and J. B. Shortall, *Plast. Rubber Process.*, 4, 108 (1979).

42. J. M. Lunt and J. B. Shortall, *Plast. Rubber Process.*, 5, 37 (1980).

43. R. P. Hegler and G. Mennig, *Polym. Eng. Sci.*, 25, 395 (1985).

44. J. Kubat and A. Szalanczi, *Polym. Eng. Sci.*, 14, 873 (1974).

45. L. Borocz and J. Kubat, *Plast. Rubber Process.*, 4, 82 (1979).

46. S. H. Emermann, *Polym. Eng. Sci.*, 27, 1101 (1987).
47. H. Bangert, *Kunstst. -Ger. Plast.*, 75, 325 (1985).
48. E. C. Bernhardt, Ed., *CAE: Computer Aided Engineering for Injection Molding*, Hanser Publishers, Munich, 1983.
49. L. T. Manzione, Ed., *Applications of Computer Aided Engineering in Injection Molding*, Hanser Publishers, Munich, 1987.
50. H. Alles, J. F. Agassant, M. Vincent, G. Dehay, P. Lerebours and G. Ginglinger, *Int. Polym. Process.*, 2, 191 (1988).
51. Z. Tadmor, *J. Appl. Polym. Sci.*, 18, 1753 (1974).
52. L. R. Schmidt, *Adv. Chem. Ser.*, 142, 415 (1974).
53. L. R. Schmidt, *Polym. Eng. Sci.*, 17, 666 (1977).
54. M. W. Darlington, A. J. Scott, and A. C. Smith, *Plast. Rubber Process. Appl.*, 3, 189 (1983).
55. J. C. Malzahn and J. M. Schultz, *Compos. Sci. Technol.*, 25, 187 (1986).
56. H. Hamada, Z. Maekawa, T. Horino, K. Lee, and K. Tomari, *Int. Polym. Process.*, 2, 131 (1988).
57. W. Woebcken, *Kunstst. Ger. Plast.*, 71, 229, 1981.
58. G. Akay, in J. C. Seferis and P. S. Theocaris, Eds., *Interrelations Between Processing Structure and Properties of Polymeric Materials*, Elsevier, Amsterdam, 1984.
59. U. Delpy and G. Fisher, *Adv. Polym. Technol.*, 19 (1986).
60. P. F. Bright and M. W. Darlington, *Plast. Rubber Process. Appl.*, 1, 139 (1981).
61. G. Fisher and P. Eyerer, *SPE ANTEC Tech. Pap.*, 32, 532 (1986).
62. D. Vigneron, *Mécanique du renforcement et mécanismes de rupture dans les polyamides renforcés par fibres courtes*, Doctoral thesis, Université de Technologie de Compiègne, 1983.
63. M. W. Darlington and A. C. Smith, *Polym. Compos.*, 8, 16 (1977).
64. M. Takeshima and N. Funakoshi, *J. Appl. Polym. Sci.*, 32, 3457 (1986).
65. R. P. Hegler, *Kunstst. Ger. Plast.*, 74, 271 (1984).
66. P. J. Wright and A. Whelan, *SPE ANTEC Tech. Pap.*, 33, 773 (1987).
67. G. Menges and P. Geibüsh, *Colloid Polym. Sci.*, 260, 73 (1982).
68. M. Sanou and C. Cohen, *SPE ANTEC Tech. Pap.*, 29, 734 (1983).
69. M. Akay and D. Barkley, *Compos. Struct.*, 3, 269 (1985).
70. M. J. Folkes and D. A. M. Russell, *Polymer*, 21, 1252 (1980).
71. M. R. Kamal, L. Song, and P. Singh, *SPE ANTEC Tech. Pap.*, 32, 133 (1986).
72. S. Fakirov and C. Fakirova, *Polym. Compos.*, 6, 41 (1986).
73. P. F. Bright, R. J. Crowson, and M. J. Folkes, *J. Mater. Sci.*, 13, 2497 (1978).
74. M. W. Darlington, A. J. Scott, and A. C. Smith, *Polym. Commun.*, 27, 109 (1986).
75. A. E. Hirsch, *SPE ANTEC Tech. Pap.*, 33, 297 (1987).
76. T. -S. Chung, *Polym. Eng. Sci.*, 25, 772 (1985).
77. T. -S. Chung, and M. E. Ryan, *Polym. Eng. Sci.*, 21, 271 (1981).
78. W. H. Harland, P. Kantas and T. Habipis, *Plast. Rubber Process.*, 20, 67 (1980).
79. A. I. Isayev, *Polym. Eng. Sci.*, 23, 271 (1983).
80. W. Dietz, J. L. White, and E. S. Clark, *Polym. Eng. Sci.*, 18, 273 (1978).
81. H. Janeshitz-Kriegl, in J. C. Seferis and P. S. Theocaris, Eds., *Interrelations Between Processing Structure and Properties of Polymeric Materials*, Elsevier, Amsterdam, 1984.
82. D. Kalyon, A. Wagner, and S. Dey, *SPE ANTEC Tech. Pap.*, 34, 605 (1988).
83. M. R. Kamal and V. Tan, *Polym. Eng. Sci.*, 19, 558 (1979).
84. K. P. Großkurth, in J. C. Seferis and P. S. Theocaris, Eds., *Interrelations Between Processing Structure and Properties of Polymeric Materials*, Elsevier, Amsterdam, 1984.
85. L. Hoare and D. Hull, *Polym. Eng. Sci.*, 17, 204 (1977).
86. P. H. Geil, *Polymer Single Crystals*, Wiley-Interscience, New York, 1963.
87. R. L. Cormia, F. P. Price, and D. Turnbull, *J. Chem. Phys.*, 37, 1333 (1962).

88. B. Wunderlich, *Macromolecular Physics*, Vol. 2, *Crystal Nucleation, Growth, Annealing*, Academic Press, New York, 1976.
89. J. P. Trotignon and J. Verdu, *J. Appl. Polym. Sci.*, 34, 1 (1987).
90. B. Heise, L. Klostermann, and W. Woebcken, *Colloid Polym. Sci.*, 260, 487 (1982).
91. J. P. Trotignon, J. L. Lebrun, and J. Verdu, *Plast. Rubber Process. Appl.*, 2, 247 (1982).
92. B. Z. Jang, D. R. Uhlman, and J. B. Vander Sande, *J. Appl. Polym. Sci.*, 29, 4377 (1984).
93. D. R. Fitchmun and Z. Mencik, *J. Polym. Sci., Polym. Phys. Ed.*, 11, 951 (1973).
94. M. R. Kamal and F. H. Moy, *J. Appl. Polym. Sci.*, 28, 1787 (1983).
95. V. Tan and M. R. Kamal, *J. Appl. Polym. Sci.*, 22, 2341 (1978).
96. F. H. Moy and M. R. Kamal, *Polym. Eng. Sci.*, 20, 957 (1980).
97. S. S. Katti and J. M. Schultz, *Polym. Eng. Sci.*, 22, 1001 (1982).
98. D. P. Russel and P. W. R. Beaumont, *J. Mater. Sci.*, 15, 197 (1980).
99. J. Bowman, *J. Mater. Sci.*, 16, 1151 (1981).
100. M. Guo and J. Bowman, *J. Appl. Polym. Sci.*, 28, 2341 (1983).
101. J. Bowman, N. Harris, and M. Bevis, *J. Mater. Sci.*, 10, 63 (1975).
102. C. M. Hsiung, M. Cakmak, and J. L. White, *SPE ANTEC Tech. Pap.*, 32, 128 (1986).
103. F. Altendorfer and E. Seitl, *Kunstst. Ger. Plast.*, 76, 47 (1986).
104. M. Fleissner, *Kolloid.-Z.*, 251, 1006 (1973).
105. J. P. Trotignon, and J. Verdu, *J. Appl. Polym. Sci.*, 34, 19 (1987).
106. R. H. Burton, and M. J. Folkes, in D. W. Clegg and A. A. Collyer, Eds., *Mechanical Properties of Reinforced Thermoplastics*, Elsevier Applied Science Publishers, London, 1986.
107. A. Misra, B. L. Deopura, S. F. Xavier, F. D. Hartley, and P. H. Peters, *Agnew. Makromol. Chem.*, 113, 113 (1983).
108. R. H. Burton, and M. J. Folkes, *Plast. Rubber Process. Appl.*, 3, 129 (1983).
109. S. Y. Hobbs, *Nature (London) Phys. Sci.*, 234, 12 (1971).
110. C. Lhymn, and J. M. Schultz, *Polym. Compos.*, 6, 87 (1985).
111. D. H. Burton, T. M. Day, and M. J. Folkes, *Polym. Commun.*, 25, 361 (1984).
112. D. Campbell, and J. R. White, *Agnew. Makromol. Chem.*, 122, 61 (1984).
113. S. Wu, *Polym. Eng. Sci.*, 19, 638 (1979).
114. H. W. Cox, and C. C. Mentzer, *Polym. Eng. Sci.*, 26, 488 (1986).
115. P. Friel, *Kunstst. Ger. Plast.*, 76, 23 (1986).
116. B. F. Blumentritt, B. T. Vu, and S. L. Cooper, *Polym. Eng. Sci.*, 14, 633 (1974).
117. B. F. Blumentritt, B. T. Vu, and S. L. Cooper, *Polym. Eng. Sci.*, 15, 428 (1975).
118. L. Dilandro, A. T. Dibenedetto, and J. H. Groeger, *Polym. Compos.*, 9, 209 (1988).
119. M. G. Bader, and W. H. Bowyer, *Composites*, 4, 150 (1973).
120. M. W. Darlington, B. K. Gladwell, and G. R. Smith, *Polymer*, 18, 1269 (1977).
121. S. Yamashiro, *Kunstst. Ger. Plast.*, 78, 231 (1988).
122. H. Krenchel, *Fiber Reinforcement, Theoretical and Practical Investigation of the Elasticity and Strength of Fiber Reinforced Materials*, Akademisk Forlag, Copenhagen, 1964.
123. H. Fukuda, and K. Kawata, *Fiber Sci. Technol.*, 7, 207 (1974).
124. J. C. Halpin, N. I Pagano, and J. M. Whitney, *J. Compos. Mater.*, 2, 154 (1968).
125. H. H. Winter, in J. C. Seferis and P. S. Theocaris, Eds., *Interrelations Between Processing Structure and Properties of Polymeric Materials*, Elsevier, Amsterdam, 1984.
126. S. Piccarolo, A. Rallis, and G. Titomanlio, *Int. Polym. Process.*, 2, 131 (1988).
127. C. Austin, in E. C. Bernhardt, Ed., *CAE: Computer Aided Engineering for Injection Molding*, Hanser Publishers, Munich, 1983.
128. S. C. Kim and N. P. Suh, *Polym. Eng. Sci.*, 26, 1200 (1986).
129. C. B. Bucknall, *Pure Appl. Chem.*, 58, 985 (1986).
130. S. C. Malguarnera, *Polym. Plast. Technol. Eng.*, 18, 1 (1982).
131. H. G. Moslé, R. M. Criens, and H. Dick, *SPE ANTEC Tech. Pap.*, 30, 772 (1984).

132. S. C. Malguarnera and A. Manisali, *Polym. Eng. Sci.*, 21, 586 (1981).

133. E. Nolley, J. W. Barlow, and D. R. Paul, *Polym. Eng. Sci.*, 20, 364 (1980).

134. A. Savadori, A. Pelliconi, and D. Romanini, *Plast. Rubber Process Appl.*, 3, 215 (1983).

135. P. J. Cloud, F. McDowell, and S. Gerakaris, *Plast. Technol.*, 22(9), 48 (1976).

136. M. J. Bozarth, and J. L. Hamill, *SPE ANTEC Tech. Pap.*, 30, 1091 (1984).

137. B. Fisa, J. Dufour, and T. Vu-Khanh, *Polym. Compos.*, 8, 408 (1987).

138. P. S. Allan, and J. J. Bevis, *Short Fiber Reinforced Thermoplastics*, International Conference, Uxbridge, U.K., June 17–18, 1985, preprints published by Plastics and Rubber Institute, London, 1985.

139. P. S. Allan, and J. J. Bevis, *Polymer Processing Machinery*, International Conference, Bradford, U.K., July 3–4, 1985, preprints published by Plastics and Rubber Institute, London, 1985.

140. J. M. Dealy, in D. V. Rosato, and D. V. Rosato, Eds., *Injection Molding Handbook*, Van Nostrand Reinhold, New York, 1986.

141. A. I. Isayev, and D. L. Crouthamel, *Polym. Plast. Technol. Eng.*, 22, 177 (1984).

142. G. Menges, A. Troost, J. Backhaus, A. Boue, W. Feser, and O. Krestschmar, *Kunstst. Ger. Plast.*, 75, 244 (1985).

143. M. St-Jacques, *Polym. Eng. Sci.*, 22, 241 (1982).

144. G. Menges, A. Dierkes, L. Schmidt, and E. Winkel, *SPE ANTEC Tech. Pap.*, 26, 300 (1980).

145. J. Kubat, and M. Rigdahl, *Polym. Eng. Sci.*, 16, 792 (1976).

146. P. J. Cloud, and M. A. Wolverton, *Plast. Technol.*, 24(12), 107 (1978).

147. A. Boldizar, and J. Kubat, *Polym. Eng. Sci.*, 26, 877 (1986).

148. L. E. Nielsen, *Mechanical Properties of Polymers and Composites*, Dekker, New York, 1974.

149. N. C. Baldwin, *Plast. Design Proces.*, 14(1), 25 (1974).

150. Anon., *Design Handbook for Dupont Engineering Plastics*, Module IV, Rynite, E.I. duPont de Nemours Co., Polymer Products Department, publication E-62620, Wilmington.

151. A. Bouti, *Composites thermoplastiques: Polypropylène renforcé avec des paillettes de verre*, M. Sc. A. thesis, Ecole Polytechnique de Montréal, 1988.

152. N. Burditt, A. King, and A. Scheibelhoffer,*Plast. Comp.*, 8(4), 62 (1985).

153. C. S. Hindle, J. R. White, D. Dawson, W. J. Greenwood, and K. Thomas, *SPE ANTEC Tech. Pap.*, 27, 783 (1981).

154. M. Thompson, and J. R. White, *Polym. Eng. Sci.*, 24, 227 (1984).

155. L. D. Coxon, and J. R. White, *Polym. Eng. Sci.*, 20, 230 (1980).

156. M. M. Qayyum, and J. R. White, *J. Mater. Sci.*, 20, 2557 (1985).

157. A. Ram. O. Zilber, and S. Kenig, *Polym. Eng. Sci.*, 25, 577 (1985).

158. A. V. Iacopi, and J. R. White, *J. Appl. Polym. Sci.*, 33, 577 (1987).

159. A. V. Iacopi, and J. R. White, *J. Appl. Polym. Sci.*, 33, 607 (1987).

160. D. P. Russell, and P. W. R. Beaumont, *J. Mater. Sci.*, 15, 208 (1980).

161. P. C. Powell, *Engineering with Polymers*, Chapman & Hall, London, 1983.

162. R. C. Crawford, *Plastics Engineering*, Pergamon Press, London, 1987.

163. P. C. Powell, in R. M. Ogorkiewicz, Ed., *Thermoplastics, Properties and Design*, Wiley, London, 1974.

164. G. J. Foote, and T. T. Lawson, *Short Fiber Reinforced Thermoplastics*, International Conference, Uxbridge, U.K., June 17–18, 1985, preprints published by Plastics and Rubber Institute, London, 1985.

165. L. C. E. Struik, *Plast. Rubber Process. Appl.*, 2, 41 (1982).

166. J. P. Trotignon, and J. Verdu, in J. C. Seferis, and P. S. Theocaris, Eds., *Interrelations Between Processing Structure and Properties of Polymeric Materials*, Elsevier, Amsterdam, 1984.

167. M. W. Darlington, and P. H. Upperton, in D. W. Clegg, and A. A. Collyer, Eds., *Mechanical Properties of Reinforced Thermoplastics*, Elsevier Applied Science Publishers, London, 1986.

168. M. W. Darlington, and P. H. Upperton, *Fiber Reinforced Composites 1986*, Second International Conference, Liverpool, U.K., April 8–10, 1986, preprints published by Mechanical Engineering Publications Ltd, London, 1986.

169. M. W. Darlington, and P. H. Upperton, *Short Fiber Reinforced Thermoplastics*, International Conference, Uxbridge, U.K., June 17–18, 1985, preprints published by Plastics and Rubber Institute, London, 1985.

170. C. J. Hooley, and J. Maxwell, *Plast. Rubber Process. Appl.*, 5, 19 (1985).

171. H. D. Vondenhagen, at 42nd Annual SPI Technical Conference, Composites Institute, 1987, 19-A.

172. J. Chaput, and J. Steward, *Plastics on the Road '84*, International Conference, London, December 5–6, 1984, preprints published by Plastics and Rubber Institute, London 1984.

173. G. Lehnert, *Kunstst. Ger. Plast.*, 78, 248 (1988).

174. H. Schmidt, and R. Izguierdo, *Kunstst. Ger. Plast.*, 78, 149 (1988).

175. M. J. Folkes, *Short Fiber Reinforced Thermoplastics*, Research Studies Press, Chichester, 1982.

176. K. O'Brian, H. F. Crincoli, and R. L. Kauffman, at 43rd Annual SPI Technical Conference, Composites Institute, 1988, 3-D.

177. D. F. Marshall, *Mater. Design*, 8(2), 77 (1987).

178. D. A. Cianelli, J. E. Travis, and R. S. Bailey, at 43rd Annual SPI Technical Conference, Composites Institute, 1988, 3-B.

179. R. Bailey and H. Kraft, *Int. Polym. Process.*, 2, 94 (1987).

180. A. G. Gibson, S. P. Corscaden, and A. N. McClelland, at 43rd Annual SPI Technical Conference, Composites Institute, 1988, 3-C.

181. C. R. Gore, G. Cuff, and D. A. Cianelli, *Mater. Eng. (Cleveland)*, 103(3), 47 (1986).

182. A. G. Gibson, *Plast. Rubber Process. Appl.*, 5, 95 (1985).

183. A. C. Gibson, and A. N. McClelland, *Fiber Reinforced Composites 86*, Second International Conference, Liverpool, U.K., April 8–10, 1986, preprints published by Mechanical Engineering Publications Ltd, London, 1986.

184. J. E. Travis, D. A. Cianelli, C. R. Gore, and O. H. Brunner, at 42nd Annual SPI Technical Conference, Composites Institute, 1987, 19-B.

185. Anon., *Plast. Eng.*, 43(10), 9 (1987).

186. J. J. Duska, *Plast. Eng.*, 42(12), 39 (1986).

187. L. E. English, *Mater. Eng. (Cleveland)*, 103(3), 36 (1986).

188. G. Menges, T. Schacht, H. Becker, and S. Ott, *Int. Polym. Process.*, 2, 77 (1987).

189. Z, Ophir, and Y. Ide, *Polym. Eng. Sci.*, 23, 792 (1983).

190. T. Weng, A. Hiltner, and E. Baer, *J. Mater. Sci.*, 21, 744 (1986).

191. F. C. Jaarsma, *SPE ANTEC Tech. Pap.*, 32, 726 (1986).

192. D. G. Baird, C. D. Nguyen, and E. G. Joseph, *SPE ANTEC Tech. Pap.*, 29, 705 (1983).

193. P. R. Lantos, *Polym. Plast. Technol. Eng.*, 26, 313 (1987).

194. W. Brinegar, in G. M. Kline, and J. F. Carley, Eds., *Modern Plastics Encyclopedia*, Vol. 63, McGraw-Hill, New York, 1986, p. 42.

195. D. H. Morton-Jones, and J. W. Ellis, *Polymer Products, Design Materials and Processing*, Chapman & Hall, London, 1986.

196. R. C. Progelhof, and J. L. Throne, *Polym. Eng. Sci.*, 19, 493 (1979).

197. P. R. Hornsby, and D. A. M. Russell, *J. Mater. Sci.*, 21, 3274 (1986).

198. P. R. Hornsby, I. R. Head, and D. A. M. Russell, *J. Mater. Sci.*, 21, 3279 (1986).

199. G. C. McGrath, and D. W. Clegg, in D. W. Clegg, and A. A. Collyer, Eds., *Mechanical*

Properties of Reinforced Thermoplastics, Elsevier Science Publishers, London, 1986.
200. J. Sneller, *Modern Plast.*, 58(10), 50 (1981).
201. H. Echardt, and H. Munschek, *Short Fiber Reinforced Thermoplastics*, International Conference, Uxbridge, U.K., June 17–18, 1985, preprints published by Plastics and Rubber Institute, London, 1985.
202. G. Akay, *Polym. Compos.*, 4, 256 (1983).
203. W. Knappe, and A. Lampl, *Kunstst. Ger. Plast.*, 74, 79 (1984).
204. B. R. K. Pain, *Fiber Reinforced Composites 86*, Second International Conference, Liverpool, U. K., April 8–10, 1986, preprints published by Mechanical Engineering Publications Ltd, London, 1986.
205. R. C. Haines, *Plast. Rubber Process. Appl.*, 5, 79 (1985).
206. J. F. Yardley, *Proceedings of the Third European Conference on Automated Manufacturing*, IFS (Publications), Kempston, Bedford, U.K., 1985.

9 · Injection Molding of Thermoset Composites

Michael E. Ryan
Musa R. Kamal

Contents

Michael E. Ryan, Department of Chemical Engineering, State University of New York at Buffalo, Amherst, New York, USA
Musa R. Kamal, Department of Chemical Engineering, McGill University, Montreal, Quebec, Canada

9.1 INTRODUCTION

Thermosetting molding compounds typically consist of a base resin and either a fibrous reinforcement or a filler. In addition, the molding compound may contain a variety of other components such as catalysts, colorants, and processing aids. The thermosetting resin is capable of undergoing an irreversible cross-linking reaction, thereby resulting in an infusible network structure. The process and extent of this cross-linking reaction depend on temperature, and therefore care must be exercised in controlling the thermal history of the material during processing. As a consequence of their three-dimensional network structure, thermosets inherently possess excellent mechanical, heat resistance, and chemical resistance properties. Their ability to retain their performance characteristics and dimensional stability at elevated temperatures makes them particularly competitive and attractive over thermoplastic polymers for some applications. The thermal, electrical, and mechanical properties are also governed by the nature of the filler or reinforcement.

Various types of thermosetting molding compound are available commercially including allyls, aminos, epoxies, phenolics, thermoset polyesters, thermoset polyimides, and silicones. The consumption of the major thermosetting resins in the United States is shown in Table 9.1. Thermosets currently command approximately 17% of the total plastics market in the United States.

TABLE 9.1   Consumption of Major Thermosetting Resins in the United States (1)

Resin	Sales ($\times 10^{-6}$ kg)	
	1986	1987
Alkyd	132	138
Epoxy	169	183
Phenolic	1,207	1,254
Unsaturated polyester	560	596
Polyurethane	1,153	1,216
Urea and melamine	662	710
Total	3,883	4,097
Cumulative total for all plastics	22,451	24,379

9.2 THERMOSET INJECTION MOLDING

The injection molding of thermosets is similar in many respects to the injection molding of thermoplastics, with the notable difference relating to the solidification of the product. In the case of thermoplastics, solidification is accomplished by means of cooling, whereas for thermosets solidification occurs by means of the cure reaction, which is achieved in a heated mold cavity.

Historically, thermosetting materials were successfully injection molded on plunger-driven machines during the 1940's. Automation of plunger and ram transfer processes led to the implementation of a hopper-fed screw plasticizer for accomplishing preforming and preheating operations. The in-line reciprocating-screw injection

323

molding machine was developed during the 1950's. The reactivity and changing rheological characteristics of thermosetting systems made processing difficult initially and hampered the widespread development and application of the screw injection process for thermosets. Nevertheless, the injection molding process permits great flexibility for complex part geometry and design, facilitates good control of the process and product, obviates the need for special preheating equipment or procedures, and is more amenable for short cycles. Recent developments in material handling, process control, robots, and automatic mold change equipment have resulted in improved production efficiency, scrap reduction, part consistency, and conformity to tight tolerances. Reliable data acquisition and automated part removal, sorting, and finishing are becoming more commonplace in the industry, with concomitant improvements in productivity.

In a conventional injection molding process, the molding compound is typically in the form of pellets or granules, which are usually gravity fed from a hopper. The material is conveyed forward by the flights of a rotating screw. Figure 9.1 is a schematic cross-sectional diagram of a typical in-line reciprocating-screw injection molding machine. The temperature of the barrel in various zones is usually maintained by circulating water or oil through a jacket surrounding the barrel. This externally supplied heat in conjunction with the mechanical action of the rotating screw causes the molding compound to be transformed to a viscous molten state. The throat end is maintained at room temperature or less, to prevent clogging or bridging of the hopper feed. The temperature along the barrel rises to approximately 93°C (200 °F) at the front end of the barrel for most thermoset molding compounds. It is critically important to avoid excessive temperatures in the barrel to minimize the possibility of premature curing of the material. Screw back pressures are typically of the order 689–1379 kPa (100–200 psi)(2).

As the screw rotates it retracts, permitting a predetermined amount of material to be accumulated in front of the screw. The extent of retraction is set by limit switches. Screw rotation ceases and the screw then translates forward, forcing the heated viscous material to flow through the nozzle, sprue, and runner system into the mold cavity. Filling times may be as short as a couple of seconds for small parts to

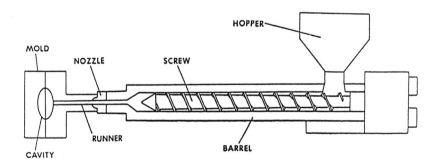

FIGURE 9.1   Schematic cross-sectional diagram of a reciprocating-screw injection molding machine.

10 seconds or more for large parts. There is considerable viscous dissipation of energy to heat as the material flows through the sprue, runner, and gate, and the temperature of the material may increase considerably as a result.

Once the mold cavity has been completely filled, an elevated hold pressure is maintained for several seconds to minimize any dimensional changes that might occur upon solidification and to ensure proper conformity to mold dimensions and design tolerances. Solidification is accomplished by means of a heated mold, which causes the material to react, cross-link, and transform to an infusible solid. Heating of the mold tool may be achieved through the use of hot circulating oil or steam but is most commonly done using electrical cartridge heaters connected to temperature control units. When the part is sufficiently rigid, the mold is opened, the hot molded part is ejected, and the molding cycle is repeated. The ejected part may continue to undergo cure as a result of the thermal inertia and residual heat present in the ejected part.

The clamping unit provides the requisite force to keep the mold closed during the filling and holding stages. The clamp opens to permit ejection of the molded part. Occasionally the clamp may be partially opened just after filling to vent gaseous reaction products from the mold at the parting line. This process is referred to as a breathe–open and breathe–dwell step and is typically of the order of 1 or 2 seconds(2).

9.3 RECENT DEVELOPMENTS

Recently, more productive molding techniques have been developed including hot cone molding, runnerless injection–compression (RIC), and lost core hollow part injection (3). These innovations have resulted in dramatically shorter molding cycles, reduced scrap, and improved finished part properties.

Hot cone or hollow sprue molding simply involves placing a mandrel or core into the sprue. This results in a hollow sprue and a consequent reduction in sprue weight. Since the sprue often is one of the thickest sections, the overall cure time may be substantially reduced. The technique was developed by General Electric and has been licensed to Rogers Corporation and the Durez Resins and Molding Materials division of Occidental Chemical Corporation (4).

Durez has developed and licensed a patented runnerless injection–compression molding process. A cold manifold plate and cooled bushing are employed with modified injection–compression tooling to minimize the production of scrap material in the form of cured sprues and runners. Plasticized material is injected into a partially closed mold without creating back pressure on the cold manifold sprue. The mold is typically held open by 6.35–12.7 mm (0.25–0.5 in.) often with the aid of mechanical stops. Once the charge has been injected, the mold is closed to compression mold the part without the accompanying sprue and runner system. The cavity is completely enclosed by telescoping the cavity from half of the mold into the matching cavity in the other half of the tool. This process is becoming increasingly popular with fiber-reinforced thermoset composites because the severity of fiber orientation is reduced. Parts that have good impact properties and are free of weldlines can be produced by this process.

325

Lost core molding involves the molding of a part around a eutectic alloy insert, typically tin–bismuth. This insert is melted out and recovered during the part postbake operation. The process is used when the hollow sections and undercuts that are produced with conventional injection molding practice are not feasible. Ford Motor Company is using this process to mold a one-piece water pump from a glass-reinforced phenolic.

New machinery for the production of thermoset parts possessing excellent mechanical properties has been developed by incorporating long reinforcing fibers directly into the machine. New multimaterial injection presses can produce parts having given surface finish while maintaining the required rigidity without a significant increase in part weight or operating cost.

9.4 THERMOSET RESINS AND MOLDING COMPOUNDS

9.4.1 Allyls

Allyl esters based on monobasic and dibasic acids are available as both monomers and prepolymers for applications as cross-linking agents for unsaturated polyester resins and in the preparation of thermoset molding compounds. Diallyl orthophthalate and diallyl isophthalate molding compounds have achieved the widest commercial usage and retain their electrical insulating properties under severe environmental conditions involving elevated temperature and humidity. They also exhibit other superior electrical properties under these adverse conditions, such as stable low loss factors, good dielectric strength, and high arc and track resistance. Molded parts also possess excellent dimensional stability, chemical resistance, mechanical strength, and heat resistance. These materials are used in such demanding electrical and electronic applications as connectors, housings, switches, relays, circuit breakers, and terminal strips for communication, computer, and aerospace systems.

Allylic monomers are widely used as cross-linking agents for unsaturated polyester resins. These applications include preform or mat bindings, laminating prepregs, rope, granular, or premix gunk molding compounds, and glass cloth and decorative laminates. Their lower volatility compared with styrene at elevated molding temperatures makes them suitable for faster molding cycles. Allylic resins are typically preimpregnated in glass cloth and roving. The molding grades are available in mineral, organic fiber, and glass fiber filled formulations.

9.4.2 Aminos

Amino resins are produced through a controlled reaction of formaldehyde with various compounds containing an amino (NH_2) reactive group. Urea–formaldehyde and melamine–formaldehyde are the two most widely known and commercially used examples. α-Cellulose fibers and a variety of other fillers may be used to enhance strength, promote moldability, improve dimensional stability, and minimize molded-in stresses. The α-cellulose filled compounds are typically used in granular form for wiring devices, closures, control housings, knobs, handles, etc. The α-cellulose melamine grades are used for dinnerware, knobs, buttons, ashtrays, appliance components, etc. Properly cured moldings will be resistant to attack by weak acids, alkalies, solvents, or water. The molding compounds may contain minerals, glass fiber, chopped cotton flock, or wood flour as filler materials. The urea molding compounds exhibit good surface finish, durability, and hardness. The resins are available in a wide variety of colors ranging from light to dark and pastel to high chromatic. Molded parts are resistant to breaking, chipping, and heat; they possess good rigidity and have good electrical arc and track resistance (5).

9.4.3 Epoxies

Epoxies possess a reactive functional group (oxirane ring) in their molecular structure and undergo very little shrinkage during cure. They are very frequently used for the encapsulation of capacitors and semiconductor devices for the electronics industry. They are also used in molding applications where high thermal stability may be required. Epoxy resins contain a hardener, catalyst, or cross-linking agent such as an aliphatic amine, anhydride, polyamide, or phenol/urea/melamine formaldehyde. Molding compounds may be filled with glass fibers, graphite fibers, organic fibers, silica, and other minerals. Reinforced epoxy composites have excellent strength-to-weight ratios along with good thermal and electrical properties. Carbon fiber and aramid epoxy composites are widely used for aircraft and aerospace applications.

9.4.4 Phenolics

Phenolic resins are produced by the reaction of phenol and formaldehyde. Single-stage resins (resols) consist of an excess of formaldehyde and phenol with an alkaline catalyst. Two-stage resins (novolacs) consist of an excess of phenol, formaldehyde, an acid catalyst, and a cross-linking agent such as hexamethylenetetramine. Good physical strength, creep resistance, heat resistance, dimensional stability, and moderate cost have made phenolics attractive in a variety of transportation, appliance, and electrical component applications.

General-purpose molding compounds are primarily two-stage resins that may be filled with mica, clay, wood flour, cotton flock, chopped fabric, and mineral fibers. General-purpose wood flour filled compounds are extensively used for knobs, handles, and appliance housings. Other applications of general-purpose compounds include ashtrays, switch gears, and pulleys. High performance engineering molding compounds are also typically two-stage resins that are reinforced with glass fibers or organic fibers. Teflon, nylon, and graphite may also be incorporated to improve toughness and self-lubricating characteristics. Mineral and glass fiber filled engineering molding compounds are used in high performance products such as wiring devices, connectors, commutators, and automotive components.

Phenolics are commonly processed by means of injection molding, which results in faster cycles, reduced scrap, and lower piece prices. New injection-moldable phenolic alloys have been developed having much greater heat deflection resistance, impact strength, and thermal stability than conventional phenolics. These materials are successfully penetrating metals replacement markets for under-the-hood and other automotive applications. One limitation of phenolic materials is color. Most molding compounds are either black or brown. However, current efforts are directed toward the development of coloring or color coating techniques.

9.4.5 Thermoset Polyesters

Unsaturated polyester resins are reaction products of organic acids and alcohols. Various monomers and catalysts are used to promote the cross-linking reaction. Dry pelletized thermoset polyester and alkyd molding compounds generally employ nonstyrenated systems (6). Various monomers and peroxide catalysts may be used to cross-link these systems. These materials typically possess good high temperature dielectric strength, high arc and track resistance, and good dimensional stability.

The injection molding of thermoset polyesters and alkyds necessitates relatively low barrel or preheat temperatures to avoid premature cure of the resin. Mold surfaces should be highly polished and chrome plated to minimize sticking problems that may result from undercure of the molded part. These materials cure rapidly at elevated temperatures. Applications include electrical and electronic parts as a result of their excellent mechanical strength and dielectric properties.

Unsaturated polyesters also include bulk molding compounds (BMC), sheet molding compounds (SMC), and thick molding compounds (TMC). Bulk molding compounds consist of unsaturated polyester resins, styrene monomer, low profile thermoplastic additives such as polyethylene or polystyrene, inorganic fillers, fibrous reinforcements, catalysts, release agents, and pigments. The diversity of resin formulations provides great flexibility in tailoring specific properties. BMC formulations typically contain 3.18 or 6.35 mm (0.125 or 0.25 in.) glass fibers before molding. High performance applications usually necessitate the use of SMC or TMC.

Low profile SMC formulations contain thermoplastic additives that interrupt and significantly reduce the shrinkage associated with the polymerization of the polyester. This consequently yields near-zero shrinkage and an ability to control surface waviness in large surface area parts. SMC possesses excellent strength-to-weight ratio, and the strength properties can be varied by changing the length, orientation, and amount of the glass fibers in the resin matrix. Typical formulations contain 25–30% chopped glass fiber by weight. Other high performance fibers can also be used. Statistical process control of incoming materials has led to improved composition uniformity. New rapid cure compounds capable of smooth Class A surface finishes have been developed.

TMC was introduced commercially in the late 1970s. The process technology of TMC provides high filler loadings and combines the lower cost economies of BMC compounding with the higher strength characteristics of SMC. Table 9.2 compares the mechanical properties of injection-molded BMC, SMC, and TMC with conventional compression-molded SMC(7). The distribution of variable length fibers yields mechanical properties that are equal to or higher than comparable SMC formulations,

TABLE 9.2   Comparison of Mechanical Properties of Injection-Molded BMC, SMC, and TMC Formulations (7)

Property	BMC	SMC	TMC	Compression Molded SMC
Tensile strength, MPa	15.2	20.0	24.1	24.8
Flexural strength, MPa	52.4	63.4	96.5	8.21
Flexural modulus, GPa	9.24	8.07	10.34	8.21
Unnotched Izod impact strength, J/m	262	267	320	342

and significantly higher than for BMC systems. Typical TMC formulations contain 15–30% glass fibers by weight.

These various compounds are used in a myriad of applications including large circuit breakers, machine housings, automotive panels, appliance housings, and components for portable hand tools. A more detailed discussion of these materials can be found in Chapters 2 and 3.

9.4.6 Polyimides

Thermoset polyimide molding compounds are particularly attractive in demanding applications requiring high quality and superior performance. Glass fiber reinforced molded parts may typically possess a flexural strength of 345 MPa (50,000 psi) and a flexural modulus in excess of 20.7 GPa (3×10^6 psi) at room temperature (8). Properties are retained at the 70% level at temperatures of 250 °C (480 °F), and the moldings may perform in an adequate manner for intermittent temperatures up to 482 °C (900 °F). Glass fiber reinforced polyimides exhibit negligible creep even at elevated temperatures. Graphite powder reinforced polyimides are self-lubricating, possess low friction, and are easily machinable. Flexural strengths greater than 69 MPa (10,000 psi) are typical for these particular molded parts. Polyimides exhibit high dielectric strength, good corona resistance, low dissipation, good chemical resistance, and excellent property retention characteristics under cryogenic conditions. Polyimides require elevated molding pressures (69 MPa) and mold temperatures [238 °C (460 °F)]. The major commercial markets are aerospace and electronics applications.

9.5   FILLERS

Thermoset molding compounds typcially contain a second discrete, discontinuous, rigid phase that has been intentionally blended or compounded into the resin matrix. If the aspect ratio of the solid particulate phase is close to unity, the material is referred to as a filler. Fillers permit easy modification of the base resin properties and a cost-effective means for modifying resin performance. Fillers include hollow and solid glass microspheres, minerals, metallic powders, and organic materials and are generally present in high percentage, usually more than 5% by volume.

329

9.5.1 Glass Fillers

Both hollow and solid glass microspheres are widely used in conjunction with thermosetting resins. The spherical shape enhances the dispersion uniformity and provides for a more uniform distribution of stress in the molded part. Hollow spheres may be used as extenders to lower costs and to provide for a reduction in density without substantial loss of mechanical performance. Solid spheres are used to improve modulus. Glass spheres are easily wetted and may be treated with special surface coatings to improve the bonding characteristics to the matrix. Compared with other fillers, spheres possess a minimum surface-to-volume ratio, and thus high filler loadings can be achieved without unacceptable increases in viscosity. For the same loading level, glass spheres yield improved mold shrinkage, warpage, and cycle time. Glass microspheres are inert, nontoxic, and stable, and they yield improved abrasion and corrosion resistance. They are easily mixed by low shear processes into epoxies, polyesters, and other thermosetting systems.

9.5.2 Mineral Fillers

Although commodity mineral fillers are used as extenders to reduce the resin cost, the compounding operation is expensive. Thus the selection of an appropriate mineral filler is generally justified by the ability of the filler to provide specific performance advantages or characteristics such as increased modulus, increased heat distortion temperature, or a lower thermal expansion coefficient. Clay, mica, talc and wollastonite have high ratios of length to thickness or length to width and thereby perform a reinforcing function. Specialty grades consist of fine particle sizes or chemically modified surfaces. Chemical treatment or modification results in improved wetting, dispersion, and processing characteristics. The bonding characteristics between the mineral and the base resin can be enhanced, along with the mechanical and electrical properties of the composite.

Alumina trihydrate is used in conjunction with unsaturated polyesters, epoxies, and phenolics because of its flame-retardant and smoke-suppressant properties. Calcium carbonate (limestone) is commonly used in polyester systems because of low cost, low oil absorption characteristics, dimensional stability during cure, and smooth surface appearance of the finished part. Loading levels of calcium carbonate are approximately 30–35% by weight in SMC materials but may be even higher in some systems. Other mineral fillers include barium sulfate, calcium sulfate, silica, and feldspar. In general, mineral fillers are readily available and comparatively inexpensive.

9.5.3 Metallic Powders

Metallic powders (e.g., aluminum flake, aluminum-coated glass, bronze, zinc, nickel) are used principally to enhance the thermal and electrical conductivity of the material and to provide shielding from electromagnetic and radiofrequency interference (EMI/RFI). Metallic powders are a generally more expensive class of filler material.

9.5.4 Organic Fillers

Organic fillers include wood flour, ground nutshells, starches, carbohydrate by-products, and various synthetic organic materials. They are frequently used as fillers and reinforcing agents with ureas, melamines, and phenolics. Their major attribute is low cost, and their major disadvantage is water absorption.

9.6   FIBROUS REINFORCEMENTS

Thermoset molding compounds may contain fibrous reinforcements that are strong and inert and form an effective bond with the matrix resin. Fibers yield dramatic improvements in both tensile and flexural strength in the fiber direction. Some of the common fibrous reinforcements are aramid fibers, carbon fibers, cellulosic fibers, ceramic fibers, glass fibers, metallic fibers, and thermoplastic fibers. The influence of fibers on the performance of thermoplastic composites has received widespread attention and is discussed in some detail elsewhere in this volume (9), as well as in other sources (10–12).

9.6.1 Aramids

Aromatic polyamide fibers, such as Kevlar, have low density, high tensile strength, excellent toughness, outstanding heat resistance, low dielectric properties, and fairly good chemical resistance. They also possess excellent vibration damping, low thermal expansion coefficients, good frictional properties at elevated temperatures, and exceptional wear resistance. Chopped aramid fibers ranging in length from 6.35 to 25.4 mm (0.25–1 in.) are commonly used to reinforce thermosets. Aramid fibers are used in applications requiring critical weight savings such as components for aircraft, space vehicles, and missiles. Ballistic and structural performance considerations have led to armor applications. Aramid–carbon hybrid composites are used with epoxies, polyimides, phenolics, and polyesters.

9.6.2 Carbon Fibers

Carbon Fibers possess a number of inherent advantages including low density, outstanding mechanical properties, excellent electrical conductivity, good chemical inertness and corrosion resistance, excellent high temperature properties, and good frictional and wear characteristics. The highest composite mechanical properties are obtained by employing 60% by volume or more of a unidirectional continuous fiber in a high strength epoxy or similar matrix. Metal-coated carbon fibers may be used to enhance electrical conductivity and provide shielding from EMI/RFI and from electrostatic discharge (ESD). Carbon–glass hybrids are encountered with SMC systems.

9.6.3 Cellulosics

Cellulosic fibers are being used as low cost extenders and reinforcements with a variety of thermosetting resins including ureas, melamines, polyesters, phenolics, and elastomers. The tensile strength of cellulosic fibers is approximately 545 MPa compared to 3.5 GPa for E-glass and 4.5 GPa for S-glass and R-glass fibers.

9.6.4 Glass Fibers

Glass fibers are the most commonly encountered reinforcement for injection-molded composite parts because of their outstanding cost/performance characteristics. Glass fibers are commercially available in a myriad of different forms and possess good dimensional stability, high strength-to-weight ratio, good corrosion resistance, and ease of fabrication. Glass fibers are treated with coupling agents or other surface modifiers, or they may be coated with metals. Continuous strand roving is fed to a chopper assembly and the chopped fibers are randomly arranged in typical SMC systems.

9.6.5 Metallic Fibers

A wide variety of metallic and metal alloy fibers are commercially available, including aluminum, nickel, and stainless steel fibers. At low loading levels they essentially maintain the mechanical properties and processing characteristics of the base resin yet impart EMI shielding or ESD protection to the molded part. Unsized chopped fibers are typically used with thermosetting compounds.

9.6.6 Thermoplastic Fibers

Thermoplastic fibers, such as polyester and nylon, are used to impart better fatigue life and impact strength and to reduce the brittleness or propensity for microcracking of the thermoset matrix. Thermoplastic fibers are inherently nonabrasive and are readily used in conjunction with BMC, SMC, and TMC applications. The main disadvantages relate to compatibility with the resin for proper wetout and adhesion, surface finish, and adequate dimensional stability over the temperature range for processing the thermoset composite.

9.7 CURING CONSIDERATIONS

The most notable differentiating characteristic of thermosetting systems is their ability to undergo an irreversible cross-linking reaction that transforms the material to an infusible, insoluble network structure. This curing process is dependent on the thermal history experienced by the material during processing. The degree of cure directly influences the ultimate mechanical, thermal, electrical, and chemical properties of the final molded part. Thus, a knowledge of the kinetics governing the cure reaction and the variation of the degree of cure with the time–temperature history of the material is essential for understanding the structural and physical property changes that accompany the cure reaction. However, it should be recognized that the degree of cure as well as the ultimate part properties are also affected by numerous other considerations, including the loading level of fillers and reinforcing agents, the degree of orientation of reinforcements or the resin matrix, the presence of other additives (e.g., stabilizers, lubricants, fire retardants, processing aids), the extent of regrind, and the specific process conditions. Improper or inadequate cure may manifest itself in terms of readily apparent visual defects including a lack of rigidity when ejected from the mold, cracks, porosity, warpage, swelling, and surface blisters.

A number of standardized tests to detect undercure conditions in various thermoset systems have been adopted by the industry. For diallyl phthalate and diallyl isophthalate, sections of a molded specimen are refluxed in boiling chloroform for 3 hours and visually inspected for evidence of serious attack such as cracking, swelling, or fissuring (13). A comparison of the Barcol surface hardness (ASTM D 2583) before and after boiling in chloroform is a measure of the degree of undercure. The surface hardness should not be measured until at least 12–14 hours after removal from the boiling chloroform. The Barcol surface hardness test can also be used to ascertain undercure conditions in polyester systems by boiling sections of a molded specimen in water for one hour. For epoxies and phenolics, the acetone extraction test (ASTM D 494) may be used for cure comparisons. However, the American Society for Testing and Materials discontinued D 494 as a standardized test in 1985. ASTM specifications and test methods for various mechanical and electrical properties are summarized in Table 9.3.

Unfortunately, none of the above-mentioned tests yields a precise quantitative measure of the degree of cure in a molded part. Consequently, a variety of experimental methods and analytical techniques have been developed for monitoring or characterizing the cure reaction. These techniques involve the analysis of changes in the physical and chemical properties as a function of time and temperature during the course of the cross-linking reaction. These methods include chemical analysis, sol–gel analysis, spectroscopic techniques (IR, NMR, ESR, EPR, Raman, etc.), physical property changes (density, electrical conductivity, refractive index, viscosity, dynamic mechanical properties, etc.), thermogravimetric analysis (TGA), differential thermal analysis (DTA), and differential scanning calorimetry (DSC). The relative merits and limitations of these techniques are discussed elsewhere (4). In general, these methods

TABLE 9.3 Standardized Specifications and Test Methods for Thermosetting Materials

Material	ASTM test method
Allyl molding compounds	D 1636–81
Epoxy molding compounds	D 3013–81
Epoxy resins	D 1763–81
Melamine–formaldehyde molding compounds	D 704–82
Phenolic molding compounds	D 700–81
Thermosetting polyester molding compounds	D 1201–81
Urea–formaldehyde molding compounds	D 705–81

are not well suited for routine production or quality control testing and are more commonly used in research and development studies.

The cumulative exothermic heat evolved during the course of an isothermal cross-linking reaction increases with increasing temperature. The kinetics of the cross-linking reaction also increases dramatically with increasing temperature. The simplest kinetic model for describing the isothermal cure kinetics of a cross-linking thermoset system is the familiar nth-order kinetic rate expression:

$$\left(\frac{\partial \alpha}{\partial t}\right)_T = k(1 - \alpha)^n \tag{9.1}$$

where α is the relative degree of cure at temperature T, k is the kinetic rate constant, and n is the kinetic exponent. The rate constant k is temperature dependent and is generally assumed to follow an Arrhenius expression:

$$k = k_0 \exp\left(-\frac{\Delta E_k}{RT}\right) \tag{9.2}$$

where k_0 is a pre-exponential constant, ΔE_k is the kinetic activation energy, R is the familiar gas constant, and T is the absolute temperature. For a variety of thermosetting systems and molding compounds including epoxies, unsaturated polyesters, and phenolics, the following generalization of Eq. (9.1) has been found to correlate isothermal kinetic data to a reasonable degree of accuracy (14–18):

$$\left(\frac{\partial \alpha}{\partial t}\right)_T = (k_1 + k_2 \alpha^m)(1 - \alpha)^n \tag{9.3}$$

Here the kinetic exponents m and n are presumed to be independent of temperature and the rate constants k_1 and k_2 are assumed to have an Arrhenius temperature dependence. The details of determining the kinetic parameters from the reaction exotherms are described elsewhere (4,19).Nonisothermal characterization techniques have also been discussed in the literature (4, 20–23).

A general kinetic model based on a free radical copolymerization mechanism has recently been developed by Stevenson (24–26). The kinetic description includes multiple noninteracting initiators, inhibition, and termination. The initial monomer concentration ratios and consumption rate ratios are assumed to be equal and constant. The rate constants are presumed to be independent of conversion, and the propagation rates are assumed to be much greater than any other rate processes during the polymerization. The relevant governing material balance expressions become:

Initiators:

$$I_j = I_{j_0} \exp\left(-k_{d_j} t\right) \qquad j = 1, 2, \ldots, N \tag{9.4}$$

Radicals:

$$\frac{dR^{\cdot}}{dt} = 2 \sum_{j=1}^{N} k_{d_j} I_j - k_z(R^{\cdot})z - k_t(R^{\cdot})^2 \tag{9.5}$$

334

Inhibitor:

$$\frac{dz}{dt} = k_z z (R^{\cdot}) \tag{9.6}$$

Monomer:

$$\frac{dM}{dt} = k_p M (R^{\cdot}) \tag{9.7}$$

where I_j denotes the concentration of initiator j, $R^{\cdot}$ denotes the total concentration of free radicals including all reacting species and initiator radicals, M is the total monomer concentration, z is the inhibitor concentration, k_p and k_{d_j} are the polymerization and decomposition rate constants, respectively, and k_z and k_t are the rate constants associated with the free radical combining with the inhibitor and another free radical, respectively. The rate constants are assumed to have an Arrhenius dependence on temperature, and the termination rate constants are considered to be equal.

The general model has good predictive ability, but evaluation of the rate constants is laborious and occasionally leads to unusual values for the individual rate constants. Some of these difficulties may be circumvented by making additional assumptions, such as neglecting the chain termination reaction. The simplified model provides a reasonable compromise between predictive ability and computational ease. The model has been used for both isothermal and nonisothermal simulations of SMC pastes and glass fiber reinforced SMC systems.

Integration of Eq. (9.1) or Eq. (9.3) can yield an estimate of the cumulative heat evolved or the ultimate level of cure of the system. Clearly the final extent of cure is dependent on the detailed thermal history of the material. Not all reactive sites or species may participate in the cross-linking reaction at some temperature T, and the material may retain some residual reactivity or cure. Sidi (27) found that for a glass and mineral filled, injection molding grade, unsaturated polyester, the residual extent of cure decreased exponentially with increasing cure temperature. If the initial cure temperature is sufficiently low, prolonged postcure does not necessarily force the cross-linking reaction to ultimate completion. Thus, to some extent the initial thermal history of the system determines the network structure and cross-link density and controls the level of accessibility of reactive segments during the later stages of reaction. Calorimetry is a useful tool for determining the residual level of cure and the cure distribution in molded parts (28, 29).

Clearly, barrel temperature, injection speed, mold temperature, cycle time, and other operating conditions play a critical role in defining the extent of cure and ultimate properties of a molded part. The optimization and control of the thermoset injection molding process requires a careful assessment of these factors. Frequently molded thermoset parts are postcured by afterbaking to enhance the mechanical properties. This results in improved strength, creep resistance, and dimensional stability and a reduction of volatiles and residual stresses. Recommended postcuring cycles for various molding compounds are given in Table 9.4 (13). Multistage

TABLE 9.4   Recommended Postcuring Cycles for Thermoset Molding Compounds (13)

Compounds	Recommended Postcuring Cycles
Diallyl Orthophthalate	7 hours over a range commencing at 275 °F (135 °C) and increasing the temperature by 20 °F/hour to 415 °F (213 °C).
Diallyl Isophthalate	8 hours over a range commencing at 275 °F (135 °C) and increasing the temperature by 20 °F/hour to 415 °F (213 °C).
Epoxy molding compounds	8 hours over a range commencing at 275 °F (135 °C) and increasing the temperature by 20 °F/hour to 415 °F (213 °C).
Glass and/or mineral reinforced phenolic molding compounds	
One-step compounds 3.175 mm (0.125 in.) thick or less	2 hours at 280 °F (138 °C) 4 hours at 330 °F (166 °C) 4 hours at 375 °F (191 °C)
Two-step compounds 3.175 mm (0.125 in.) thick or less	2 hours at 300 °F (149 °C) 4 hours at 350 °F (177 °C) 4 hours at 375 °F (191 °C)

For parts exceeding 3.175 mm (0.125 in.) in thickness it is recommended that the time be doubled for each additional 1.5875 mm (0.0625 in.) of thickness.

temperature cycles are generally employed for best results, with the starting temperature below the actual molding temperature. The nature and level of the reinforcement will determine the time–temperature cycle. Organic reinforcements must be postcured at lower temperatures than for glass or mineral reinforcements. Parts of uneven cross-sectional thickness will typically exhibit uneven shrinkage.

9.8 RHEOLOGICAL CONSIDERATIONS

Measurement of the rheological or flow behavior of thermosetting materials is rendered difficult as a result of the changes accompanying the cross-linking reaction. A variety of standardized tests have been developed to characterize the flow characteristics of thermoset compounds. These tests include ASTM cup closing test, disc flow tests, orifice flow tests, instrumented spiral flow tests (EMMI and Mesa), and measurements obtained from a Brabender Plasti-Corder.

The ASTM cup closing test (ASTM D 731) consists of a flash-type compression mold for a cup of specific geometrical configuration (13). The time required to close the mold to a specified flash thickness is reported as the closing time. The stiffer the material, the longer the closing time will be. Although this test is quite useful for comparative evaluations of materials intended for compression molding applications, the test is of limited value for injection molding compounds.

The disc flow tests involve the compression of a measured quantity of material between two heated die plates at predetermined temperature and pressure. The diameter or thickness of the disc is then used as a measure of the flow behavior. The disc flow tests have the same limitations as the ASTM cup closing test.

Various orifice flow tests have been devised. One of the orifice flow tests involves placing a measured charge of material into a cavity and compressing the material under a standard set of conditions. Two small grooves or orifices are cut in the edge of the plunger, permitting the material to flow out of the mold cavity. The amount of residual material remaining in the cavity is weighed and is used as a measure of the flow behavior of the sample. This test also has limited applicability to injection molding applications.

Another form of orifice flow test involves placing a known quantity of material (90 g) into a heated mold cavity [152 °C(305 °F)]. Pressure (25 tons) is applied to the material, causing it to flow through a runner and out of the mold cavity. The weight of material displaced from the cavity as well as the time required for flow to cease is recorded.

The spiral flow of thermoset molding compounds involves an instrumented spiral flow mold in conjunction with a transfer press. A standardized procedure is followed (ASTM 3123-72), and the material is forced to flow in a channel that has the shape of an Archimedean spiral. The channel length is marked in inches from the center of the mold, where the material enters and the spiral flow length of the material is measured (30). The Mesa spiral test employs a mold runner that is 6.35 mm wide, 0.84 mm deep, and 1270 mm long (0.25 in. × 0.033 in. × 50 in.) and molding pressures of either 5 or 25 tons. The Mesa spiral test is generally used with conventional molding compounds. The EMMI spiral test uses only 6.9 MPa (1000 psi) molding pressure and a 3.18 mm (0.125 in.) half-round runner that is 2540 mm (100 in.) long. The EMMI test is used primarily for very soft epoxy molding compounds or encapsulating compounds (13).

In the Brabender torque rheometer the temperature of the mixing head is controlled at specific levels (ASTM D 3795-79). The material is compacted, melted, and cross-linked, and the torque is measured as a function of time. The maximum loading torque, the maximum torque, the final torque peak, the time required to reach the minimum torque, and the time required to reach the final torque peak are all reported.

The major limitation of all these methods is that since they do not measure any fundamental rheological property, they are virtually useless for purposes of analysis or design. A more basic understanding of the influence of time, temperature, and shear rate on the rheological or flow behavior of thermosetting systems requires the use of carefully defined and controlled flows.

A commonly used experimental technique involves the steady or oscillatory shear of a sample between concentric cone-and-plate fixtures (14, 31–38). The two most widely used commercial instruments are the Weissenberg rheogoniometer and the Rheometrics Mechanical Spectrometer. An environmental chamber surrounds the test fixtures to control the temperature. The samples are sheared in a steady manner, to

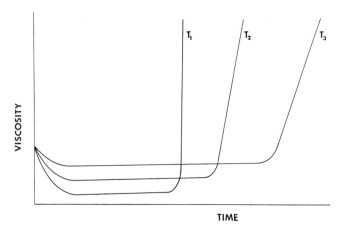

FIGURE 9.2   Variation of viscosity as a function of time at various temperatures.

permit the determination of the variation of viscosity with time at a particular temperature and shear rate. Typical viscosity–time isotherms are shown schematically in Figure 9.2. After the sample has been introduced between the test fixtures, a short transient period is required for the sample to equilibrate to the desired preset temperature. During this initial transient period the measured viscosity decreases with time as the sample is heated to the test temperature. The viscosity remains effectively constant for some extended period of time, then increases rapidly as the material transforms to an infusible solid. At higher temperatures, the viscosity is initially lower because of the conventional temperature-dependent reduction of viscosity. However, at the higher temperature the cross-linking reaction proceeds more quickly and the viscosity increases more rapidly at an earlier time, as can be seen in Figure 9.2.

In addition to measuring viscosity, these instruments may be used to determine the first normal stress difference by measuring the normal force or thrust on the platens. The dynamic storage modulus G' and the loss modulus G'' may also be measured as a function of frequency using an oscillatory shear deformation.

The cone-and-plate geometry is generally restricted to shear rates below 50 s^{-1} because of problems associated with viscous heating, flow instabilities, and loss of the sample because of centrifugal forces. A capillary rheometer may be employed to characterize the material at high shear rates. However, for curing systems two serious limitations of the capillary rheometer are the relatively short measurement interval (dictated by the capacity of the reservoir) and the relatively long time interval required to reach thermal equilibrium.

Other methods for determining the rheological behavior of thermosetting systems include falling sphere (39), parallel plate (40, 41), squeezing flow (42), and eccentric rotating discs (43). Clearly meaningful measurements are restricted to systems that do not polymerize or cross-link very rapidly or are otherwise conducted at low temperatures.

Various relationships have been proposed to describe the variation of viscosity with time and temperature for a curing thermoset system. One expression that has been used with some success for a nonisothermal curing history is:

$$\ln \eta = C_1 + \frac{\Delta E_\eta}{RT} + \int_0^t C_2 \exp\left(-\frac{\Delta E_k}{RT}\right) dt \qquad (9.8)$$

where η is the viscosity, ΔE_η denotes the flow activation energy of the uncured system, and C_1 and C_2 are constants. More recently, the Williams–Landel–Ferry (WLF) expression has been found to accurately correlate the variation of viscosity with temperature as follows (85–87):

$$\ln\left(\frac{\eta_T}{\eta_{T_g}}\right) = \frac{-C_1'(T - T_g)}{C_2' + (T - T_g)} \qquad (9.9)$$

where η_T is the viscosity of the reactive system at temperature T, η_{T_g} is the viscosity at the glass transition temperature T_g, and C_1' and C_2' are the WLF parameters. An expression for the variation of the glass transition temperature with degree of cure must be introduced, and the parameter C_1' is also permitted to depend on the glass transition temperature.

9.9 MOLD FILLING CONSIDERATIONS

The most important consideration for the successful injection molding of thermoset composites relates to the level of cure of the material during the filling stage. If the level of cure is too advanced, unnecessarily high pressures will be required to fill the mold, or the material may prematurely cure in the cavity or runner system, giving rise to incomplete filling or short shots. Thus, an understanding of the conditions under which the viscosity becomes dominated by the curing reaction is essential to ensure that such conditions are avoided. The molding conditions must be selected to achieve an appropriate balance between ease of filling, cycle time, and part quality. Recommended molding conditions are generally provided by the resin suppliers.

One particular problem commonly encountered with highly filled thermoset molding compounds is "jetting". Jetting causes the incoming material to enter the cavity as a filament or finger. Additional material may subsequently flow into the cavity in a conventional spreading radial flow pattern. However, the presence of the jetted material within the cavity causes the formation of weldlines, which generally lead to a reduction of mechanical properties and may even remain visible in the finished part. Jetting may often be avoided by proper gate design, gating against a mold wall, or changing operating conditions.

In the vicinity of the advancing flow front the actual flow pattern exhibits the so-called fountain flow effect, in which material elements at the midplane are accelerated to the melt front. The moving front divides as the material elements flow away from the midplane and toward the mold walls. The material coming into contact with the heated mold surfaces begins to cure more rapidly, forming a thin cured layer or skin analogous to the frozen skin that is formed in thermoplastic injection molding. These considerations play an important role in determining the orientation of reinforcing fibers and the cure distribution in the molded part.

Consequently, the ultimate properties and part quality are influenced by this "fountain flow" phenomenon.

The presence of inserts or multiple gates inherently results in the formation of weldlines. Proper gate location in designing the tool can prevent the location of weldlines in critical areas of the molded part. The potential for air entrapment must also be considered in such cases; to avoid a deleterious effect on the finished product, air entrapment must not be allowed to occur.

Another important factor that has received relatively little attention in conjunction with the injection molding of reinforced thermoplastic or thermosetting systems is the flow-induced migration of particles or fibers. Particle migration is generally undesirable, since it leads to density variations and inhomogeneities in the part. Some particle migration is a direct consequence of the presence of walls that bound a channel or cavity. Solid particles are necessarily excluded from the region in the immediate vicinity of a wall because of their finite size. For concentrated suspensions or highly filled systems, particle interactions and collisions may be shielded by the walls. These factors result in radial concentration gradients. Furthermore, the particles that are forced away from the solid boundaries enter a faster moving stream, leading to an axial concentration gradient (44). It has been found experimentally that larger particles migrate faster than smaller ones leading to segregation effects when a size distribution is present.

Theoretically, no particle migration is predicted for the inertialess flow of a suspension of solids in a Newtonian liquid. However, the Reynolds number is never zero in practical flow situations. Segre and Silberberg (45, 46) observed that spherical particles collected into a thin annular region when an initially uniform dilute suspension was caused to flow under laminar conditions in a straight tube. The neutrally buoyant spheres tended to reach an equilibrium radial position of 0.6 times the tube radius. The particle migration became more pronounced on increasing the tube length, the average flow velocity, and especially the ratio of particle to tube diameter.

Particle migration is found to occur as a result of elasticity, buoyancy, the presence of secondary flows, and a spatially nonhomogeneous rate of deformation, among other causes. The actual trajectory of a particle will obviously depend on the interplay of these various factors, particle–particle interactions, and the reactivity of the system. A variety of different phenomena have been observed (47), and many of the different studies are summarized by Leal (48) in his recent review of the subject. However, most of the experimental studies and theoretical analyses are restricted to dilute suspensions or single particles and neglect all interparticle effects.

Kubat and coworkers (49, 50) considered the injection molding, into a spiral mold, of polyethylene and nylon containing 5–25% glass spheres having diameters ranging from 4 to 200 μm. In general, a gradient in concentration and a pronounced fractionation effect along the length of the spiral were observed. The separation effect was found to increase with increasing sphere size, and for large spheres the concentration at the tip of the spiral for the long (2 m) mold was 6 times the inlet concentration. Concentration gradients were also observed in the sprue. The molding was done according to conventional practice.

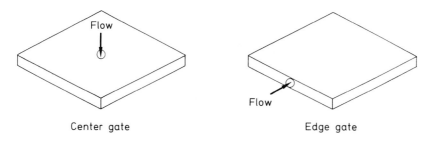

FIGURE 9.3   Gate configurations for mold-filling studies by Schmidt (51).

A more complex flow situation was examined by Schmidt (51), who conducted mold filling studies using five glass bead filled polypropylene composites having 0–26% by volume of solids. The two basic flows studied were shear flow in an edge-gated rectangular cavity and stagnation flow leading to a diverging radial flow pattern in a center-gated rectangular cavity. These two gate configurations are shown in Figure 9.3. The most remarkable finding was that resin-rich layers resulting from bead migration were observed in molded plaques formed by radial flow but were not observed in any of the edge-gated plaques. These findings illustrate the importance of flow geometry and indicate that particle migration can be either reduced or enhanced by changes in the placement of the gate.

Fiber orientation strongly influences the ultimate properties of the molded part. The orientation is clearly influenced by the mold and gate geometry, the operating conditions, and the properties of the matrix resin. It is important to realize that the kinematics and stress distribution during both filling and packing stages will influence the orientation distribution. Most investigators tacitly assume that the orientation is principally determined during filling.

Numerous studies have been undertaken with regard to the orientation of short fibers in relatively thin, flat, thermoplastic molded parts. A proper treatment of the subject is beyond the scope of the present discussion, and the interested reader is referred elsewhere in this volume (9) or to any of the recent reviews of the subject (82, 83). The determination of fiber orientation poses a formidable problem for several reasons. The kinematics is extremely complex even for simple mold geometries because of the presence of the advancing flow front and the associated "fountain flow" phenomenon. In addition, for most commercial molding compounds the loading levels of reinforcing fibers are high and consequently interaction between the fibers is significant. These high loading levels also interact with and perturb the kinematics of the flow field.

9.10 FIBER BREAKDOWN

Chiu and Shyu (90) studied the effect of various processing conditions on the fiber length of injection-molded composites. They found that the fiber content decreased from the surface to the core of the sample. The number average fiber length of twice-through extruded material was shorter (0.96 mm) than the once-through material (2.66 mm), while the original length before processing was 6 mm. Larger fiber content produced more broken fibers. The average fiber length increased with higher melt temperature. Similar observations were made by von Turkovich and Erwin (91), who compared the effects of extrusion and injection molding on fiber length. They found that changing within the limits allowed for conventional processing did not significantly alter the length of surviving fibers. They also found that extrusion and injection molding produced essentially the same results. Some breakage occurred in the mold and runners, but most of the breakage occurred in the injection unit. The melting zone in extrusion had the greatest influence on fiber breakage.

Fiber length reduction lowers the efficiency of the reinforcement thus affecting the final mechanical properties (91). Fiber damage may be due to one or more of the following processes (92):

1. Fiber–fiber interactions: surface abrasion as well as fiber overlap leads to stress concentration and bending stresses that produce fiber fracture.
2. Fiber–machine surface interactions: screw action; abrasion at the runner and mold surfaces.
3. Fiber–polymer interactions: viscous forces imparted by the polymer may cause fracture.

Theoretical treatments of suspensions of thin rigid rods in shear flow are generally inadequate in accounting for the observed fiber fracture in concentrated or highly loaded systems for several reasons. The nonuniform distribution of fibers due to interaction or exclusion from mold surfaces or due to migration has been discussed. Concentrated fiber suspensions also exhibit plug flow velocity profiles, as opposed to parabolic profiles, thereby altering the hydrodynamic interactions between fibers. In addition, many commercial reinforced materials have fiber loading levels exceeding the maximum packing volume fraction. This suggests that the fibers are aligned in clumps or are bent or broken. The maximum packing volume fraction for randomly oriented fibers as a function of average fiber aspect ratio is shown in Figure 9.4 (93).

In their study of the fiber length distribution along the length of the barrel, Chiu and Shyu (90) found that fiber bending due to solid wall and interfiber abrasion did not play a major role in fiber fraction but that during melting there is a rapid reduction of fiber length due to viscous loading of the fiber by the polymer in shear flow.

Biggs (94) showed that there was substantial fiber breakage during the compounding of carbon fiber composites. The fibers that were initially 3 mm long were reduced to between 100 and 200 μm in length. The highest reduction in fiber length was found in the composites with the highest melting matrix. The critical fiber aspect ratio was in the range 16:1 to 25:1 for the polymer carbon composites.

Recently Billion, a French company, introduced the ZMC process, which is claimed to produce both thermoplastic and thermosetting injection-molded products with minimum fiber breakdown.

The integrated injection molding system, involving both the machine and mold design, incorporates the following main components:

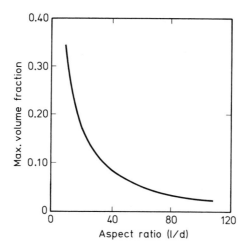

FIGURE 9.4 Maximum packing volume fraction for randomly oriented fibers as a function of average fiber aspect ratio (93).

- A heating cylinder for melting the charge.
- A feed piston for feeding the material to the extrusion unit.
- A screw melt conveying and injection unit.
- A specially designed nozzle.

The preprocessing and melting of the material in the first cylinder reduces fiber breakdown during the melting and solids conveying stages. Besides producing articles without substantial reduction in fiber length, the system contributes to the improvement of part homogeneity and quality.

9.11 COMPUTER SIMULATION

For thermoset injection molding, the pathway from the original part concept or design to the successful production of the finished part can be time-consuming and economically risky. Computer simulation of the injection molding of thermoplastic systems has made enormous strides during the past two decades. Commercially available CAD/CAM packages incorporating three-dimensional color graphics and extensive material property data bases are now commonly used as an engineering tool to expedite, analyze, and optimize many facets of the process. These simulation packages typically employ a finite element numerical scheme to analyze complex mold cavity configurations. In addition to their predictive capability for pressure, temperature, and stress distributions throughout the molded part, these simulation programs may be used to ascertain the filling time, determine appropriate locations of gates and weldlines, and permit the balancing of runners for multicavity or family molds.

Computer simulation provides a cost-effective means for the diagnosis of molding problems such as short shots or overpacking and obviates the need of costly tooling

343

changes or modifications. A commercial mold filling simulation program for thermosets has been recently developed (52, 53). The analysis neglects the fountain flow at the moving front, which could have a significant influence on the predicted temperature and degree of cure distributions across the part.

The basis of the computer simulation is a mathematical model of the injection molding process that is typically based on the basic equations of change, namely, continuity, the equation of motion, and the conservation of energy. The mathematical description is simplified by seeking to achieve a reasonable compromise between a rigorous description of the physics and computational tractability. The application of CAD/CAM packages to thermosetting systems has been limited primarily by the difficulties associated with material characterization and the incorporation of the dependence of the degree of cure on the physical properties of the material. The state of the art has been limited to the use of user-friendly CAD/CAM packages to qualitatively assess trends with regard to changes in material properties, process conditions, or mold geometry.

Recent research efforts (27, 54–57) have indicated that a mathematical analysis of the injection molding of thermosets may be developed by means of an approach similar to that for thermoplastics, provided adequate material properties and constitutive relations for reactive systems are included. In these cases, reasonably good quantitative comparisons have been achieved.

9.12 ULTIMATE PROPERTIES

Numerous studies on the interrelationships between material properties, process conditions, moldability, and the ultimate properties of injection-molded thermoset parts have been reviewed elsewhere (57, 61–64).

The influence of the design of the delivery system (runners and gates) on the ultimate properties of the part, including blistering and shrinkage, has been considered by Vaill (65, 66). He discussed the determination of the optimum cross-sectional area of runners and gates based on a knowledge of the weight of material being conveyed through the delivery system and the weight of the molded part. An increase in the runner and gate size has been found to give some improvement in impact strength and a reduction in mold shrinkage (67). In addition, the rate of cure was found to be independent of runner size but dependent on gate size, with the highest cure rate associated with the smallest gate. Numerous other mold design guidelines have been suggested (68–72).

The effect of molding conditions on the ultimate properties of a molded thermoset part has been the subject of numerous investigations (27–29, 73–80). The mechanical properties of the molded part are directly dependent on the level of cure attained during the molding cycle. The methodology suggested by Barone and Caulk (81) for thermal balancing could be employed with injection mold tooling to provide for a more uniform distribution of cure and property levels throughout the part.

Further improvements in equipment design, process control, and composite formulations will undoubtedly lead to additional high volume applications of thermosetting systems. However, a more detailed understanding of the complex inter-relationships between resin properties, processing history, and part performance is required before new innovations can be fully exploited.

9.13 RECYCLING

It is primarily because of the lack of economic incentives that relatively little effort has been directed to the recovery, reuse, and recycling of postconsumer plastic materials. However, there has recently arisen enormous environmental concern over the generation and disposal of solid wastes, particularly tires, plastic containers, and various plastic packaging materials. There is increasing concern in many states and communities about the necessity for significantly reducing the solid waste stream going to landfills. Numerous guidelines, regulations, and legislative proposals relating to plastics disposal have been passed or are under consideration. Any economic incentive for the private sector to recycle is dependent on the relative cost of recycling versus the cost of disposal.

There are four major categories of plastics recycling technology. Primary recycling refers to the processing of waste into a product having characteristics similar to those of the original product. Although the plastics industry has historically recycled a great deal of the waste that occurs during resin production, conversion, and product assembly, primary recycling does not allow for any significant waste contamination or alteration and is therefore not applicable to postconsumer thermoset injection-molded products. Secondary recycling refers to the processing of plastics waste into products having characteristics somewhat inferior to those of the original product. Many thermosetting materials are recycled in this sense by grinding the defective part and reusing the ground material as a filler. However, the regrind is not significantly different from other inexpensive fillers and does not offer dramatic benefits as such. Tertiary recycling refers to the recovery of basic chemicals or fuels from the waste resin by processes such as hydrolysis or pyrolysis. Quaternary recycling refers to the retrieval of the heat content by means of incineration. The incineration of wastes containing plastics is environmentally precarious because toxic fumes are generated when most plastics are burned. Standard incinerators cannot be used, since the combustion of plastics requires 3–5 times the amount of oxygen required by conventional incineration.

It is estimated that thermoplastics comprise 87% and thermosets 13% of postconsumer plastics wastes (84). Plastics compose approximately 4–5% of total solid wastes. The separation of plastics from the municipal waste stream poses a formidable technological and economic challenge. Thus, an important element in the recycling of plastics waste in a relatively uncontaminated form is the separation and collection of the waste before it enters or is commingled with the municipal waste stream. A representative example that lends itself to plastics recycling as a relatively uncontaminated waste is the shredder residue from automobile recycling operations. The separation of specific individual resins from one another is not expected to become economically viable.

The cross-linked nature of thermosetting resins prohibits their being remelted and formed again in any conventional sense. Reground thermosetting waste can be added as a filler up to 15% without a dramatic change in physical properties such as impact, tensile, or flexural strength (88, 89). However, impact strength and flexural strength sharply decrease when the loading level exceeds 20%, and there may be some irregularities or cosmetic defects in the appearance of the finished part.

9.14 CONCLUSION

Although many high performance engineering thermoplastics have been developed for use at elevated temperatures or for parts requiring demanding ultimate properties, thermoset composites can often favorably compete because of their inherent rigidity and lower cost. Injection molding permits greater flexibility for part geometry or design and better control of the process; there is no necessity for special preheating equipment or procedures, cycle times are short, and adaptability for automation is greater. Appropriate screw design enables fiber-reinforced materials to be molded without serious damage to the fibers, thereby enabling the achievement of ultimate properties comparable to those available from compression molding.

ACKNOWLEDGEMENTS

M. E. Ryan expresses his sincere appreciation to Professor R. K. Gupta for valuable discussions in relation to flow-induced particle migration.

REFERENCES

1. *Mod. Plast,* 65 (1), 95 (1988).
2. J. L. Hull, "Injection Molding Thermosets," *in Modern Plastics Encyclopedia 1988*, 64:10 A, McGraw-Hill, New York, 1987, P.256.
3. J. A. Sneller, *Mod. Plast,* 63 (7), 48 (1986).
4. M. R. Kamal and M. E. Ryan, "Thermoset Injection Molding," in A. I. Isayev, Ed, *Injection and Compression Molding Fundamentals,* Dekker, New York, 1987, Ch.4.
5. R. J. Schupp, "Amino," in *Modern Plastics Encyclopedia 1988,* 64:10 A, McGraw-Hill, New York, 1987, P.110.
6. T. E. Steiner, "Polyester, Alkyd," in *Modern Plastics Encyclopedia 1989,* 65:11, McGraw-Hill, New York, 1988, p.134.
7. D. J. Clavadetscher, "Thick Molding Compound (TMC)," in *Modern Plastics Encyclopedia 1989*, 65:11, McGraw-Hill, New York, 1988, p.212.
8. N. A. McGarry, "Polyimide, Thermoset," *in Modern Plastics Encyclopedia 1989,* 65:11 McGraw-Hill, New York, 1988, p.135.
9. B. Fisa, "Injection Molding of Thermoplastic Composites," this volume, Ch. 8.
10. M. R. Kamal, *Plast Eng,* 38(12), 31 (1982).
11. J. R. Fried, *Plast Eng,* 38(12), 21 (1982).
12. A. J. Gwinn and J. M. Castagno, *Plast Eng,* 41(12), 27 (1985).

13. R. E. Wright, *Thermoset Molding Manual*, Rogers Corporation, Manchester, CT, 1984.
14. M. R. Kamal, S. Sourour, and M. Ryan, *SPE. Tech. Pap* , 19, 187 (1973).
15. M. R. Kamal and S. Sourour, *Polym. Eng. Sci.,* 13, 59 (1973).
16. A. Thakar, A. C. Hettinger, and K. C. Guyler, *SPE. Tech. Pap*, 24, 295 (1978).
17. A. Dutta and M. E. Ryan, *J. Appl. Polym. Sci.*, 24, 635 (1979).
18. S. Y. Pusatcioglu, A. L. Fricke, and J. C. Hassler, *J. Appl. Polym. Sci.,* 24, 937 (1979).
19. M. E. Ryan and A. Dutta, *Polymer*, 20, 203 (1979).
20. J. F. Flynn and L. A. Wall, *J. Res. Natl. Bur. Stand.,* 70A, 487 (1966).
21. R. N. Rogers and L. C. Smith, *Thermochim. Acta*, 1, 1 (1970).
22. B. Carroll and E. P. Manche, *Thermochim. Acta*, 3, 449 (1972).
23. R. B. Prime, *Polym. Eng. Sci.*, 13, 365 (1973).
24. J. F. Stevenson, *SPE. Tech. Pap.*, 26, 452 (1980).
25. J. F. Stevenson, "Industrial Processing of Materials Which React by the Free Radical Mechanism: Kinetic Models and Isothermal Simulation," *Proceedings of the First International Conference on Reactive Processing of Polymers*, Pittsburgh, Oct. 28–30, 1980.
26. J. F. Stevenson, *Proc. 2nd World Congr. Chem. Eng.*, 6, 447 (1981).
27. S. Sidi, "The Dynamics of Thermoset Injection Molding and the Anisotropies of Molded Parts,"M. Eng. thesis, McGill University, Montreal, Canada, 1980.
28. S. Sidi and M R. Kamal, *Polym. Eng. Sci.,* 22, 349 (1982).
29. M. R. Kamal, V. Tan, and S. Sidi, *Proc. 2nd World Congr. Chem. Eng.,* 6, 471 (1981).
30. D. L. Kerr and A. J. Dontje, *Mod Plast*, 44(2), 147 (1966).
31. F. G. Mussatti and C. W. Macosko, *Polym. Eng. Sci.,* 13, 236 (1973).
32. R. P. White, Jr., *Polym. Eng. Sci.*, 14, 50 (1974).
33. L. T. Kale and K. F. O'Driscoll, *Polym. Eng. Sci.*, 22, 402 (1982).
34. Y. A. Tajima and D. Crozier, *Polym. Eng. Sci.*, 23, 186 (1983).
35. J. S. Osinski, *Polym. Eng. Sci.*, 23, 756 (1983).
36. C. D. Han and K.-W. Lem, *J. Appl. Polym. Sci.*, 28, 743 (1983).
37. C. D. Han and K.-W. Lem, *J. Appl. Polym. Sci.*, 28, 779 (1983).
38. K.-W. Lem and C. D. Han, *J. Appl. Polym. Sci.*, 28, 79 (1983).
39. R. E. Cuthrell, *J. Appl. Polym. Sci .*, 12, 955 (1968).
40. M. D. Hartley and M. L. Williams, *Polym. Eng. Sci.*, 21, 135 (1981).
41. R. J. Farris and C. Lee, *Polym. Eng. Sci.*, 23, 586 (1983).
42. R. J. Silva-Nieto, B. C. Fisher, and A. W. Birley, *Polym. Eng. Sci.*, 21, 499 (1981).
43. D. W. Sundstrom and S. J. Burkett, *Polym. Eng. Sci.*, 21, 1108 (1981).
44. A. B. Metzner, *J. Rheol.*, 29, 739 (1985).
45. G. Segre and A. Silberberg, *Nature*, 189, 209 (1961).
46. G. Segre and A. Silberberg, *J. Fluid Mech.*, 14, 136 (1962).
47. C. D. Denson, E. B. Christiansen, and D. L. Salt, *AIChE J.*, 12, 589 (1966).
48. L. G. Leal, *Ann., Rev. Fluid Mech.*, 12, 435 (1980).
49. J. Kubat and A. Szalanczi, *Polym. Eng. Sci.*, 14, 873 (1974).
50. L. Borocz and J. Kubat, *Plast. Rubber Process.*, 82 (June 1979).
51. L. R. Schmidt, *Polym. Eng. Sci.*, 17, 666 (1977).
52. T. Smith, *SPE. Tech. Pap.*, 31,1260 (1985).
53. R. Thomas, *SPE. Tech. Pap.*, 33, 706 (1987).
54. M. E. Ryan, "The Injection Molding of Thermosets," Ph.D. thesis, McGill University, Montreal, Canada, 1978.
55. T. S. Chung, "The Injection Molding of Phenolic Resin," Ph.D. thesis, Suny/Buffalo, Amherst, NY, 1980.
56. M. R. Kamal and M. E. Ryan, *Polym. Eng. Sci.*, 20, 859 (1980).
57. M. R. Kamal and M. E. Ryan, *Adv. Polym. Technol.,* 4, 323 (1984).

347

58. J. D. Domine and C. G. Gogos, *SPE. Tech. Pap.,* 22, 274 (1976).

59. J. D. Domine and C. G. Gogos, *Polym. Eng. Sci.,* 20, 847 (1980).

60. M. R. Kamal and M. E. Ryan, *SPE. Tech. Pap.,* 23, 521 (1977).

61. M. R. Kamal, *Polym. Eng. Sci.,* 14, 231 (1974).

62. J. J. Fleischmann, *Polym. Eng. Sci.,* 16, 235 (1976).

63. M. R. Kamal. V. Tan, and M. E. Ryan, in N.P. Suh and N.-H. Sung, Eds., *Science and Technology of Polymer Processing,* MIT Press, Cambridge, MA, 1979, P.34.

64. J. A. Biesenberger and C. G. Gogos, *Polym. Eng. Sci.,* 20, 838 (1980).

65. E. W. Vaill, *SPE. Tech. Pap.,* 13, 585 (1967).

66. E. W. Vaill, *SPE J.,* 24(3), 61 (1968).

67. D. L. Messenger and R. S. Papoojian, *SPE. Tech. Pap.,* 20, 657 (1974).

68. T. E. Marchewka and J. A. McAlear, *SPE. Tech. Pap.,* 13, 1036 (1967).

69. J. M. Grigor, *SPE J.,* 24(1), 63 (1968).

70. R. S. Papoojian, *SPE J.,* 24(8), 67 (1968).

71. J. C. O'Brien, *Plast Eng,* 32(2), 23 (1976).

72. T. J. Wolff, *Polym. Plast. Tech. Eng.,* 11, 117 (1978).

73. D. W. Sundstrom and L. A. Walters, *SPE J.,* 27:4, 58 (1971).

74. H. J. Barth and J. J. Robertson, *Mod. Plast.,* 47(11), 142 (1970).

75. S. Y. Choi, *SPE J.,* 26(6), 51 (1970).

76. L. D. Fishberg, D. C. Longstreet, and A. R. Olivo, *SPE J.,* 26(3), 34 (1970).

77. R. H. Beck, *SPE. Tech. Pap.,* 17, 392 (1971).

78. R. H. Beck, *SPE J.,* 27(5), 43 (1971).

79. R. S. Papoojian and R. Lynn, *Mod. Plast,* 50(11), 78 (1973).

80. B. Kleinemeier and G. Menges, *SPE. Tech. Pap.,* 23, 7 (1977).

81. M. R. Barone and D. A. Caulk, *Polym. Eng. Sci.,* 21, 1139 (1981).

82. A. E. Hirsch, *SPE. Tech. Pap.,* 33, 297 (1987).

83. C. L. Tucker III, in T. G. Gutowski, Ed., *The Manufacturing Science of Composites,* Vol. IV, ASME, New York, 1988, P.95.

84. T. R. Curlee, *The Economic Feasibility of Recycling—A Case Study of Plastics Wastes,* Praeger, New York, 1986.

85. M. L. Williams, R. F. Landel, and J. D. Ferry, *J. Am. Chem. Soc.,* 77, 3701 (1955).

86. Y. A. Tajima and D. G. Crozier, *Polym. Eng. Sci.,* 23, 186 (1983).

87. T. H. Hou, *SPE. Tech. Pap.,* 31, 1253 (1985).

88. S. H. Bauer, *SPE. Tech. Pap.,* 22, 650 (1976).

89. S. H. Bauer, *Plast Eng,* 33(3), 44 (1977).

90. W. Y. Chiu and G. D. Shyu, *J. Appl. Polym. Sci.,* 34 1493 (1987).

91. R. von Turkovich and L. Erwin, *Polym. Eng. Sci.,* 23, 743 (1983).

92. D. R. Fitchum and Z. Mencik, *J. Polym. Sci. Polym. Phys. Ed.,* 11, 951 (1974).

93. J. Milewski, *Polym. Plast. Technol. Eng.,* 3, 101 (1974).

94. D. M. Biggs, *SPE. Tech. Pap.,* 30, 668 (1984).

10 · Quality Systems for Automotive Plastics

Gilbert B. Chapman II

Contents

Gilbert B. Chapman II, Chrysler Motors Corporation, Highland Park, Michigan, USA

10.1 INTRODUCTION

10.1.1 Composites: Advantages and Challenges to Quality

The obvious advantage of composite materials is in their high strength and stiffness-to-weight ratios. Moreover, the anisotropy available to the material allows for additional strength to be utilized in a given design direction without the addition of significant extra weight resulting from the additional mass required, as is usually the case in isotropic materials. These engineering materials have, by design, a wide range of chemical, physical, and mechanical characteristics. They often contain organic and inorganic, fibrous and particulate materials, and endless combinations of these, which not only give enhanced strength characteristics but deliver them in very specific design locations and directions.

Composites are, therefore, desirable engineered materials because they can be fabricated to meet the specific mechanical requirements of an application. This application-specific characteristic is derived from the strong dependence of composites properties on the ingredients and processes by which they are formed. This high degree of dependency, a notable advantage, is also responsible for the unintentional wide variations that occur in material properties when input and process parameters are not optimally controlled. Suboptimal control in these areas can present a plethora of problems in controlling quality.

This is especially true when anomalies such as inconsistencies or undesired anisotropy occur during fabrication, or when the part is plagued by local resin richness, reinforcement anomalies, delamination, or porosity in critical areas. Furthermore, composites and plastic components are often assembled by welding thermoplastics or by adhesively bonding thermoplastics and thermosets. These processes often need inspection methods for quality control or assurance during process development, optimization, and proof of capability.

The importance of these materials, both present and future, demand that unique, comprehensive quality methods be developed for their design, fabrication, testing, and evaluation. Their inherent anisotropy and nonhomogenity create problems for many conventional design, production, and testing procedures; however, through the use of new and emerging techniques, these difficulties can be overcome.

10.1.2 Quality and Cost

Quality and cost are two ever-present issues in designing and manufacturing composite material products to meet customer requirements involving expectations of value, performance, durability, reliability, fit, finish, and style. The concern for quality, the conformity to requirements, is necessary not only for the safe and effective performance of the product, but for the economic survival and prosperity of its producer. This concern and how best to acquire and keep it, is the reason for this chapter.

This chapter deals with issues that impact the quality of plastics and polymer composites in the context of a comprehensive quality system (CQS). This comprehensive system addresses quality control as it is defined in the Japanese Industrial Standards, "A system of production methods which economically produces quality

351

goods or services meeting the requirements of consumers. Modern quality control utilizes statistical methods and is often called statistical quality control." It also incorporates the definition of a well-known quality champion, Dr. K. Ishikawa, which states "To practice control is to develop, design, produce, and service a quality product, which is most economical, most useful, and always satisfactory to the consumer" (1).

The three major components for this CQS are strategic concepts, quality improvement mangement, and inspection technology. The first two are discussed only briefly, because they have been quite thoroughly discussed elsewhere (1–4) by several quality champions. Nevertheless, these two essential components of a CQS approach to quality are mentioned to underscore their importance as prerequisites for the third, inspection technology.

Inspection, however, is but one small part of an effective quality system. Quality systems, as used here, include elements in the product development cycle that have a measureable positive impact on process or product quality. Some of these are soft elements, such as concepts, philosophies, simultaneous engineering practicies, quality function deployment (QFD), design review methods, operating practices, and data acquisition and reduction software for statistical process control (SPC). Design review also includes attempts to assure a suitable relationship among customer requirements, design, material, function, and manufacturing process elements. Quality function deployment, a system championed first by Mizuno and the Japanese in 1972, provides a comprehensive methodology for focusing resources on meeting customer requirements in the development phase of a product's evolution. Others can be considered to be hard elements, such as design, material, testing, development, and manufacturing systems hardware, including computers for automation and SPC data acquisition and reduction.

Quality can be very expensive when the impact of soft elements is overlooked or ignored. When this is the case, delayed attempts at quality often focus on testing the product rather than assuring that a good and proper relationship exists among design, material, function, and manufacturing and assembly processes. The suitability of this relationship is essential to providing quality products at reasonable costs. This is especially true in high volume, cost-sensitive manufacturing environments.

Fast cycle times (desirable) and/or multiple tooling (undesirable and capital intensive) that provide synchronous production and its resulting just-in-time (JIT) inventories, satisfy the high volume production requirement. The high volume requirement for the quality system must be satisfied as well, and this is why upsteam strategic planning in this area is essential.

This chapter discusses a quality system based on such strategic planning and focuses on inspection technology to support it. It provides some examples of inspection technologies that have been developed and used in applications to support polymer composites quality in a high volume industry. Utilizing a quality system that brings concepts, philosophy, management, and technology to bear on all elements of the product life cycle will provide quality composite components at costs that meet customer requirements.

10.2 STRATEGIC CONCEPTS

It is prohibitively expensive to meet quality objectives by resorting to quality screening to remove defective or substandard products resulting from a faulty process or an improper match among design, material, application, and manufacturing processes. Unfortunately, early quality control concepts focused on screening out substandard products to prevent them from reaching the consumer. This produced a negative impact on cost. Figure 10.1 illustrates these early concepts, where non-destructive testing (NDT) or nondestructive evaluation (NDE) methods were used primarily as a quality control window to screen out substandard or defective units. This linear scheme protects the customer from faulty products, but it often fails to provide feedback to correct the process. Feedback is essential in a high volume production environment, where scrap and rework costs can destroy the small profit margins provided by each part. Feedback begins with implementation of a CQS and utilization of the output from inspection methodology in SPC. Information derived can be applied immediately to control the process, whether it is a design, material selection, testing, engineering development, or manufacturing process.

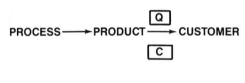

FIGURE 10.1   Linear quality control scheme.

Issues impacting quality in plastics and polymer composite components arise throughout the range of the product development cycle. If cost-conscious quality is to be realized, these issues must be dealt with in design, material selection, manufacturing, and assembly processes, as well as in cost-effective, reliable approaches to in-process and post-process inspection.

Cost-sensitive materials requirements must be satisfied by materials with high performance-to-price ratios and cost-efficient production methodologies that are fast and not labor intensive. These two requirements place uncompromisng demands on quality systems that focus on defect prevention and on inspection methodologies for quality assurance and control. These demands can be met only by systems that extend from design through production, and by methods that are fast and cost effective. This is especially true in plastics and composite components, where optimizing the match between material, processes, design, and intended performance is so important.

Inspection costs must be minimized; nondestructive inspection (NDI) must be fast and not labor intensive, and it must contribute to improvement of the process to reduce scrap, rework, and warranty costs. Thus NDI, as well as the more extensive CQS of which it is a subset, must support quality and cost objectives, as well as offering a cost advantage over processes used previously.

10.3 QUALITY IMPROVEMENT MANAGEMENT

As strategy for the quality system is formulated and put in place, appropriate attention must be given to its management. The system must be managed so that process capability and control are established early on by measurements on parameters that are coupled to quality.

For example, it is said that the material characteristics of components made from sheet molding compound (SMC) depend on nearly 150 material and process variables (5). Among them, only a few significantly impact quality in an operation and should, therefore, be identified and controlled. The control and capability requirements of the process are important quality issues, which should be defined early during design and material selection, and resolved before process development is completed.

The process development must also include a process capability study. This means the systematic study of a process by means of statistical control charts, to discover whether it is behaving naturally or unnaturally; also necessary are the investigation of any unnatural behavior to determine its cause and action to get rid of any of the unnatural behaviors that should be eliminated for economic or quality reasons. The natural behavior of the process after unnatural disturbances are eliminated is called the "process capability" (6).

This study can be supported by effective mechanical, chemical, physical, and nondestructive test procedures. The nondestructive tests have the potential, calibrating with destructive test results, to provide fast and inexpensive process/product quality monitoring to support the capability study.

10.4 INSPECTION METHODS

Because the quality of composite components is quite sensitive to the manufacturing process, a marginal process will produce a significant number of very poor parts. This will have a negative impact on cost and quality. Because of the industry's drive to implement composites in applications where cost or performance advantages can be realized, cost-effective methods of producing and assuring the quality of these materials is important. Quality, the conformity to requirements, is here being applied not only to the appropriateness of the design for its intended function and appearance, and the capability of the manufacturing and assembly processes, but also to the ability or provision of measuring the process output quickly and cost effectively by nondestructive inspection.

The ability to use NDI as a part of a quality system is important, especially when new materials and/or processes are introduced. The main uses of NDI in such a case are to help establish process control and to ensure that the desired process capability is established and maintained. The NDI system should remain in place until process optimization yields a process capability index that exceeds that required for "defect-free" operation.

In addition to well-known NDI technology, certain testing and inspection methods are often used to support process and product quality. These include

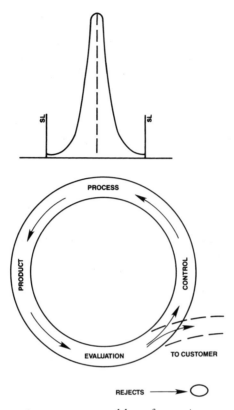

FIGURE 10.2   Quality-sensing process control loop for causing process optimization.

chemical analysis, physical testing, various methods of materials characterization, and mechanical testing. Sometimes these methods are used to support SPC and sometimes they are used to investigate the process or failure of the product. Some of these methods have been selected for discussion in the sections that follow.

Process development engineers need production-compatible testing and inspection methods that detect such anomalies rapidly and cost-effectively to ensure that the loop in Figure 10.2 has a short cycle time. Thus, process development can proceed to optimization, and anomalies can be eliminated before production is launched. The quality-sensing process control loop (Fig. 10.2) illustrates the concept under which a CQS, including NDI methodology, must be implemented to address concerns of both quality and cost.

10.4.1 Methods for Determining Fiber Orientation and Concentration

These NDI methods provide strength–quality assurance for fiber-reinforced plastic (FRP) parts made from sheet molding compound and other similar mass-produced composites. The methods can be employed at the mold site, before further processing or assembly is undertaken. The immediate proximity of inspection to production, in both time and distance, provides for more effective process control by helping to achieve and maintain process optimization through immediate feedback of quality information to molding process personnel.

Plant and field experience has shown that SMC parts tend to crack because of common process-related material anomalies which occur (a) along the prevailing direction of resin and/or reinforcing fiber orientation, or flow lines, (b) in regions where fibers are virtually absent, or resin-rich regions, and (c) in regions where insufficient knitting of fibers and resin occurs at SMC molding charge–flow confluence or knit lines. The tensile and flexural strength values in such regions approach that of the resin, when virtually all fibers either are aligned perpendicular to the load axis, are absent from the region, or do not knit across it.

To assure the strength of FRP material, an effective inspection method must detect and evaluate characteristics that significantly affect strength. These include fiber orientation, fiber concentration, and material thickness. The inspection for fiber orientation determines whether the fibers are random or aligned, as specified, and whether, where, and to what degree and direction parallel fiber flow lines and knit lines exist. The inspection for fiber concentration and wetting determines whether, where, and to what degree resin-rich regions and poorly wet regions exist. The inspection for thickness determines whether specified thickness requirements have been met, and in addition, detects delamination in the material.

Three separate, production-compatible NDT techniques can be used for general SMC strength–quality assurance: (a) a 25-kHz acoustic method for determining fiber orientation, thus detecting flow lines, (b) a 5 MHz ultrasonic method for determining fiber concentration, thus detecting resin-rich regions, and (c) a mechanical proof-test procedure for checking suspected and/or critical regions for weakness caused by flow lines, knit lines, or resin-rich regions.

X-Ray inspection has proved to be a very effective NDI method for detecting all three of the aforementioned SMC flaws, in addition to others. It has been used at the production site and in the laboratory to support process development (see Section 10.4.4).

10.4.1.1 Determining Fiber Orientation

The methods described next were developed to meet inspection requirements through utilization of commercially available instruments and production-compatible techniques. Advantages and limitations of each method are discussed.

Fiber orientation anomalies can usually occur in regions where material must flow over relatively long distances, or where two flowing charges meet during the molding process. Both phenomena are generally responsible for weak regions in SMC material. Parallel fiber alignment characterizes flow lines. Regions where fibers are curled by the collision of two flowing charges and/or do not provide interwoven reinforcement characterize knit lines.

Instrument A strength-indicating NDI method for determining fiber orientation has been developed using a 25 kHz, commercially available bond tester. The bond tester is pictured in Figure 10.3, with its dual-transducer, hand-held probe in the

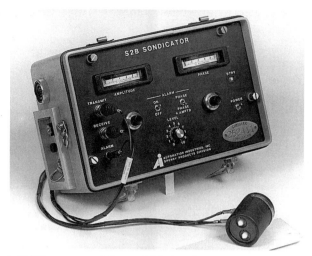

FIGURE 10.3   A 25 kHz commercially available bond tester.

foreground. A sectioned view of the dual transducer probe on a part is shown in Figure 10.4.

The bond tester operates by exciting a series of 25 kHz wave trains in the material with the transmitting transducer; and monitoring these waves with an identical receiving transducer 1.8 cm away from the sender.

In a bond testing application, the amplitude and phase of the received waves are electronically compared to previously established reference levels to determine the adhesive bond integrity of the bond joint under inspection. Details of this inspection method, and an evaluation of its effectiveness in several inspection applications, are presented in References 7 and 8.

The fiber orientation NDI method uses this bond tester to measure the Lamb wave velocity of a 25 kHz acoustic wave train propagated in the material. The transducers, 1.8 cm apart, interrogate circular regions about 2 cm in diameter when the probe is held against the part and rotated through about 180° (π radians).

Changes in wave velocity occurring during probe rotation are indicated by a phase shift in the received signal. Oscillographs showing this phase shift appear in

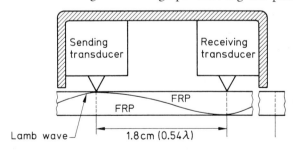

Sectioned view of dual-transducer probe

FIGURE 10.4   Sectional view of 25 kHz dual-transducer bond tester probe.

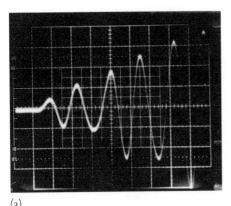
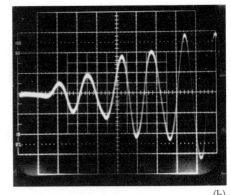

(a)   (b)

FIGURE 10.5   Oscillographs showing phase shift caused by change in acoustic velocity with fiber direction: (a) along fiber direction and (b) across fiber direction.

Figure 10.5. Figure 10.5a indicates a relatively high velocity for the wave traveling along the prevailing direction of fiber orientation. Figure 10.5b indicates a significantly lower wave velocity obtained when the major axis of the probe (a line joining the two transducers) is rotated 90° and the velocity is measured across the fibers.

The difference in the velocity occurs because the reinforcing fibers have a high modulus, therefore carry the acoustic energy much faster than the resin. In a fiber-reinforced composite, however, neither of the constituents solely constitutes the energy path; rather, both contribute to it, as they do to the modulus of the material. Hence, the extremes in velocity are not expected to be as large as the differences in individual moduli, but these differences in velocity are expected to indicate anisotropies in the mechanical properties of the material.

Lamb wave velocity measurements were made on chopped glass fiber reinforced polyester sheet molding compound, a common automotive fiber-reinforced plastic material. Although velocity differences were the measurements of interest, nevertheless, a reliable estimate of the velocity is needed to determine the relative magnitude of velocity change. The value obtained was about 800 m/s.

The velocity difference is obtained by measuring the phase shift during probe rotation. The probe location on the surface of the part is fixed by templates that defined positions to be inspected. The probe, once fixed, is rotated 180° in about 4 seconds. The phase shift observed during rotation is recorded as a velocity indicator reading (VIR) from the bond tester.

A small or insignificant change Δ in VIR is interpreted as a random fiber orientation or a resin-rich region. A significant change in VIR during rotation is interpreted as an indicator of a prevailing fiber orientation, Θ. This Θ value is related to the angle of fiber orientation at which maximum VIR was observed.

To summarize, the magnitude of the change in VIR, ΔVIR is used as an indicator of the degree to which the fibers are preferentially oriented. The direction of the change is taken as positive when the fibers are aligned parallel to the axis along which the mechanical test load is applied (0°) and negative when the fibers are aligned perpendicular to that axis (90°).

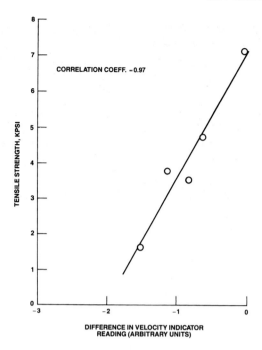

FIGURE 10.6   Tensile strength versus difference in Lamb wave velocity indicator readings.

Tensile strengths are plotted against ΔVIR in Figure 10.6. The correlation coefficients for these data generally range from 0.74 to 0.97, well above the level needed for practical use. Note that the values of the tensile strengths at ΔVIR = 0, the intercept, are within 2 standard deviations of the nominal value of 76 MPa (11,000 psi) for this material. A summary of the curve fit statistics in presented in Table 10.1,

The data in Table 10.1 indicate that the differences between the means are highly significant, especially when measured in terms of the standard deviation of the means (standard deviation/$\sqrt{N}$). This high level of singificance indicates how much the mechanical characteristics of SMC material can vary with location and orientation in a given SMC part. These variations are far greater than variations from part to part

TABLE 10.1   Summary of Tensile Strength Data from Different Regions of Similar SMC Parts

Sample		Tensile Strength				Correlation with Δ VIR
No.	Size (N)	Mean (kpsi)	Intercept (kpsi)	Standard deviation (kpsi)	Coefficient of variation (%)	
1	40	11.4		1.48	13	
2	11	4.8	10.0	2.15	44	0.86
3	5	4.1	6.9	1.98	48	0.97
4	30	5.8	7.9	3.85	66	0.74

in a well-controlled region, as measured in sample 1. These data show the importance of specifying, as well as testing for minimum mechanical requirements in certain critical regions where, because of flow and pressure complications during molding, those minimum requirements are difficult to maintain. A more complete discussion of these data and results is found in Reference 8.

Similar dispersions in these mechanical strength data have been reported by other investigators (9, 10). These data represent realistic variations to be expected in ordinary production SMC parts. Although specimens taken from flat-molded plaques generally yield mechanical strength data with narrow dispersions, data from certain regions of actual parts have wide dispersions in mechanical strength, which tend to emphasize the necessity of providing more effective process control. The application of an effective process NDI procedure immediately after a part is molded (or bonded, in case of adhesive bond inspection) can significantly contribute to accomplishing that objective. The immediate feedback of quality information provides the process operators with a readily available correspondence between slight unintentional deviations from optimum process parameters and critical product characteristics. Even when critical process parameters are not known, this information will help to identify and control them.

In an actual QC application, these values should not be read, but an alarm level can be set to be activated when ΔVIR falls outside a predetermined critical acceptable range.

A computer program has been developed (in BASIC) to compute the maximum tensile strength expected in a region from ΔVIR and angle Θ data from both inner and outer surfaces. A listing of the program is given in the appendix. The maximum expected tensile strength is computed from the equations listed in lines 620–680 of the program. The definitions of the terms in the equations are documented in lines 150–370. Note the provisions in the program for plotting the maximum tensile strength expected versus the actual tensile strengths of the specimens, when tensile tests are made.

Tensile strengths computed by the program are considered to be the maximum value expected for the fiber orientation indication detected, because the program does not take into account the deleterious effect of resin-rich regions and knit lines on SMC strength. These defects, not normally detected by this method, cause only downward deviations in strength from the maximum values computed, as shown in Figure 10.7, but they can be detected by the two NDT methods covered next.

Flexural Strength   A poor correlation coefficient of 0.43 was computed for a linear regression of flexural strength on ΔVIR for 14 specimens tested. Higher degree polynomial fits yielded no significant improvement in correlation.

The mean of these data was 18,030 psi (124.3 MPa) with a coefficient of variation of 50%. Data from sample set 1 yielded means of about 25,200 psi (173.7 MPa) and coefficient of variation of about 10%.

The wide dispersion and poor correlation of these results were not unexpected. The discrepancies agree with results from flex–peel strength tests of adhesive bonds undertaken in previous work (11). There too, the flex–peel strength correlation with

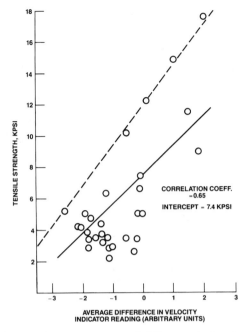

FIGURE 10.7   Tensile strength versus average of differences in Lamb wave velocity indicator readings taken from opposite sides of several regions; some were resin-rich.

the NDI indicator was poor. Therefore, this method is not a reliable indicator of flexural strength.

Limitation   There are occasions when regions of resin richness occur in a molded part. These regions are usually weaker than regions where flow lines or knit lines occur, because in some cases virtually no glass fiber reinforcement is present. These resin-rich regions, readily seen by X-ray because of their lower density, often cannot be detected by the method developed for detecting fiber orientation. This is because insufficient fibers are present in the regions. This is especially true if the region is more than 2 cm wide; in such a case, the inspection probe can remain completely inside the region during its rotation, and no significant change in the VIR will be detected. These resin-rich regions can be detected by other acoustic methods. One is the measurement of ultrasonic longitudinal wave velocity, which appears to have the required effectiveness.

10.4.1.2 Ultrasonic Pulse–Echo Inspection for Fiber Concentration and Thickness

The ultrasonic pulse–echo inspection of FRP components serves two purposes: to assure the integrity of the material and to determine the thickness. The first purpose is accomplished if the ultrasonic pulse passes through the material with the velocity and attenuation expected for that material. Further ultrasonic assurance of compositional integrity is discussed later. The second purpose, ultrasonic thickness gauging, is accomplished by measuring the time t required for the pulse to travel a given

361

distance through the entire thickness of the material. If the velocity v of the pulse is known for the material, the thickness d can be computed from the transit time t, by $2d = vt$. This common, well-known NDI procedure can be studied further in Reference 12.

Method The pulse–echo method utilizes a 5 MHz transducer 1.27 cm (0.5 in.) in diameter to transmit an ultrasonic pulse of longitudinal waves through the thickness of the part; the transducer then "listens" for the echo returning from the opposite side. Figure 10.8 is a sketch of this arrangement.

The first amplitude peak at the left-hand edge of the oscillograph is produced when the pulse leaves the transducer face. The second peak from the left is produced when the first returning echo excites the transducer after time t, represented by the distance between the two peaks. This is the time required for the pulse to make one round trip through the material.

A routine, commercially available embodiment of this procedure is the ultrasonic thickness gage, used in a pulse–echo mode. The gage must be calibrated by setting the velocity to that determined for the material to be gauged. This is usually accomplished by calibration with a similar material of known thickness.

Velocity Measurement In an example of the operation, the gage was calibrated with an SMC plaque containing about 27% glass and used to read acoustical thicknesses of several suspected resin-rich regions of an SMC specimen. The mechanical thicknesses of these regions were also measured. The ratio of mechanical thickness

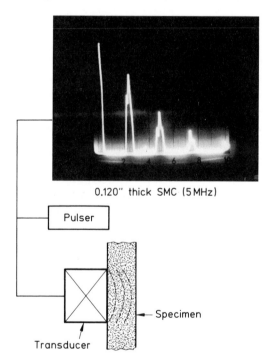

0.120″ thick SMC (5 MHz)

FIGURE 10.8   Ultrasonic pulse–echo thickness gauging.

TABLE 10.2 Regional Compositions of an SMC Part

Specimen location	Specimen weight (g)	Concentration (wt %)			
		Glass	Resin	Filler	Ash
G-1	4.661	17.0	33.3	49.7	66.7
G-2	4.250	19.0	32.7	48.3	67.3
H-1[a]	3.935	27.1	30.5	42.4	69.5
H-2[a]	3.196	22.7	30.8	46.5	69.1
I-1	4.221	17.4	33.4	49.2	66.6
I-2	4.199	19.9	32.3	47.8	67.7
J-1	3.200	12.7	35.7	51.6	64.3
J-2	3.754	19.3	35.6	45.1	64.4
K-1[a]	3.807	21.8	32.0	46.2	68.0
K-2[a]	4.506	26.4	30.5	43.1	69.5
M-1	3.797	18.7	33.8	47.5	66.2
M-2	3.555	15.8	34.4	49.7	65.6
Specified composition		25 (min.)		50 (max.)	

[a] From region where nominal composition expected.

to acoustical thickness, simply another convenient way of expressing relative ultrasonic velocity, was determined for six regions. The velocity was also determined.

Glass Analysis Glass fiber concentrations of these regions were determined on small square specimens. Two specimens from each of the six regions were chemically analyzed. The analytical data are tabulated in Table 10.2. The first of each pair was taken as the specimen through which the ultrasonic velocity was determined. Although the specimens were about four times larger than the half-inch-diameter circular regions interrogated by the transducer, they nevertheless yielded glass concentrations that correlated well with the ultrasonic velocities obtained by thickness gauging. The data plotted in Figure 10.9 used the glass concentrations from the first of each of the pairs of data in Table 10.2 and have a correlation coefficient of 0.97 for a linear regression on velocity.

The scatter among the three points at the lower left of Figure 10.9, for plaques containing no glass, is indicative of the velocity measurement error. These data have a coefficient of variation of 2%. The thickness measurement error was about 1%. The errors in the other data shown in Figure 10.9 occur primarily because the specimens used for glass analyses were about four times larger than the region interrogated ultrasonically for thickness. The pulse–echo ultrasonic inspection was performed on regions directly underneath the 5 MHz, half-inch-diameter transducer. The glass analyses were performed on sections measuring about one inch square. These sections, having areas much larger than the area of the transducer face, generally extended well into regions where the glass concentration was higher than in the region directly beneath the transducer. This measurement problem, which was confirmed by radiographic examination, is assessed to be the major cause of the scatter in Figure 10.9.

The data in this example provide a useful relationship by which glass fiber concentration can be determined nondestructively, if actual thickness is known to

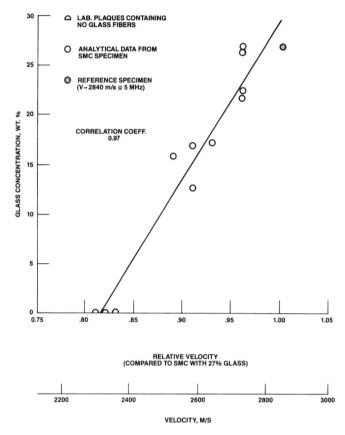

FIGURE 10.9   The effect of glass fiber concentration on ultrasonic longitudinal wave velocity of propagation through the laminate.

Low attenuation
0.097 in thick resin at 5 MHz

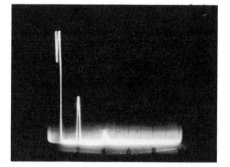

High attenuation
0.097 in thick HSMc at 5 MHz

FIGURE 10.10   CRT traces showing main bang and ultrasonic pulse echoes from two vastly different laboratory specimens. The lower trace shows higher attenuation and velocity because of increased fiberglass concentration.

within $\pm 5\%$. The curve shows that a 10% change in velocity, or indicated thickness, is indicative of a change in glass concentration by about 10%. This is about a 40% change in glass concentration at the 25% glass level. Because composite strength is related to glass fiber concentration, this ultrasonic method is an effective tool in identifying areas that are weak due to high resin-to-glass ratios.

Other indications   Alternatively, the thickness can be determined if a constant material composition is maintained, thereby maintaining a constant ultrasonic velocity. This method is also effective in detecting delaminations and, when embodied in a CRT video display flaw detector, can detect insufficient wetting of the glass by the resin. This condition is indicated by increased attenuation, such as that observed in the lower, as compared to the upper, oscillograph of Figure 10.10. the use of ultrasonic attenuation to measure porosity is discussed in Section 10.4.1.6.

Overcoming a Limitation with the Pulse–Echo Technique   The inability to determine thickness and velocity independently is a serious limitation of the pulse–echo inspection method, especially in applications where neither constant thickness nor constant composition can be relied on. This limitation can be overcome by a combination through-transmission, pulse–echo ultrasonic inspection method, which is capable of measuring velocity indpendent of thickness variations. Such a commercially available inspection system and methodology can be applied to the glass concentration inspection problem, based on the relationship between glass concentration and ultrasonic velocity established with data obtained by pulse–echo inspection. Some minor adjustments in the curve may be necessary, however, for each material or inspection system.

This through-transmission velocity method is well suited for use on small submersible parts, but it may be used on large bulky moldings as well. A full description of the method may be found in Reference 13. This method can also be effective in detecting delaminations and insufficient wetting.

Ultrasonic thickness gauging of FRP should be undertaken only with an awareness of the sensitivity of the relationship between ultrasonic velocity and glass fiber concentration, as shown by the curve in Figure 10.9. Changes in the glass-to-resin ratio can cause changes in velocity, which result in erroneous thickness readings unless the thickness gage is recalibrated to accommodate the new velocity.

10.4.1.3 Radiometric Inspection for Glass Fiber Concentration

A selective radiation technique for measuring glass concentration in plastics has been embodied in an easy-to-use quantitative system called the Compuglass Analyzer. The system, developed and marketed by Radiation Monitoring Devices, Inc., of Watertown, Mass., is a fast, nondestructive test for glass content that allows on-line process control. The Compuglass Analyzer employs a selective radiation technique and functions independent of resin type, fillers, addditives, and pigments. Samples can be assayed for glass content in 2 minutes with accuracy better then 1%. This type of nondestructive test can be utilized to support improved quality and productivity especially during process development (14).

10.4.1.4 Analysis of Analytical Chemistry Data

The self-consistency of the analytical data is demonstrated by plotting the glass fiber concentration against ash content of the 12 specimens analyzed. There is a good linear fit to the data, with a correlation coefficient of 0.97 if only one of the data points is excluded. This close distribution of the data about the curve in Figure 10.11 supports the self-consistent reliability of the analytical data. Only one analysis is suspect.

The curve also shows that a reasonably reliable and routine estimate of glass concentration can be obtained by merely ashing the SMC specimens at 650 °C, chemically leaching out the inorganic filler, and weighing the dry residue. This curve must be established or confirmed for each batch of SMC analyzed.

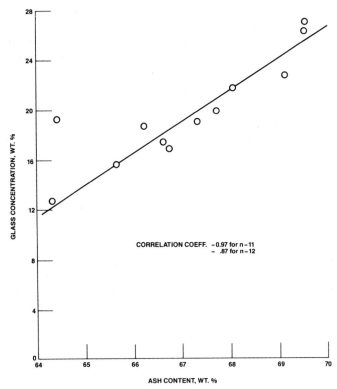

FIGURE 10.11   Glass fiber content versus total ash content for a set of SMC specimens.

10.4.1.5 Inspection for Interlaminar Integrity

The 25 kHz acoustic inspection method, as well as the pulse–echo and the through-transmission ultrasonic methods, have been proved to be effective indicators of material characteristics that adversely affect interlaminar strength.

The 25 kHz inspection method has been used to detect microscopically thin delaminations, more than 2 cm in diameter, in graphite–glass fiber reinforced plastic structural parts. The parts were inspected, any suspected delaminations detected, and their locations identified. Parts were inspected again after fatigue testing, and the suspected delaminations were observed to have increased in area. The parts were then sectioned and examined by optical microscopy, confirming the presence of the delaminations. Delaminations in these parts could not be detected or their presence confirmed by pulse–echo or through-transmission ultrasonic inspection because of the high attenuation of the material at frequencies above 1.0 MHz.

The 25 kHz method has also proved to be an effective indicator of interlaminar strength in SMC. It was used in an adhesive bond inspection application, and when the inspected bond joints were tensile tested for shear strength, many failed by delamination. Correlations between indicator values were good (11). This indicated the effectiveness of the method as a nondestructive indicator of interlaminar shear strength in a bond joint NDI applciation. Interlaminar strength is of the highest importance in such locations.

Pulse–echo and through-transmission inspection were effective in detecting delaminations and have been used by some investigators to determine material strength (10). Generally, these inspection methods should be evaluated for effectiveness and efficiency in each inspection application proposed.

QC Applications: Inspection Criteria   Both magnitude and direction of fiber orientation must be considered to determine the acceptability of a region of a part to perform in service. There may be some regions in which local loads experienced in one direction far exceed those experienced in another. There may also be regions in which loads are directed along flow lines (strong), and others in which loads are directed across flow lines (weak). Each of these three conditions must be taken into account in establishing accept/reject fiber orientation criteria. Therefore, the critical accept/reject ΔVIR values must be established for each critical region of the part, based on the performance requirements of that region in terms of both magnitude and direction of the expected load.

Once the critical regions have been identified, and the minimum strength requirements established, a curve such as that plotted in Figure 10.7 can be used to establish the minimum required ΔVIR. The VIR values need not be read by the operator, but an alarm can be set to activate when ΔVIR falls outside a predetermined acceptable range, orientation, and probe angular position.

Parts Inspected   The fiber orientation NDI method has been utilized in several prototype and production applications.

Fiber orientation studies have also been performed on a reinforced reaction injection molding (RRIM) fender and on crossmembers made of fiber-reinforced plastic molding compound with continuous glass fibers that are nearly unidirectional (X pattern). In both cases, feasibility was established.

The 25 kHz ultrasonic method has also been used to detect delaminations in FRP air conditioner compressor brackets, to eliminate substandard prototype parts from service, and to suggest process improvement needs.

In summary, the quality assurance of FRP components can be achieved by the nondestructive inspection of critical regions of the part with the three inspection procedures. The first procedure, a 25 kHz acoustic method, determines the direction and degree of fiber orientation in the region inspected, by measuring the Lamb wave velocity of waves propagated along the region. Tensile test results have shown good correlations of strength with NDI results from this inspection method. The second and third procedures indicate the concentration of glass fiber reinforcement in a region by measuring the ultrasonic longitudinal wave velocity of pulses propagated through the region, or by using a quantitative, through-transmission radiometric method. Analyses of these regions have shown good correlation of glass concentration with velocity.

The first two of these methods are capable of detecting delaminations and, when used together in a production QC application, can provide quality information essential to quality assurance and helpful in process control.

10.4.1.6 Ultrasonic Methods for Detecting Porosity

Ultrasonic attenuation can be used to detect and, to some degree, to quantify porosity in composite materials (15–18). The ultrasonic attenuation of a solid containing a distribution of voids depends on the number of voids per unit volume and the ultrasonic scattering cross section of the voids. This physical picture has been shown to provide an accurate description of the attenuation behavior of ultrasound in metals containing low level porosity (19).

Fiber-reinforced composites are structurally more complex than a homogeneous metal, and consequently the morphology of the porosity in composites is also characteristically different. Voids in composite laminates tend to occur at the interface between the plies and are generally flattened and elongated along the axial direction of adjacent fibers (20). In composites, the distribution of the geometric dimensions of the pores spans a much wider range than pores in metals. The frequency dependence of the ultrasonic attenuation also appears to be different in the two cases. The presence of porosity in composites increases the attenuation level, which can be observed as in Figure 10.10 or determined quantitatively by establishing the correlation between the porosity volume fraction and the frequency dependence of the attenuation as in Reference 15.

10.4.1.7 Mechanical Proof Test for Weak Regions

A crude, inexpensive, fast, and effective mechanical test that can be considered to be nondestructive if the part is good, but destructive if the part is defective in the region tested, is a hand-loaded proof test. This procedure can be applied to any SMC part where quantitative tension and/or flexural loading can be done manually.

Instrument A Chatillon force gage, model CFG100, or equivalent, is recommended for proof testing. The device may be used in either tension or compression mode, and it has a "hold at maximum" provision that allows the operator easy post test reading of the maximum force exerted during the test.

The force gage attachment used to couple the gage to the specimen should be configured to provide a safe, reliable, non damaging coupling to the specimen. The disc attachment, covered with a hard rubber pad, is recommended for compression testing. A hook attachment, designed to accommodate each specific part, is recommended for tension testing.

Test Procedure The critical region to be tested is identified by the engineer, and the test conditions are specified. Normally, these include, but are not limited to, a specification of test loads (including load direction) at each specified location.

An example of the use of this test procedure follows. In an earlier application of this QA technique to an SMC part experiencing sporadically high rates of flow line failures during assembly, a flexural load of 290 N (65 lb) was applied to the critical area. The load was applied by hooking the tensile test attachment to a nearby boss and pulling outward, perpendicular to the part surface, with a force that was gradually increased until fracture was detected audibly or visually, or until the force reached 290 N. A 2% failure rate was observed during testing on two batches totaling slightly more than 100 parts. Of the 98% passing the 290 N proof test, none failed during assembly, and no premature in-service failures have been reported. The 290 N critical value was established for the proof test to accomplish this result.

The proof test is, therefore, recommended as a cost-effective and practical test method, but it yields little specific quantitative information for process control because it does not indicate whether the region failed because it was a flow line, a knit line, or resin-rich.

10.4.2 Methods for Evaluating Adhesive Bonds

An inspection method for use in a high volume, low cost production environment must be fast, convenient, cost-effective, and able to effectively assure some minimum acceptable level of bond integrity. It must also be operationally simple and virtually maintenance free. An effective method must provide inspection for adhesion within the bond joint, rather than for the mere presence of adhesive material. Moreover, the prevailing interest here is in a method of inspecting for adhesion rather than for the cohesive strength of the adhesive layer, because less than 1% of the failures observed in typical automotive bond joints are cohesive. Therefore, proven methods of measuring bond cohesive strength—for example, by ultrasonic Fokker bond testing —are far more effective in applications where cohesive failure is a significant failure mode. Furthermore, liquid couplant (generally required for ultrasonic inspection) is not preferred because of its possible adverse effect on subsequent product processibility.

Several approaches to adhesive bond nondestructive evaluation meet some of the requirements for production quality control application. Among those surveyed, a 25 kHz bond test method was determined to hold the highest potential for meeting most of the production QC requirements. Such a method needs no liquid couplant and utilizes one of several simple instruments that are commercially available and portable. Because of its operational simplicity and speed, the method has a high potential for cost effectiveness.

Its flaw-finding effectiveness is dependent on two key parameters that provide the basis for a practical semiquantitative NDE method, namely local bond integrity

369

(LBI) measurements, which are based on the use of a statistically selected reference specimen for "calibrating" the bond tester, and bond merit factor (BMF) values, which are estimates of regional bond integrity computed from LBI data. The effectiveness of these two parameters as indicators of bond strength in FRP lap joints has proved valuable in high volume QC applications.

10.4.2.1 Bond Test Instrument

Choice of Instrument An Automation Industries Sondicator Model S2B is a simple but effective tool for bond testing. Its combined phase and amplitude detection circuit is a desirable feature for detecting and quantifying certain disbond conditions and marginal adhesive strength, as described later.

Principles of Operation The bond tester operates by exciting the material under inspection with a series of acoustic pulses from the transmitting transducer. Adhesive bond defects are detected by comparing the wave train of the pulses received from the specimen to the wave train of pulses received during calibration from a reference specimen of known bond integrity. The contrast between these pulses can be seen in oscillograms of the entire pulse envelopes shown in Figure 10.12. The significant difference between the amplitudes from the good reference bond and the unbonded regions can easily be seen by comparing the three oscillograms in Figure 10.12.

The mode of pulse transmission is predominantly by plate-type Lamb waves, because the specimen thickness, in this application, is only a fraction of the wavelength. The vibrations are detected by the receiving transducer and electronically compared with those received from the reference bond. The leading portion of the wave train from a reference bond vibration is shown in Figure 10.13b. That resulting from an unbonded specimen is shown Figure 10.13c. Note the increase in amplitude and shift in phase apparent in comparing the wave trains from bonded and unbonded specimens.

This change in phase and amplitude occurs because the pulse transmission is along a layer of unbonded material. The unbonded layer is free to vibrate at higher amplitude, with less energy dissipation and lower wave velocity than a bonded reference specimen. Evidence that the phase shift is due to lower wave velocity rather than to a change in resonant frequency appears in Reference 11.

The velocity decreased by about 35 m/s, or 4%, from a bonded to an unbonded specimen. The velocity was computed as 833 m/s. The wavelength was measured with the same transducer and also computed from velocity and frequency data. Both methods yielded 3.3 cm.

Modes of Operation

Alarm Mode To make a valid comparison of pulses from parts of unknown integrity and pulses from a reference sample of known integrity, the bond tester is first referenced using the required reference specimen. Amplitude and phase levels are adjusted to prescribed values while monitoring the reference specimen. Then the alarm mode is selected, and its level set. There are two options for the alarm mode.

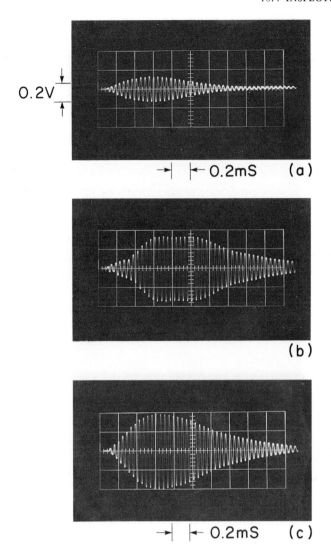

FIGURE 10.12 Oscillographs showing 25 kHz signals from three specimens interrogated by the bond tester. (*a*) Good bond; (*b*) disbond at edge of adhesive void; (*c*) single thickness FRP adherend or extensive void.

The first provides for flaw detection by sensing deviations of the phase from its reference value. The second provides for flaw detection by sensing the combined deviations of the phase and amplitude. The latter is the recommended mode. The alarm activation level (AAL) adjustment, made after the amplitude and phase levels have been adjusted, determines the turn-on point of a Schmitt trigger, which activates the alarm. Its setting determines the magnitude of increase in amplitude and shift in phase required for flaw indication by alarm activation. Ordinarily, when inspecting bond joints by this mode, AAL, amplitude, and phase adjustments remain set after

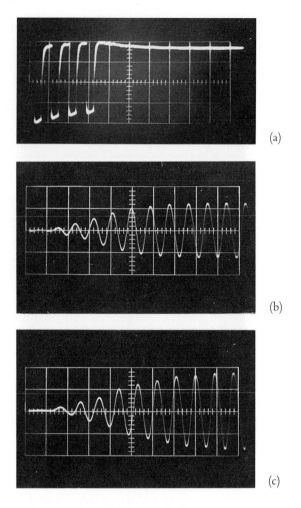

(a)

(b)

(c)

FIGURE 10.13 Oscillograph showing 25 kHz signals from bonded and unbonded specimens interrogated by the bond tester. (*a*) Input pulse to transmitting transducer; (*b*) pulse received from reference specimen; (*c*) phase-shifted pulse received from unbonded specimen.

referencing. However, the AAL setting can be used as a quantitative indicator of bond joint integrity, under certain limiting circumstances.

Metering Mode In addition to the alarm mode for which the bond tester was designed, the procedure can be quantified by reading the instrument response to the acoustical property of the local bond region. This attempt to quantify bond tester response allows correlation of these responses with bond strength, at least for certain failure modes.

The bond tester alarm circuit (11), a portion of which is shown in Figure 10.14, was altered so that the potential difference between the base of transducer Q19 and ground could be measured with digital voltmeter M4. The AAL potentiometer, R87, was set to read zero (at 200 Ω maximum resistance) during the NDI monitoring, and the voltage readings were recorded as LBI readings for each bond locale.

Bond Reference Specimens

Need The proper preparation and selection of a bond reference specimen is a prerequisite to obtaining meaningful and consistent results in the sonic NDI of bond joints. A reference specimen is required for use as a calibration standard because the bond tester essentially operates as a difference-detecting device. It compares the acoustic characteristics of unknown parts with those of a reference of known, acceptable integrity. It is, therefore, essential that the reference specimen resemble the parts to be inspected in material composition and bond joint geometry. It should also possess a level of bond integrity equal to or only slightly better than that required in the final product.

Selection A reference specimen is required for each material type from sample sets that are similar in geometry and composition to the corresponding test specimens. Each must be prepared under adhesive bonding conditions that closely simulate accepted and expected production practice. These specimens contained no intentional adhesive flaws.

Each reference specimen is selected from the sample set by the following procedure.

1. Use one arbitrary bonded specimen as an interim reference for initial adjustment of the bond tester according to instructions in the instrument manual. During this referencing, and all subsequent measurements in the procedure, each specimen should be supported by a material of low acoustic impedance, such as soft plastic foam sheet about 2 cm thick, to reduce acoustic interferences from the supporting structure.

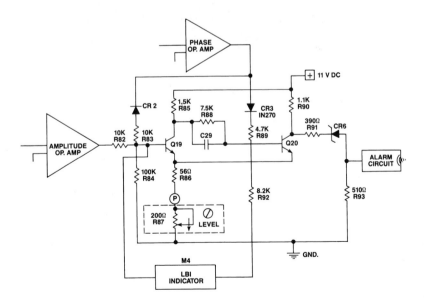

FIGURE 10.14 Circuit schematic for local bond integrity indicator and alarm.

2. Monitor several bond joint locations on each of the specimens by placing the probe down firmly on a location along the bond line. A 1 kg mass is recommended for use as a probe hold-down weight, to assure the application of firm, evenly distributed, reproducible, and operator–independent pressure on the probe. (About four locations per specimen are suggested.) Avoid locations where sharp changes in the readings are observed for small changes in probe position. Record phase, amplitude, and AAL values for each monitored location. (LBI readings may be recorded instead of AAL.)

3. Permanently identify (mark) each location and indicate the transducer probe orientation for future reference and duplication.

4. Rank each monitored location according to its AAL value. Phase and amplitude values are recorded to verify that the arbitrary interim reference specimen was indeed bonded. AAL values are the combined results of phase and amplitude values and are used here to indicate bond integrity.

5. Select bond locations ranking at or near the 40th percentile as reference candidates. The references candidates should have bond integrities, as indicated by their AAL values, higher than 35% but no higher than 45% of the AAL values at bond locations monitored in the sample set. Each specimen containing one or more of these locations is a reference specimen candidate.

6. Reference the bond tester on the reference specimen candidate nearest the 40th percentile according to the described procedure. Then inspect each specimen ranked below the 40th percentile, according to the described procedure, to find the highest ranking specimen causing bond tester alarm activation.

7. Destructively test the highest ranked specimen causing alarm activation, and its two nearest neighbors, according to a prescribed mechanical test or durability test procedure, to determine whether they met engineering bond specifications.

8. If the bond strength meets minimum requirements for the intended adhesive system application, the candidate specimens from the 40th percentile are confirmed as valid initial reference specimens for the subject adhesive–adherend assembly application.

10.4.2.2 Nondestructive Inspection of Specimens

Instrument Referencing   Three adhesively bonded FRP lap joint specimens are used to perform the referencing adjustments. Two of these—reference specimens of known, acceptable bond integrity—were selected from a sample set of bonded specimens according to the eight-step procedure just described. The third contained an unbonded area about 1 in.^{2} within the lap joint. One of the two reference specimens of sound integrity was kept with the instrument for regular use as a control. The other was retained as a "primary standard" and kept in a protected place for daily comparison to the first. The third specimen was used to confirm disbond detection after completion of the referencing adjustments. These specimens were used to reference the bond tester according to instructions in the instrument operator's manual.

Inspection for Regional Bond Integrity   In a practical production QC application, disbond and weak bond data are used to accept or reject a bonded assembly or region thereof. It is impractical to assume that any minor substandard bonding in an assembly would warrant rejection of the part. Thus a formula has been devised to quantify the bond integrity of a bonded region, say 25–40 cm long, in terms of local bond inspection data. Disbond dimensions are defined in terms of the operating sensitivity level of the bond tester. These disbond data, which may include areas of weak or substandard bonding, are then used to compute a bond merit factor (BMF) for the region by the equation,

$$\text{BMF} = \frac{L - C - P/2}{L} \tag{10.1}$$

where L is the total length of bond joint region inspected, C is the total length of all complete disbonds within the region, and P is the total length of all partial disbonds within the region.

Partial and complete disbonds are defined, and the inspection process for distinguishing between them is described, in Reference 11. Figure 10.15 shows examples of a partial and a complete disbond.

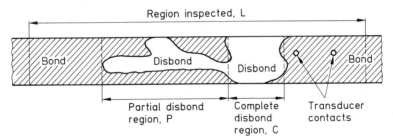

FIGURE 10.15   Plan view into an adhesive bond joint showing partial and complete disbonds.

Mechanical Testing   When mechanical testing is required for NDT calibration or process audit, the specimens are prepared by cutting the plaques into segments 2.5 cm wide with a diamond-tooth saw. Each such segment is then cut to an overall length of 18 cm. The resulting specimens are then subjected to the following tests to determine bond joint strength.

Shear Strength by Tension Loading   Recommended test methods are based on procedures described in the American Society for Testing and Materials (ASTM): Recommended Practice for Determining the Strength of Adhesively Bonded Rigid Plastic Lap–Shear Joints in Shear by Tension Loading [D 3163-73(1984)] and Test for Strength Properties of Adhesives in Shear by Tension Loading of Laminated Assemblies [D 3165-73(1979)].

This test procedure is recommended over the neutral axis test to stimulate the tension-shear load conditions experienced by the lap joints while in service. This test also conformed to mechanical QC tests required by typical engineering specifications, with which the NDI method must correlate. It did not measure the pure shear strength of the bond, because immediately upon loading, the lap joint specimen deformed to the configuration shown in Figure 10.16. This places the joint under a combination of shear and peeling forces, as well as a flexural moment, enhancing tearing of the FRP adherend.

An attempt can be made to measure a more accurate bond shear strength. This test procedure, based on ASTM test D 3165-73(1979), is used to increase the shear component at the adhesive–adherend interface by loading along the neutral axis. It reduces distortion-induced tearing and reduces the fraction of failures by delamination within the adherend. Specimens for this test procedure are fabricated to have a dual thickness over the entire span of the specimen, except for a notched indentation, on alternate sides to define the lap joint test area.

Peel Strength by Flexural Testing A test method based on the procedure described in ASTM Test for Flexural Properties of Plastics and Electrical Insulating Materials [D 790-71(1978)] is recommended. The bonded specimens, with previously described conventional lap joint geometry, are loaded at the rate of 5 cm/min at the center of the lap joint by a loading nose of 3 mm radius, while the specimens are supported over a 5 cm span.

Disbonds Detected The results from inspecting the sample group of 60 bonded specimens for disbonds are summarized in Figure 10.17. They show that the effect of inherent operator and instrument variability was virtually insignificant, compared with the size of a typical disbond. The agreement on location of the lineal center of the disbonds was within 0.7 cm (0.3 in.) pooled standard deviation. The agreement on the length of these disbonds was within 2.2 cm (0.9 in.) pooled standard deviation.

This desirable result was highly dependent on using a common or identical reference specimen. Use of the same reference specimen provided uniform control of the sensitivity of the instruments, hence nearly uniform disbond detection and definition. Use of a slightly dissimilar reference specimen (the same in composition and geometry, but chosen from the 60th rather than 40th percentile rank in LBI), resulted in causing the area detected as disbonded to nearly double.

This bond NDI mode detects unbonded regions whether caused by lack of adhesive or lack of adhesion. It does not, by the recommended alarm mode, distinguish between disbonds due to adhesive voids and disbonds due to unbonded adhesive. The operator can, however, distinguish between adhesive voids and unbonded adhesive by observing the deflections of the phase and amplitude meters. The phase meter is usually deflected to the left when the disbond is caused by unbonded adhesive. The amplitude meter is usually deflected to the right when disbonds are due to adhesive voids. Furthermore, disbonds detected by the bond tester cannot always be confirmed by X-ray, because many of them are not adhesive voids.

The probability of detecting regions of unbonded adhesive and adhesive voids increased, as expected, with disbond size. The probability of detecting adhesive voids varied from about 0.25 at the 1 cm limit of detection, to about 0.5 at 1.5 cm, to 0.95 at 2.0 cm. The probability of detecting unbonded adhesive ranged from about 0.25

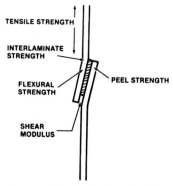

FIGURE 10.16   Lap joint under actual load showing the five components of strength.

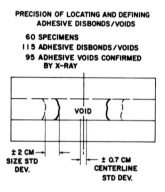

FIGURE 10.17   Typical adhesive bond joint flaw.

at the same 1 cm limit of detection, to about 0.5 at 3.3 cm, to 0.95 at 4.0 cm. The lower detection probability for regions of unbonded adhesive, compared with that for adhesive voids of the same size, may be due to the unbonded adhesive restricting the oscillation amplitude of the excited adhered layer. This reduces the amplitude of the pulse transmission, which reduces the amplitude component in the phase-plus-amplitude disbond detection circuit.

Bond Strength   The nondestructive determination of bond strength requires a method of inspection that has been verified by destructive mechanical test results. These results were obtained for the specimens described and are discussed in Reference 11.

The specimens were tested to failure and the load at failure recorded, along with the observed failure mode. Of the four possible failure modes—adhesion, delamination in the adherend, tensile, and cohesive failure of the adhesive material—only the first three were observed. The first of these is failure of the adhesion at the interface between the adhesive and the adherend. The last, cohesive failure, is the separation of the adhesive material from itself, while remaining adhered to the adherend or substrate of the bond joint. An example of each of the first three is shown in Figure 10.18. The most frequent failure was by delamination of the adherend enhanced by distortion-induced tearing.

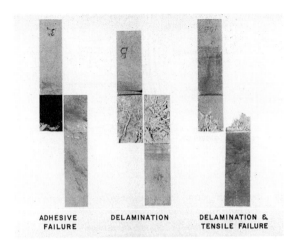

ADHESIVE   DELAMINATION   DELAMINATION &
FAILURE   TENSILE FAILURE

FIGURE 10.18   Specimen failure modes.

Bond joint strength is determined as a function of the AAL, LBI, and BMF indicators, as discussed next.

AAL indicator results revealed the potential of this inspection approach to bond strength determination. Data are plotted in Figure 10.19. A good linear fit to these data has a correlation coefficient of 0.998. But the problem that prohibits practical use of this empirical relationship to determine bond joint strength is the unplotted superpositioned scattered data from other modes of joint failure. These scattered data were within the same domain, and they had a very poor correlation coefficient. The best correlation coefficient was 0.33, obtained for a second-degree polynomial fit. The correlation coefficient of the linear fit was 0.10. These data, from all failure modes, underscore a very real problem encountered when evaluating bonds for strength. Because the inspector cannot nodestructively distinguish joints that will fail adhesively from joints that will fail by all other modes, a method that indicates adhesive strength only is an ineffective indicator of bond joint strength. This reveals the limited extent to which an unaltered commercial bond tester can be an effective indicator of bond strength.

To overcome this limitation, the instrument and inspection procedures were altered to provide a more effective NDI indicator of bond strength, regardless of whether the failure mode is adhesive or by delamination of the adherend. This alteration was described in the "Metering Mode" subsection above (see Fig. 10.14).

The LBI indicator is a more time-efficient index of local bond strength than the AAL and may provide improved correlation with bond strength for bond joint failures of all types. LBI data and corresponding destructive test results were obtained according to the previously described procedures and plotted versus bond strength. The LBI readings usually ranged from about 0.8 V for good bonds to 2.6 V for disbonds. These values, like AAL values, depend on the reference specimen used to adjust the operating sensitivity of the instrument. This reemphasizes the importance of the reference specimen in determining the absolute LBI values.

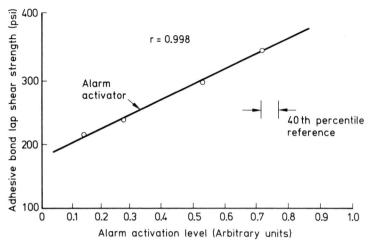

FIGURE 10.19   Bond shear strength versus alarm activation level setting of AAL potentiometer. (Zero is at maximum resistance.)

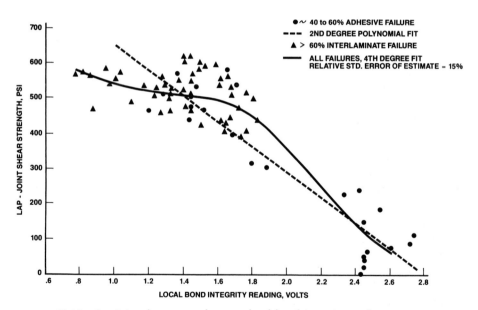

FIGURE 10.20   Lap joint shear strength versus local bond integrity reading.

Correlation of LBI with bond shear strength is shown by the data plotted in Figure 10.20. The distribution of these data is such that all failure modes observed appear to belong to the same data set. Nevertheless, a second–degree polynomial curve was fit to data from specimens manifesting predominantly adhesive failure. The 25 circled points in the figure are from specimens manifesting 40–60% adhesive failure. These data deviate from the broken-line fit by a relative standard error of estimate of 28%. The correlation coefficient is 0.92.

379

The relative standard error of estimate is less for all fits to the data from all failure modes. The optimum fit was obtained for fourth-degree polynomials, with a correlation coefficient of 0.91. The reduced correlation appears to result from the reduced slope in the LBI region < 1.7 V. The reduced slope in this region should not be a serious detriment to the use of the NDI method, because the critical decision region for identifying marginal or substandard bonds in this population is that region where shear strength is less than 2.8 MPa (400 psi). Only about 14% percent of the specimens tested were weaker. Both curves have sufficient slope to indicate usable sensitivity in that critical region. Figure 10.19, where shear strength is plotted against AAL, also shows sufficient slope to indicate useful sensitivity in the linear region below 2.8 MPa (400 psi).

These data form the experimental basis for the NDI method recommended herein. They do not provide physical insights into the empirical relationships established. These relations were established, and their correlations evaluated, to provide quantitative interpretation of adhesive bond NDI results in quality assurance applications.

The correlation of bond peel strength by flexural testing with LBI was poor. No good correlation or curve fit was found for any of the observed failure modes, or combination thereof. This agrees with results from the Fokker bond tester reported in Reference 21.

When the data points for adhesive failures are plotted (Fig. 10.21), the scatter of the data about the best polynomial fit can be seen. The standard deviation of all data shown is approximately equal to the standard error of estimate of the curve fit. This implies a curve fit so poor that about one-third of the LBI readings will be outside

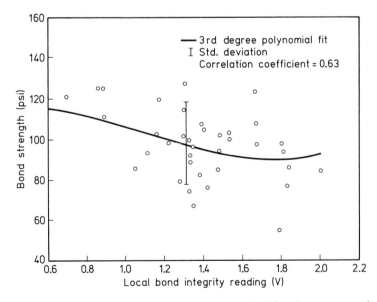

FIGURE 10.21   Flex–peel adhesive bond strength versus local bond integrity reading.

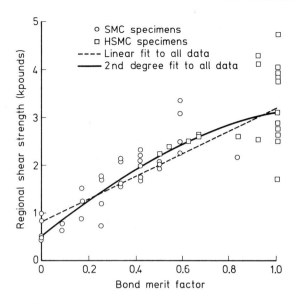

FIGURE 10.22   Bond shear strength versus bond merit factor for six-in. lap joint regions of bonded HSMC plaques.

the bond strength range over which the regression function is defined. The regression function is therefore virtually useless for practical applications.

In routine QC applications, the bond tester is most effectively used as a disbond detector. In such use, the AAL potentiometer is adjusted so that alarm activation occurs whenever the local bond joint strength is below a predetermined critical level. This level must be within the region of acceptable sensitivity, indicated by linear or near-linear curve fits in Figures 10.19 and 10.20. The domain of this region may be, different for each specimen population, and it must be established experimentally before reliable use of the method can be assured. Past experience indicates that the linear region will generally extend through bond strength data ranking from 1 to about the 30th percentile.

Once the AAL has been adjusted to the desired level, areas that cause alarm activation during bond joint inspection are identified as regions of substandard bond strength. The extent of these regions is determined as described in connection with Eq. (10.1), and a *BMF* is computed by that equation. The degree to which the BMF correlates with regional bond strength is indicated in Figure 10.22. These data represent 37 regions. The correlation coefficient of each curve fit is nearly 0.9. These data, and the resulting curve fits, indicate a useful correlation of regional bond strength with BMF for these adhesive–adherend systems.

The poor precision of the bond strength data at BMF values near unity occurs mainly because the disbond definition data, from which the BMF values are computed, contain no information on the bond strength of bonded regions except that it exceeds a minimum critical value. Obviously, there is a wide range of bond strengths that exceed this value, especially among these inspection data, which were

381

obtained at a low bond tester sensitivity. If, on the other hand, the operating sensitivity of the bond tester were increased, the sizes and frequency of regions defined as disbonds would increase, and the range of bond strengths existing in the subcritical or disbonded regions would consequently increase. Lower range strength data at each low sensitivity BMF would be shifted to a lower BMF at the high sensitivity, and the poor precision would then shift to lower values of BMF, accumulating at zero, where a higher intercept of the curve-fit would be observed.

The nonzero intercept in Figure 10.22 is a result of two simple disbond data acquisition features. First, recall that a disbond is called complete even though there may be narrow bonded regions along its edges. Second, poorly bonded regions with bond integrity below the critical level, defined when the bond tester is referenced, will be identified as disbonds. Obviously, these regions always have bond strength of zero or greater. The maximum level of bond strength existing in such disbonds is determined by the referenced operating sensitivity of the bond tester.

The correlation of regional bond flex–peel strength with BMF was investigated by inspecting 11 bonded HSMC plaques and cutting them into 110 specimens for destructive testing. This provided 66 regions for BMF computation for correlation with strength. The BMF of these 66 regions ranged from 0.833 to 1.0, an insufficient range to establish a useful correlation between regional flex–peel strength and BMF. Of these 66 regions, 55 had BMF values of unity. The average strength of these 55 regions, each 15.2 cm long, was 2366 N (532 lb), with a coefficient of variation of 24%. The coefficient of variation for the strength of all five regions with BMF of 0.833 was 5.3%, about a mean of 2491 N (560 lb). These data tend to show again the poor strength precision at BMF values at or near unity.

Bond Geometry   These correlations, established for flat specimens, are assumed to hold for typical bond joints in various automotive FRP assemblies with a variety of geometrical configurations. The validity of this assumption is based on the small size of the 2.5 cm (1 in.) bond length over which the measurements were made. In most actual assemblies encountered thus far, few bond joints deviate significantly from flatness over such a short distance.

10.4.2.3 Conclusions

Quantitative NDE of adhesively bonded FRP lap joints for shear strength can be routinely accomplished by using an electrically modified, commercially available bond tester to obtain two NDI indicators. Obtained by a simple cost-effective inspection procedure, these indicators provide bond inspection data that show fairly good correlations with bond joint strength, as well as good sensitivity over the range of expected use.

Each of these two indicators has a specific bond NDI application. The first, LBI, indicates local bond integrity over a region approximately $2\,cm^2$. The LBI is read from a meter wired into the bond tester circuitry. The second, BMF, is bond merit factor. It indicates regional bond integrity over a bond line ranging in length from about 10 cm to perhaps 40 cm. The BMF is computed from bond inspection data defining unbonded regions of the bond line. The definition of these unbonds is based on the operational sensitivity of the bond tester, which is initially adjusted during calibration with a bond reference specimen. Ultimately, both reliability and validity of the method are based on the reference specimen selection and verification procedure.

10.4.3 Instrumented Impact Testing

Redundant joining systems are sometimes used in automotive composite applications to improve load distribution and joint durability. An example of this is a composite leaf spring that is both adhesively bonded and riveted onto the steel end plate by which it is fastened to the vehicle. In this case, the adhesive was applied to the joint to provide better load distribution in the highly stressed region where the load is transferred between the steel plate and composite spring.

This joining system could not be easily and cost-effectively inspected ultrasonically for several reasons.

1. The effectiveness of pulse–echo inspection was limited by the extreme difference between the acoustic impedance of the adhesive and that of the steel at the adhesive–steel interface, compared with the much closer match between the adhesive and composite.
2. The effectiveness of the 25 kHz bond testing approach was severely limited by the total joint thickness.
3. Through-transmission inspection was not an attractive alternative because of the first technical reason given above (item 1) and because high attenuation in the composite prevented much of the ultrasonic energy from being transmitted, even through good bond joints.

These unusual bond joints yielded to inspection by computerized instrumented impact testing. Figure 10.23 shows significant differences in the load–deflection diagrams generated during impact of the falling instrumented tup on the bond joint. The instrumented tup provides for real-time data acquisition of impact dynamics while delivering energy to the impact zone. Such a method could provide a relatively inexpensive inspection procedure to correlate with expensive and labor-intensive destructive tests. The NDT then becomes a useful, cost-effective tool for use during process development and process capability studies.

10.4.4 X-Ray Inspection

Fiber-reinforced composites that are molded (by injection, resin transfer, reaction injection, or several other processes) may sometimes manifest anomalies in the concentration or orientation of reinforcing fibers that can be detected by X-radiography or real-time microfocus X-ray systems.

X-ray inspection for these anomalies is far more effective than the older, more conventional "burnout" method. X-ray inspection provides a view completely through the part, not just a view of fiber directions near the surfaces, as provided by

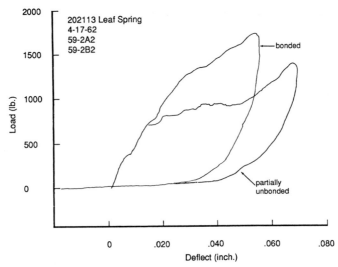

FIGURE 10.23   Load–deflection diagram obtained during instrumented impact testing of steel-to-composite bond joint.

a burnout. This is important because frequently fiber directions near the surface are not the same as those found throughout the thickness of the material. Moreover, with a modest capital investment, X-ray inspection can prove to be more cost-effective and much faster than the "burnout" procedure. X-ray inspection is therefore recommended, and has been installed in plants where new part molding patterns are developed.

The source and detector requirements are fairly simple for a radiographic system. A source with electrical X-ray tube potential ranging between 25 and 50 kV and a spot size slightly smaller than 1 mm are sufficient. The aluminum filter should be removed, using only the beryllium window, to avoid removing the soft X-rays, which are to be absorbed by the glass fibers.

A fairly simple methodology has been developed and applied, which can assist radiographic operators and other interested and involved engineers and technicians in the systematic interpretation of X-ray inspection observations. In addition, this approach can provide consistent results that can be considered to be semiquantitative in performance predictions. In X-ray inspection, the operator records numerical ratings for each of five anomalies that are frequently found in polymer composite components. These five anomalies are shown at the top of Table 10.3. The fifth, incipient crack potential (ICP), is a comprehensive rating by the observer to indicate a judgment of the probability that this region will fail prematurely in normal service.

Those who use this method must first be trained on X-ray "consensus" specimens that have been rated in the five areas by a panel of four or five observers. The "consensus" specimens are then mechanically tested to determine the desired tensile and/or flexural properties. The values from these mechanical tests are recorded with the X-ray data and ratings, with their concomitant mechanical test data, and are used not only to train new raters but to serve as reference "standards" for use during future

TABLE 10.3   Radiographic Rating of SMC Quality

Knit line	Fiber orientation	Resin rich area	Porosity	ICP

TABLE 10.4   Summary of Radiographic Rating of SMC Quality

X-Ray ID	B.A.H. rank	J.R.H. rank	A.2 rank	P.J.H. rank	G.B.C. rank	Ranking total	Std. dev.	Coeff. of Var.
16	1	2	2	2	1	8	0.5	6.8
17	2	4	1	3	2	12	1.1	9.5
15	5	1	4	4	3	17	1.5	8.9
14	3	5	8	1	4	21	2.5	12.3
37	4	6	5	5	5	25	0.7	2.8
36	8	3	3	9	6	29	2.7	9.5
38	6	8	7	6	7	34	0.8	2.4
35	7	9	8	8	9	39	1.3	3.3
40	9	7	9	7	8	40	1.0	2.5

semiquantitative work. There should be reference radiographs for each anomaly, at increments of about 7 MPa (1000 psi) through the critical region for performance.

Table 10.4 is a sample statistical summary showing the reliability of the precision of this approach. The rankings assigned by each of the five observers are shown. Note the acceptable level of the standard deviations and the coefficients of variation on the right side of the table.

Enhancement of X-ray inspection can be accomplished by adding metallized tracer fibers (22) to assist in detecting fiber orientation in short fiber composites.

10.4.5 Infrared Inspection

Thermoplastic components can be joined rapidly and cost-effectively by friction welding. Sometimes performed at low ultrasonic frequencies, these welds tend to be strong and reliable. When welding process variables are not controlled or kept constant, however the strength of these welds can vary over a very wide range.

The five circled points in Figure 10.24 indicate the wide range over which weld joint strength can vary while welding at what the operator thinks are constant weld parameters. The five points represent welds performed at the same energy, weld time, pressure, and hold time. Nevertheless these wide variations show a process out of control.

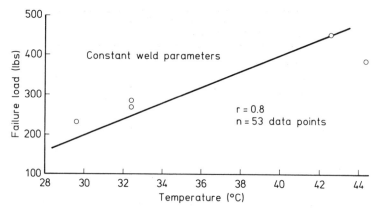

FIGURE 10.24   Weld joint strength versus joint surface temperature as determined by infrared detection about 10 seconds after welding a thermoplastic part.

Implementing infrared inspection of this thermal process provided an indicator of weld integrity with fast enough feedback to help achieve process control and optimization. With this approach, acceptable process capability was achieved and part production proceeded without 100% inspection. In another application, involving a more critical part, 100% inspection was continued during the first year of production. In both cases, the object of inspection was first to control the process and also to provide quality assurance to the product.

10.5 OTHER ADVANCED INSPECTION METHODS AND ARTIFICIAL INTELLIGENCE

Many other nondestructive inspection techniques (e.g., ultrasonic attenuation, C-scan, and acoustic microscopy) are effective in supporting quality in composite components (23–25). Some of these have been automated, and some have been very helpful in interrogating composites, and in acquiring and storing the inspection data for subsequent interpretation in various ways at desired times. The Advanced Data Acquisition, Imaging, and Storage System (ADIS), developed by McDonnell Douglas (26), is a very powerful and versatile quality inspection tool for use in the development of advanced materials.

Several artificial intelligence systems used to enhance the inspection and quality assurance of polymer composites have been under development for some time. One such project at the NDE Center at Iowa State University seeks to size and classify flaws detected ultrasonically (27). Another system, under development jointly by Chrysler Motors and the University of Michigan, seeks to provide a knowledge-based system to enhance the identification, selection and performance of NDI of composites (28). Figure 10.25 summarizes the wide variety of methods required for the various materials and processes. Additionally, there are several such systems under development, and in use, which should provide considerable aid to the composite community in its ongoing quest for quality.

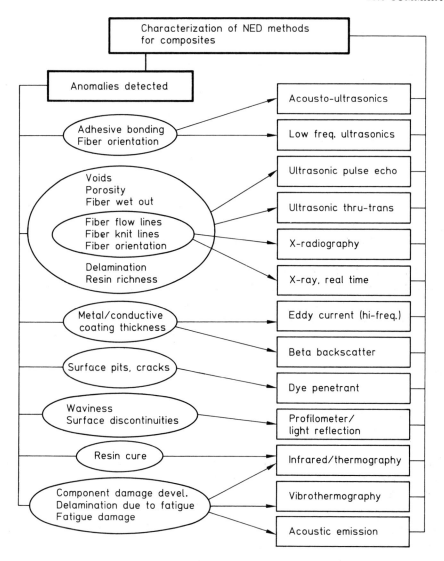

FIGURE 10.25   A variety of NDE methods.

10.6 SUMMARY

The concepts and methods discussed in this chapter seek to emphasize the importance of considering quality and productivity issues in every phase of the product development cycle: from concept through design to manufacture and performance in the field. The focus of attention should include both hard and soft elements that support composite quality throughout the product development cycle. Such a comprehensive quality approach not only pays its own way, but adds to customer satisfaction and profits, especially when its implementation leads to process optimization. This should

387

result in the virtual elimination of scrap, rework, warranty costs, customer dissatisfaction, and the minimization of inspection costs.

The methods discussed in this chapter for inspecting polymer composites are but a few of many that can be useful in providing measurements for statistical process control and/or quality assurance. These methods were selected because their speed, cost-effectiveness, and operational simplicity make them compatible with mass production manufacturing environments. Once correlation between the inspection parameters and the physical, chemical, mechanical, or performance characteristics have been established, the inspection parameters can be used as indicators, and variability in process parameters can be studied in a cost-effective manner. Thus, these inspection methods reduce the cost and labor intensity of providing data for determining and maintaining process control and capability. Determining process control and capability characteristics before mass production is launched is an essential element in the CQS, as is the philosophy of continuous quality improvement. And improvements in quality, like any other quantity, must be measured to be made.

REFERENCES

1. Kaora Ishikawa, *What is Total Quality Control? The Japanese Way*, Prentice-Hall, Englewood Cliffs, NJ, 1985.
2. W, Edwards Deming, *Quality, Productivity, and Competitive Position*; Massachusetts Institute of Technology, Center for Advanced Engineering Study, 1982.
3. J. M. Juran, F. M. Gryna, and R. S. Bingham: *Quality Control Handbook*, 3rd ed., McGraw Hill, New York, 1974.
4. Philip B. Crosby, *Quality Is Free*, McGraw-Hill, New York, 1979.
5. Mark T. Hutson, Owens Corning Fiberglas Croporation to Aaron D. Rosenstein, Chrysler Corporation, personal communication, May 8, 1985.
6. *Statistical Quality Control Handbook*, © 1956 by Western Electric Company, Inc., 1958, 11th printing, 1985.
7. Gilbert B. Chapman and Laszlo Adler, "Nondestructive Inspection Technology in Quality Systems for Automotive Plastics and Composites," SAE Technical Paper 880155, 1988.
8. G. B. Chapman, "Nondestructive Inspection for Quality Assurance of Fiber Reinforced Plastic Assemblies," *SAE Trans.*, 91 (1982).
9. J. H. Sinclair, and C. C. Chamis, "Fracture Modes in Of-Axis fiber Composites," *34th SPI/RP Annual Technology Conference*, 1979, 22-A.
10. Lie K. Djiauw, "Relation of Composite Part Strength to Ultrasonic Inspection," Automotive Plastics Durability Conference Proceedings, Society of Automotive Engineers Technical Paper 811350, December 1981.
11. G. B. Chapman, II, "A Nondestructive Method of Evaluating Adhesive Bond Strength in Fiberglass Reinforced Plastic Assemblies," in K. T. Kedward, Ed., *Joining of Compsite Materials*, ASTM STP 749, American Society for Testing and Materials, Philidelphia, 1981, pp. 32–60.
12. Julian R. Frederick, *Ultrasonic Engineering*, Wiley, New York, 1965.
13. Harvey E. Henderson, "Ultrasonic Velocity Technique for Quality Assurance of Ductile Iron Casting" *Iron Worker*, Summer 1976.

14. Elisa Redler, Gerald Entine, and Anthony J. Ratkowski, at 38th Annual SPI Technical Conference (Composite Solutions to Material Challenges), Houston, Tx, February 7–11, 1983. Published by SPI, New York, Sess. 24G.

15. M. S. Hughes, S. M. Handley, J. G. Miller, and E. I. Madaroas, "A Relationship Between Frequency Dependent Ultrasonic Attenuation and Porosity in Composite Laminates," in D. O Thompson and D. E. Chimenti, Eds., *Review of Progress in Quantitative NDE,* Vol. 7B, Plenum Press, New York, 1988, pp. 1037–1044.

16. David K. Hsu, "Porosity Evaluation in Composite Laminates by Ultrasound" Technical Paper IQ87-354, presented at AUTOCOM '87 (Society of Manufacturing Engineers), June 1–4, 1987.

17. David K. Hsu, "Ultrasonic Measurements of Porosity in Woven Graphite Polyimide Composites," in D. O. Thompson and D. E. Chimenti, Eds., *Review of Progress in Quantitative NDE,* Vol, 7B, Plenum Press, New York, 1986, pp. 1063–1068.

18. Peter B. Nagy, and Laszlo Adler, "Scattering Induced Attenuation of Ultrasonic Back-scattering," in *Review of Progress in Quantitative NDE,* D. O. Thompson and D. E. Chimenti Eds, Plenum Press, New York, 1986 Vol 7B, 1263–1271.

19. L. Adler, J. H. Rose, and C. Mobley, "Ultrasonic Method to Determine Gas Porosity in Aluminum Alloy Casting: Theory and Experiment," *J. Appl. Phys.,* 59, (2), 336–347 (1986).

20. D. K. Hsu, and K. M. Uhl, "A Morphological Study of Porosity Defects in Graphite–Epoxy Composites," *Review of Progress in Quantitative NDE,* Vol. 6B, D. O. Thompson and D. E. Chimenti, Eds., (Plenum, New York, 1986), pp. 1175–1184.

21. D. F. Smith and C. V. Cagle, "*Appl Poly Symp,* 3, 411–437 (1966).

22. D. L. Denton and S. H. Munson-McGee, "Use of X-Radiographic Tracers to Measure Fiber Orientation in Short Fiber Composites," *High Modulus Fiber Composites in Ground Transporation and High Volume Applications,* ASTM STP 873, American Society for Testing and Materials, Philadelphia, 1985, pp. 23–35.

23. J. H. Williams, Jr., and S. S. Lee, "Promising Quantitative Nondestructive Evaluation Technqiues for Composite Materials," *Mater Eval.,* 43 561 (April 1985).

24. J. H. Williams, Jr., and S. S. Lee, "An Overview of Research at MIT in the Quantitative Nondestructive Evaluation of Fiber Composite Materials and Structure," SME Technical Paper EM84103, Third Composites in Manufacturing Conference, Anaheim, CA, Jan. 10–12, 1984 (Society of Manufacturing Engineering,).

25. J. H. Williams, Jr., S. S. Lee, and T. K. Wang, "Quantitative Nondestructive Evaluation of Automotive Glass Fiber Composites" *J. Compos Mater,* 16, 20–39 (1982).

26. Joe Flaherty, "The Advanced Data Aquisition, Imaging and Storage System (ADIS)," Technical Paper MS86-366, presented at AUTOCOM '86 (Society of Manufacturing Engineers) June 9–12, 1986.

27. L. W. Schmerr, K. E. Christensen, and S. M. Nugen, "Development of an Expert System for Ultrasonic Flaw Classification," in *Review of Progress in Quantitative Nondestructive Evaluation,* ed. D. O. Thompson and D. E. Chementi, 879–887. New York: Plenum Press, 1988.

28. Gilbert B. Chapman, II, and Pamela F. Lee, "The Development of a Knowledge-Based System for the Nondestructive Inspection of Composites," SAE Technical Paper 890246, presented at International Congress and Exposition, Detroit, February 27–March 3, 1989.

APPENDIX

Listing of the BASIC Program for Computing Maximum SMC Tensile Strength from Fiber Orientation Indications

```
00100 rem       This BASIC program is written to compute the expected tensile strength predicted
00110 rem       from NDI data from sonic velocity indications (Sondicator Phase Readings).
00120 rem
00130 rem       Definitions:
00140 rem       A1 is the angle between the prevailing direction of fiber orientation and
00160 rem       the longitudinal axis of the specimen (along which the tensile strength is
00170 rem       measured) for the top side.
00180 rem
00190 rem       A2 is the same for the opposite side of the specimen.
00200 rem
00210 rem       D1 is the Delta VIR reading from side 1 of the specimen
00220 rem
00230 rem       D2 is the Delta VIR reading from side 2 of the specimen
00240 rem
00250 rem       T is the expected tensile strength of the specimen
00260 rem       T1 is the expected tensile strength component from side 1
00270 rem       T2 is the expected tensile strength component from side 2
00280 rem       T3 is the expected tensile strength component from random fiber SMC
00290 rem       T4 is the expected tensile strength component from resin-filler matrix
00300 rem       T5 is the actual tensile strength of the specimen.
00310 rem
00320 rem       K is a strength-sensitivity constant, unique to each instrument and material
00330 rem
00340 rem       Data section: Lines 900-1000.
00350 rem PROGRAM:
00360 DIM A1(100), A2(100),D1(100), D2(100)
00370 DIM S(100)
00380 DIM T(100), T1(100), T2(100), T5(100)
00390 DIM T3(100)
00400 DIM X(100),Y(100)
00410 DIM Z(100)
00420 rem
00430 K=34567
00440 T4=3333
00450 READ L$,N
00460 rem       Table Heading--------------
00470 PRINT 'K=';K
00480 PRINT 'T4=';T4
00490 PRINT
00500 PRINT
00510 PRINT '                    '; L$
00520 PRINT
00530 PRINT
00540 PRINT 'Specimen','Delta VIR','Fiber','Delta VIR','Fiber','Expected','Actual'
00550 PRINT 'Number','Side 1','Angle','Side 2','Angle','T. Strength','T. Strength'
00560 PRINT
00570 FOR I=1 TO N
00580 READ S(I), D1(I), A1(I), A2(I),T5(I)
00590 REM  CONVERT DEGREES TO RADIANS FOR COSINE FUNCTION
00600 A1(I)=A1(I)*(3.1416/180)
00610 A2(I)=A2(I)*(3.1416/180)
00620 T1(I)=K*D1(I)*ABS(COS(A1(I)))
00630 T2(I)=K*D2(I)*ABS(COS(A2(I)))
00640 T3(I)=1000
00650 T(I)=T1(I)+T2(I)+T3(I)+T4
00660 REM   CONVERT RADIANS BACK TO DEGREES
00670 A1(I)=A1(I)*(180/3.1416)
00680 A2(I)=A2(I)*(180/3.1416)
00690 GO TO 710
00700 PRINT I;S(I),D1(I),A1(I),D2(I),A2(I),T(I),T5(I)
00710 NEXT I
```

```
00720 REM      THIS SECTION, LINES 900-1900, RESERVED FOR DATA
00730 DATA 'L-SERIES TRUCK HOOD SPECIMENS - 2 NOV. 82',42
00740 DATA 71,.04,0,.02,45,6700
00750 DATA 72,.04,0,.04,0,7500
00760 DATA 73,.03,0,.04,0,5900
00770 DATA 74,.04,90,.06,90,3700
00780 DATA 75,.07,90,.06,0,4500
00790 DATA 76,.06,0,.06,0,17600
00800 DATA 77,.06,90,.04,90,4400
00810 DATA 78,.07,0,.06,0,16500
00820 DATA 79,.09,90,.08,90,4200
00830 DATA 710,.08,90,.06,0,4000
00840 DATA 711,.06,90,.05,90,3100
00850 DATA 712,.01,0,.01,0,6500
00860 DATA 713,.06,145,.05,145,12200
00870 DATA 714,.06,0,.05,0,10600
00880 DATA 81,.04,145,.04,0,6900
00890 DATA 82,.07,90,.04,0,8300
00900 DATA 83,.05,90,.06,0,7300
00910 DATA 84,.05,0,.08,0,11800
00920 DATA 85,.04,45,.06,0,6400
00930 DATA 86,.08,0,.07,0,19400
00940 DATA 87,.09,90,.06,90,3900
00950 DATA 88,.09,0,.07,0,18200
00960 DATA 89,.10,90,.09,90,4500
00970 DATA 810,.13,90,.08,0,3500
00980 DATA 811,.06,90,.08,90,2900
00990 DATA 812,.04,0,.08,90,3800
01000 DATA 813,.05,0,.04,90,7400
01010 DATA 814,.08,0,.06,0,8600
01020 DATA 91,.04,0,.06,0,12800
01030 DATA 92,.04,145,.09,90,7400
01040 DATA 93,.06,0,.05,90,4000
01050 DATA 94,.07,0,.08,0,11700
01060 DATA 95,.08,90,.06,0,4600
01070 DATA 96,.08,0,06,0,16900
01080 DATA 97,.04,90,.05,90,4000
01090 DATA 98,.05,0,.06,0,10900
01100 DATA 99,.06,90,.04,90,4100
01110 DATA 910,.08,90,.04,09,4200
01120 DATA 911,.02,90,.08,145,2200
01130 DATA 912,.08,145,.02,0,3800
01140 DATA 913,.00,45,.06,0,12400
01150 DATA 914,.03,145,.08,0,13300
01160 REM  THIS PROGRAM READS & PLOTS DATA.
01170 REM   X(I) IS ABSCISSA
01180 REM   Y(I) IS ORDINATE
01190 PRINT
01200 PRINT
01210 PRINT
01220 PRINT
01230 PRINT 'PLOT DATA                    ';
01240LINPUT Y$
01250 IF Y$<'PASS> THEN 2180
01260 PRINT 'NEW PAGE                  '
01270 LINPUT A$
01280 PRINT 'ENTER SCALE FACTOR FOR ABSCISSA.';
01290 S1=160
01300 PRINT 'ENTER SCALE FACTOR FOR ORDINATE.';
01310 S2=160
01320 PRINT 'WHAT PLOT SYMBOL';
01330 P$='*'
01340 PRINT
01350 FOR T=1 TO N
01360 X(T)=T(I)/S1
01370 Y(I)=T5(I)/S2
01380 NEXT I
01390 PRINT
01400 S8=0
```

```
01410 FOR I=1 TO N
01420 IF X(I-1)>X(I) THEN 1940
01430 NEXT I
01433 FOR I=1 TO N
01435 Z(I)=INT(X(I))
01438 NEXT I
01440 FOR I=1 TO N
01450 IF Z(I)=Z(I+1) THEN 2060
01460 NEXT I
01470 PRINT
01480 REM  PLOTTING AMPLITUDE
01490 PRINT 'K=';K
01500 PRINT 'T4=';T4
01510 PRINT
01520 PRINT
01530 PRINT 'I','Z','X','T-EXPECT','Y','T-ACTUAL'
01540 FOR I=1 TO N
01550 REM     PRINT THE REARRANGED VALUES
01560 PRINT I,Z(I),X(I),X(I)*S1,Y(I),Y(I)*S2
01570 NEXT I
01580 PRINT
01590 PRINT 'NEW PAGE';
01600LINPUT A$
01610 PRINT 'DO YOU WANT THE ORIGIN SHOWN';
01620 LINPUT A$
01630 PRINT 'THE ACTUAL TENSILE STRENGTH (Y) VS THE EXPECTED TENSILE STRENGTH (X)'
01640 PRINT '                              FOR'
01650 PRINT '                         ';L$
01660 PRINT
01670 PRINT
01680 PRINT
01690 PRINT'SCALE FACTORS','X','Y'
01700 PRINT ,S1,S2
01710 PRINT
01720 PRINT 'K=';K;'    T4=';T4
01730 PRINT
01740 PRINT'                    THE ACTUAL TENSILE STRENGTH, KPSI'
01750 PRINT
01760 PRINT 'Y:    +----+----+----+----+----+----+----+----+----+----+----+----+----+
--+----+'
01770 PRINT 'X(1)=';Z(1)*S1
01780 IF A$>'PASS' THEN 1800
01790 Z(0)=Z(1)-5
01800 FOR I=1 TO N
01810 D7=Z(I)-Z(I-1)
01820 FOR D8=1 TO D7
01830 PRINT
01840 NEXT D8
01850 PRINT TAB(7+Y(I));P$;
01860 NEXT I
01870 PRINT
01880 PRINT
01890 PRINT
01900 PRINT
01910 PRINT
01920 STOP
01930 REM   BUMP-BACK SECTION
01940 REM   REARRANGE X(I) AND Y(I) VALUES IN INCREASING ORDER.
01950 REM   THIS X(I-1) ARRANGEMENT PREVENTS A FINAL X(1)=0
01960 X=X(I-1)
01970 X(I-1)=X(I)
01980 X(I)=X
01990 Y=Y(I-1)
02000 Y(I-1)=Y(I)
02010 Y(I)=Y
02020  GO TO 1410
02030 REM    PRINT X-LOCATION
02040 PRINT X(I);
02050 GO TO 1850
```

```
02060 IF Y(I)>Y(I+1) THEN 2090
02080 GO TO 1460
02090 REM SWITCH Y
02100 REM
02110 Y=Y(I+1)
02120 Y(I+1)=Y(I)
02130 Y(I)=Y
02170 GO TO 1440
02180 END
```

INDEX